T0135737

Fluid flows around moving obstacles: Non-autonomous rotation and fluids with variable density

Vom Fachbereich Mathematik

der Technischen Universität Darmstadt

zur Erlangung des Grades eines

Doktors der Naturwissenschaften

(Dr.rer.nat.)

genehmigte

Dissertation

von

Dipl.-Math. Tobias Hansel

aus Seeheim-Jugenheim

Referenten:	Prof. Dr. Matthias Hieber
	Prof. Dr. Bálint Farkas
	Prof. Dr. Giovanni P. Galdi
	Prof. Dr. Yoshihiro Shibata
Tag der Einreichung:	20. Oktober 2011
Tag der mündlichen Prüfung:	24. November 2011

Darmstadt 2012
D 17

Bibliografische Information der Deutschen Nationalbibliothek

Die Deutsche Nationalbibliothek verzeichnet diese Publikation in der
Deutschen Nationalbibliografie; detaillierte bibliografische Daten sind
im Internet über http://dnb.d-nb.de abrufbar.

ISBN 978-3-8325-3095-2

Logos Verlag Berlin GmbH
Comeniushof, Gubener Str. 47,
10243 Berlin
Tel.: +49 (0)30 42 85 10 90
Fax: +49 (0)30 42 85 10 92
INTERNET: http://www.logos-verlag.de

Preface

This thesis deals with the analytical investigation of viscous fluid flows around moving, in particular around rotating obstacles. The main emphasis is placed on *non-autonomous rotation*, or more precisely on the case where the obstacle undergoes a rotation described by a non-autonomous equation, and on fluids with *variable density*.

In general, when dealing with moving obstacles one faces the problem that the exterior of the obstacle changes with time. Therefore, in order to use classical techniques, it is reasonable to consider the governing equations for the fluid in a frame attached to the obstacle. However, due to the rotation, the transformed fluid equations contain a drift term with linearly growing coefficients. Herein lies the particularity and the main difficulty in the analysis of fluid flows around rotating obstacles, and therefore this problem has attracted a lot of attention for the last 20 years.

Here, we consider the incompressible Navier-Stokes flow past an obstacle, rotating with a prescribed *time-dependent* angular velocity. In addition, we assume that the obstacle is also translating with a prescribed time-dependent translational velocity, or (equivalently) that the fluid flow is subject to a time-dependent outflow condition at space-infinity. After rewriting this problem on a fixed exterior domain[1] $\Omega \subset \mathbb{R}^d$, one obtains a *non-autonomous* system of equations with an *unbounded and time-dependent* drift term. The linearization of these equations leads to a family of non-autonomous operators of *Ornstein-Uhlenbeck type*. By these we mean second order differential operators with linearly growing drift coefficients. Such operators arising from fluid flow around a rotating obstacle have been studied intensively and systematically in the autonomous case, but a clomplete and systematic analysis of the non-autonomous case was still open.

Since the difficulty in the analysis of these operators is the unbounded drift, it is reasonable to first consider the model situation $\mathbb{R}^d$. In this case we show, based on an explicit solution formula, that the family of non-autonomous linear operators generates a strongly continuous evolution system on $L^p_\sigma(\mathbb{R}^d)$ for $1 < p < \infty$. The explicit formula also allows to prove L^p-L^q-estimates and gradient estimates for the evolution system. Based on a localization technique, we then show the analogous results in the case of exterior domains $\Omega \subset \mathbb{R}^d$.

These linear results are key ingredients to obtain local existence of a mild solution to the full non-linear problem for initial values $u_0 \in L^p_\sigma(\Omega)$, $p \geq d$, by using a variant of Kato's iteration scheme. Moreover, we prove long time existence in the sense that for an arbitrarily fixed $0 < T_0 < \infty$ we obtain a mild solution on $[0, T_0]$ if the given data is sufficiently small (depending on T_0). Furthermore, it is shown that for a related non-linear model problem in $\mathbb{R}^d$ one even obtains a global mild solution for small initial values $u_0 \in L^d_\sigma(\mathbb{R}^d)$.

---

[1] Here $d = 3$ is the physically realistic and interesting case, but mathematically we can allow $d \geq 2$.

Since so far, most results for fluid flows past rotating obstacles are only known to hold for "classical" Navier-Stokes fluids with a constant density, we shall prove a well-posedness result in this context for fluids with variable density which are still incompressible. The motion of such a fluid is described by the *density-dependent incompressible Navier-Stokes equations*. This system of equations consists of a non-linear momentum equation for the velocity field of the fluid and a transport equation for the density.

Here, we consider the density-dependent Navier-Stokes flow in a domain $\Omega \subset \mathbb{R}^3$ exterior to a compact obstacle that rotates with a prescribed constant angular velocity. It is shown that if the initial velocity field u_0 and the initial density ρ_0 are taken from appropriate function classes, then there exists a unique, local strong solution to the initial boundary value problem in the L^2-setting. The proof is based on the Faedo-Galerkin method and on suitable a priori estimates for the constructed sequence of approximating solutions. Here again the unbounded drift terms, as well as the coupled transport equation for the density, cause a lot of technical difficulties. To handle these, new techniques needed to be developed. In particular, we derive new elliptic L^p-estimates for the modified stationary Stokes problem with rotating effect which are crucial for deriving the a priori estimates and which are also interesting in their own right.

Moreover, the problem of deriving global decay estimates for the Stokes semigroup in exterior domains is also discussed in this thesis. Although these estimates are well-known, we give a survey on this topic and present a rather elementary and self-contained proof which combines ideas from several different approaches.

Acknowledgements

At this point, it is a great pleasure for me to thank several people who contributed to the development of this thesis:

First of all, I want to express my deep gratitude to Professor Matthias Hieber and Junior-Professor Bálint Farkas for their constant support and their encouragement during the past three years. Their guidance and many suggestions contributed invaluably to the development of this thesis. At the same time they gave me sufficient freedom to develop and pursue my own ideas. Moreover, I want to thank Bálint Farkas for his guidance and his advices outside of mathematics.

I would like to thank Professor Giovanni P. Galdi and Professor Yoshihiro Shibata for acting as co-referees to this thesis. I also thank both of them for fruitful and inspiring discussions on mathematical fluid dynamics during their various visits to Darmstadt. Moreover, I am deeply grateful for Professor Shibata's kind hospitality and his support during my six months stay at Waseda University in Tokyo. This was really a memorable time for me.

I wish to thank my co-authors, Matthias Geissert and Abdelaziz Rhandi, for the fruitful collaboration, the many interesting and stimulating discussions and for many remarks. I certainly enjoyed the joint work with both of them in every perspective.

I thank Professor Ting Zhang for fruitful discussions and the many valuable hints I received during and after his visit to Darmstadt. I certainly learned a lot from him about density-dependent fluids.

I am greatly indebted to Matthias Geissert, Robert Haller-Dintelmann, Horst Heck and Mads Kyed, as they were always ready to kindly answer my questions. In particular, I thank them for proof-reading parts of this manuscript.

I always enjoyed working at the Department of Mathematics and the International Research Traning Group 1529 "Mathematical Fluid Dynamics" in Darmstadt. I thank my colleagues Dario Götz, Karoline Götze, Christian Komo, Manuel Nesensohn, Felix Riechwald, Martin Sauer, George Schöchtel, Raphael Schulz and Christof Trunk for the amicable atmosphere and for the many discussions on mathematics and beyond. Moreover, I wish to thank all the Japanese professors and students involved in the International Research Traning Group 1529, especially my Japanese colleagues Norihisa Ikoma and Kohei Soga, for their help during my time in Japan.

I would like to thank the DFG International Research Training Group 1529 "Mathematical Fluid Dynamics" for the financial support that enabled me to work on this thesis and for the possibility of a research stay at Waseda University in Tokyo.

Last but not least, I thank my family for their support, patience and encouragement throughout my studies and the work on this thesis.

Zusammenfassung

Die vorliegende Arbeit beschäftigt sich mit der analytischen Behandlung von Strömungen viskoser Fluide um sich bewegende, insbesondere rotierende, Gegenstände. Das Hauptaugenmerk liegt dabei auf dem nicht-autonomen Fall sowie auf einer Klasse von Fluiden mit variabler Dichte.

Das grundlegende Modell für die mathematische Betrachtung viskoser, inkompressibler Fluiden sind die inkompressiblen Navier-Stokes-Gleichungen. In dieser Arbeit wird dieses System auf einem Gebiet außerhalb eines Gegenstandes betrachtet, dessen Schwerpunkt sich mit einer vorgeschriebenen Geschwindigkeit bewegt und der zusätzlich mit einer vorgeschriebenen zeitabhängigen Winkelgeschwindigkeit rotiert. Nach einer geeigneten linearen Koordinatentransformation erhält man ein System von nicht-autonomen Gleichungen auf einem festen Außenraumgebiet. Die Besonderheit und die Hauptschwierigkeit des transformierten Problems liegt in der Präsenz eines Terms erster Ordnung mit *zeitabhängigen* und *linear wachsenden* Koeffizienten.

Zunächst wird das linearisierte System im Ganzraum betrachtet und eine explizite Lösungsformel hergeleitet. Mit Hilfe dieser Formel wird gezeigt, dass das lineare Problem im Ganzraum wohlgestellt ist. Desweiteren können regularisierende Eigenschaften der Familie von Lösungsoperatoren nachgewiesen werden. Durch ein geeignetes Lokalisierungsargument werden diese Resultate anschließend auf den Außenraum-Fall überführt. Das sich hieraus ergebende Hauptresultat ist die Existenz einer eindeutigen, zeitlokalen milden Lösung zum nichtlinearen Problem, die mit Hilfe eines geeigneten Iterationsverfahrens konstruiert wird. Außerdem diskutieren wir Langzeit-Existenz und zeigen Existenz einer eindeutigen, globalen milden Lösung für ein relevantes nichtlineares Problem im Ganzraum, unter geeigneten Annahmen an die Daten.

Im Folgenden werden außerdem *inkompressible* Fluiden mit *variabler Dichte* außerhalb eines rotierenden Gegenstandes betrachtet. Die relevanten Gleichungen für solche Fluide bestehen aus einer nichtlinearen Gleichung für das Geschwindigkeitsfeld, gekoppelt mit einer Transportgleichung für die Dichte. Auch hier sorgt die Rotation dafür, dass die transformierten Gleichungen Terme erster Ordnung mit linear wachsenden Koeffizienten enthalten. In dieser Arbeit wird die Existenz einer eindeutigen, zeitlokalen starken Lösung für das gekoppelte Problem bewiesen. Der Beweis basiert auf der Faedo-Galerkin Methode und geeigneten *A Priori* Abschätzungen für die konstruierte Folge von approximierenden Lösungen. Dabei sorgen sowohl die linear wachsenden Koeffizienten als auch die gekoppelte Transportgleichung für gewisse technische Schwierigkeiten, die neue Argumente erfordern. Insbesondere werden elliptische L^p-Abschätzungen für das modifizierte stationäre Stokes-System mit Rotationseffekt bewiesen, welche eine wichtige Rolle im Beweis der *A Priori* Abschätzungen spielen.

Desweiteren wird die Stokes-Halbgruppe in einem Außenraumgebiet untersucht und es wird ein Überblick über globale L^p-L^q-Abschätzungen und Gradientenabschätzungen gegeben. Solche Abschätzungen sind aus der Literatur wohlbekannt, allerdings ist der hier präsentierte Beweis in sich geschlossenen und recht elementar.

Contents

Introduction

The *Navier-Stokes equations* are a system of non-linear partial differential equations describing the dynamics of viscous fluids. These equations were first introduced in the 19th century, see [Nav27] and [Sto45]. Nowadays the Navier-Stokes equations serve as the fundamental mathematical model in fluid dynamics, and besides applications to various disciplines of engineering and natural sciences they have a particular importance within the development of the modern mathematical theory of partial differential equations.

In this thesis problems in the framework of the incompressible Navier-Stokes equations, including the density-dependent case, are studied. As it was outlined in the preface, the main interest lies in (local) existence and uniqueness questions of Navier-Stokes flows in the exterior of a rotating obstacle. In the following, a broad overview on the topics of this thesis, including aspects of modeling, are given. Moreover, the main results are explained without technical details and references to related works are provided. The intention of this introduction is to navigate the reader through the thesis and to motivate the problems studied.

Governing equations for viscous incompressible fluids

Before we turn our attention to rigorous mathematics, we shall briefly derive the relevant equations considered in this thesis. For a more rigorous modeling and for more information on the physics behind the equations we refer to standard monographs, e.g. to Landau and Lifschitz [LL59] or Batchelor [Bat99]. A detailed derivation of the governing equations can also be found in the preliminary parts of mathematical monographs, see e.g. Lions [Lio96, Chapter 1], Antontsev, Kazhikhov and Monakhov [AKM90, Chapter 1] or Chorin and Marsden [CM93, Chapter 1].

We recall here the standard derivation of the relevant equations in eulerian form. We consider a viscous fluid moving within a domain $\Omega \subset \mathbb{R}^d$, $d \geq 2$. Let $(0, T) \subset \mathbb{R}$ be a time interval in which we follow the fluid motion. The velocity field of the fluid is denoted by $u := u(t, x)$ and the (mass) density of the fluid by $\rho := \rho(t, x)$. Here $t \in (0, T)$ is the time and $x = (x_1, \ldots, x_d)$ is a point in Ω. The particles of the fluid are moving along the integral curve $X(t)$ solving $\dot{X}(t) = u(t, X(t))$, and the mass $m(t; V)$ of the fluid contained in a fixed control volume $V \subset \Omega$ at time t is given by

$$m(t; V) = \int_V \rho \, \mathrm{d}V.$$

The principle of conservation of mass states that the variation of mass inside V must be equal to the flux of mass on the boundary $\Gamma := \partial V$. Mathematically this means that

$$\int_V \frac{\partial \rho}{\partial t} \, dV = \int_\Gamma \rho u \cdot \nu \, d\Gamma,$$

where ν is the outer normal vector at $\Gamma := \partial V$. Since the control volume V is arbitrary we conclude, by the Gauss theorem, that the evolution of the density is given by

$$\frac{\partial \rho}{\partial t} + \operatorname{div}(\rho u) = 0 \qquad \text{in } (0, T) \times \Omega. \tag{1}$$

This evolution equation for the density is the so-called *continuity equation*. Similarly, by using Newton's law of conservation of momentum, one can also derive an equation for the evolution of the velocity field $u = (u_1, \ldots, u_d)$, or equivalently for the evolution of the momentum ρu:

$$\rho u_t + \rho u \cdot \nabla u - \operatorname{div} T(u, p) = \rho f \qquad \text{in } (0, T) \times \Omega. \tag{2}$$

This is the so-called *momentum equation*. Here $f := f(t, x)$ represents external forces per unit volume acting on the fluid, such as e.g. gravity or electromagnetic forces, and $T(u, p)$ is the Cauchy stress tensor representing surface forces. Since a real, also called viscous, fluid in motion is under the influence of compression effects and *viscous* effects one writes

$$T(u, p) = -p \operatorname{Id} + S(u),$$

where Id is the identity matrix, $p := p(t, x)$ is the pressure of the fluid and $S(u)$ is the viscous stress tensor. In the case of so-called Newtonian fluids, which are exclusively studied in this thesis, the viscous stress tensor is given by

$$S(u) = \lambda(\operatorname{div} u)\operatorname{Id} + \mu \left(\frac{\partial u_i}{\partial x_j} + \frac{\partial u_j}{\partial x_i} \right)_{i,j=1}^d.$$

Here λ, μ denote the bulk and the dynamic viscosity of the fluid, respectively. In general, λ and μ are functions of ρ, but in this thesis we shall assume that $\lambda, \mu > 0$ are constants.

Combining the continuity equation (1) with the momentum equation (2), we obtain the *compressible Navier-Stokes equations*:

$$\begin{cases} \rho_t + \operatorname{div}(\rho u) = 0 & \text{in } (0, T) \times \Omega, \\ \rho u_t + \rho u \cdot \nabla u + \nabla p - (\lambda + \mu)\nabla(\operatorname{div} u) - \mu \Delta u = \rho f & \text{in } (0, T) \times \Omega. \end{cases} \tag{3}$$

Of course, these equations need to be supplemented by suitable boundary and initial conditions and the pressure p must obey some given state equation of the form $p = p(\rho)$. For more details and a rigorous treatment of the compressible Navier-Stokes equations we refer to the monographs [Fei04, NS04] and the references therein.

In this thesis we shall not consider the compressible Navier-Stokes system (3), but we simplify this model by assuming that the fluid is incompressible which mathematically means that

$$\operatorname{div} u = 0. \tag{4}$$

In order to understand the physical interpretation of the divergence-free condition (4), fix some time t and consider a control volume V at time $t + h$ for some $h > 0$, i.e. $V(t + h) = \{X(t + h; y) : y \in V(t)\}$. Here $X(\tau; y)$ denotes the solution to the equation $\dot{X}(\tau; y) = u(\tau, X(\tau)), \tau > t$, and $X(t; y) = y \in \Omega$. The volume of $V(t + h)$ is now given by

$$\int_{V(t)} J(h; y) \, dy,$$

where $J(h; y)$ is the Jacobian determinant of the map $y \mapsto X(t + h; y)$. From the fact that

$$\frac{\partial}{\partial \tau} J(\tau; y) = J(\tau; y)[\operatorname{div} u(X(\tau; y), \tau)],$$

see e.g. [CM93, page 8], we can conclude that (4) holds if and only if the volume of $V(t + h)$ is equal to the volume of $V(t)$. Thus, the incompressibility means that the volume of any part of the fluid remains constant during the motion. In the incompressible case the relevant system of equations reads as follows:

$$\begin{cases} \rho_t + u \cdot \nabla \rho &= 0 \quad \text{in } (0, T) \times \Omega, \\ \rho u_t + \rho u \cdot \nabla u + \nabla p - \mu \Delta u &= \rho f \quad \text{in } (0, T) \times \Omega, \\ \operatorname{div} u &= 0 \quad \text{in } (0, T) \times \Omega. \end{cases} \tag{5}$$

Again these equations need to be supplemented by suitable boundary and initial conditions. Equations (5) are the so-called *density-dependent incompressible Navier-Stokes equations* or *inhomogeneous incompressible Navier-Stokes equations*.

A particular relevant case consists in choosing $\rho \equiv \bar{\rho}$ for some constant $\bar{\rho} > 0$. Without loss of mathematical generality, we may always assume that $\bar{\rho} \equiv 1$. In this case we say the fluid is *homogeneous* and the equations

$$\begin{cases} u_t + u \cdot \nabla u + \nabla p - \mu \Delta u &= f \quad \text{in } (0, T) \times \Omega, \\ \operatorname{div} u &= 0 \quad \text{in } (0, T) \times \Omega, \end{cases} \tag{6}$$

are called *homogeneous incompressible Navier-Stokes equations*. Many authors just speak about the Navier-Stokes equations if they mean the homogeneous incompressible Navier-Stokes equations. In this thesis we keep this convention. If we consider the more general situation (5) with variable density ρ, then we always explicitly state this. Also in many cases we omit the word incompressible. For giving an overview of known results we restrict ourselves in this introduction to the case without external forces, i.e. $f \equiv 0$.

In general, the system of equations (5) with variable density ρ has mathematically additional and in some sense different properties compared to the homogeneous Navier-Stokes equations (6). First of all, this is due to the presence of an additional equation of first order for the density ρ. Second, the density ρ occurs in the momentum equation in front of the highest derivative u_t. Therefore, the system (5) has a non-linearity in the highest order derivative, whereas in (6) the non-linearity lies in the convection term $u \cdot \nabla u$, only.

There is an interest in the inhomogeneous model (5) due to its significance for applications e.g. in the field of oceanic and atmospheric sciences. We refer to [Yia65] for applied aspects of inhomogeneous fluids; see also the introductory part of [AKM90] and the references therein.

From a purely mathematical point of view the homogenous Navier-Stokes equations have gained much more attention than their inhomogeneous counterpart. This is probably due to the fact that the homogeneous Navier-Stokes equations are simpler than the inhomogeneous equations, but still already reflect a lot of fundamental and challenging mathematical difficulties and problems. Since the pioneering works of Leray [Ler34] and Hopf [Hop51] in the 30s and 50s of the last century the homogeneous Navier-Stokes equations have gained a particular importance within the modern mathematical theory of partial differential equations and many new ideas, concepts and theories in this field originated from the study of (6).

In the analysis of the Navier-Stokes equations different notions of solutions, e.g. weak solutions, very weak solutions, mild solutions and strong (or regular) solutions have been introduced. See e.g. the introduction in [Ama00] for a brief overview of some of these concepts. Weak solutions which exist globally in time were first constructed by Leray [Ler34] in 1934. However, since then the fundamental question, if for $d \geq 3$ and for smooth initial data such solutions are regular or smooth for all times, is still open. Mild and strong solutions can be constructed for small time intervals, or globally in time if the data is sufficiently small (see e.g. [FK64, FJR72, Sol77, Hey80, Kat84, CM95, Ama00]). Such solutions can even be shown to be unique, however it is unknown whether such solutions exist for all times. Closely related to this problem is the question whether weak solutions are unique. These fundamental open problems in the context of existence and regularity theory for the Navier-Stokes equations have recently gained even more attention since in 2000 the *Clay Mathematics Institute* in Cambridge, Massachusetts named this problem as one of the seven "Millennium Prize Problems"; see the problem description by Fefferman [Fef00]. For a comprehensive overview on the mathematical theory of the Navier-Stokes equations we refer e.g. to the monographs by Ladyzhenskaya [Lad69], Galdi [Gal94a, Gal94b], Sohr [Soh01] and Temam [Tem01].

The starting point for the analysis of the inhomogeneous Navier-Stokes equations was about 30 years after the work of Leray on the homogeneous Navier-Stokes equations. In the 60s and 70s of the last century researchers from the Russian school, in particular Kazhikhov, studied weak solutions of (5). For details we refer to the monograph [AKM90] and the references therein. Weak solutions were also studied by Lions, and there is a whole chapter in his book devoted to this problem, see [Lio96, Chapter 2]. The question of finding unique (local) strong solutions to (5) for sufficiently smooth data was first solved by Ladyzhenskaya and Solonnikov in [LS75] and then also by Okamoto [Oka84]. More recently, strong solutions were also studied in a series of papers by Danchin, see [Dan03, Dan04, Dan06]. We also refer to [BRMFC03, CK03, CK04] for related results in the context of strong solutions. Clearly, the fundamental open problems for the homogeneous Navier-Stokes equations (global existence of strong solutions or uniqueness of weak solutions) are also open for the inhomogeneous Navier-Stokes equations.

The semigroup approach to the Navier-Stokes equations

One possible way to study the Navier-Stokes equations, which is also favored in most parts of this thesis, is to use the theory of abstract evolution equations (semigroup methods). The theory of abstract evolution equations is briefly summarized in Section 1.2 to the extent needed in this thesis.

In order to describe the semigroup approach to the Navier-Stokes equations, we shall consider (6) for $\mu \equiv 1$ and $f \equiv 0$. Moreover, we assume that the equations are complemented by the following boundary and initial conditions:

$$\begin{cases} u = 0 & \text{on } (0,T) \times \partial\Omega, \\ u|_{t=0} = u_0 & \text{in } \Omega. \end{cases} \tag{7}$$

If the domain Ω is unbounded and if it is not stated otherwise, we shall always assume that the solution u to (6) satisfies

$$\lim_{|x|\to\infty} u(t,x) = 0 \qquad \text{for } t \in (0,T).$$

This condition is always hidden in finding solutions in an L^p-scale. The starting point for the semigroup approach is to consider the linearized system, consisting of the so-called *Stokes equations*, which are obtained from (6) by dropping the non-linear convection term $u \cdot \nabla u$:

$$\begin{cases} u_t - \Delta u + \nabla p = 0 & \text{in } (0,T) \times \Omega, \\ \operatorname{div} u = 0 & \text{in } (0,T) \times \Omega, \\ u = 0 & \text{on } (0,T) \times \partial\Omega, \\ u|_{t=0} = u_0 & \text{in } \Omega. \end{cases} \tag{8}$$

The basic idea of the semigroup approach is to consider the partial differential equations (8) as an ordinary differential equation on an infinite-dimensional Banach space. For this it is necessary to work in a suitable functional analytic setting: We introduce the function space $L^p_\sigma(\Omega)$, $1 < p < \infty$, of *solenoidal* (divergence-free) vector fields in $L^p(\Omega)^d$ by setting

$$L^p_\sigma(\Omega) := \overline{C^\infty_{c,\sigma}(\Omega)}^{\|\cdot\|_{L^p(\Omega)}}, \qquad C^\infty_{c,\sigma}(\Omega) := \{u \in C^\infty_c(\Omega)^d : \operatorname{div} u = 0\}.$$

In the following, we shall assume that the domain Ω admits a *Helmholtz decomposition* of $L^p(\Omega)^d$ for all $1 < p < \infty$. By this we mean that $L^p(\Omega)^d$ can be decomposed into the direct sum of $L^p_\sigma(\Omega)$ and the subspace of $L^p(\Omega)^d$ containing all gradient fields. Then there exists a linear projection

$$\mathbb{P}_{\Omega,p} : L^p(\Omega)^d \to L^p_\sigma(\Omega),$$

called the *Helmholtz projection*, which in particular maps gradient fields to zero. For $p = 2$ the validity of the Helmholtz decomposition and the existence of the Helmholtz projection follow from Hilbert space techniques, whereas for $p \neq 2$ the Helmholtz decomposition is known to hold only for special classes of domains. All necessary details concerning the Helmholtz decomposition are presented in Section 1.3.

By formally applying the projection $\mathbb{P}_{\Omega,p}$ to (8) the pressure ∇p gets eliminated and the Stokes equations may be rewritten as an abstract linear evolution equation on the Banach space $L^p_\sigma(\Omega)$:

$$\begin{cases} u'(t) - A_{\Omega,p}u(t) = 0, & t > 0, \\ u(0) = u_0. \end{cases} \tag{9}$$

Here $A_{\Omega,p}$ is the so-called *Stokes operator* on $L^p_\sigma(\Omega)$ which is defined by

$$A_{\Omega,p}u := \mathbb{P}_{\Omega,p}\Delta u.$$

The definition of the Stokes operator provides the starting point for employing operator theoretic arguments in the study of the Navier- Stokes equations. It is well-known, see e.g. the results in [Sol77, McC81, Gig81, Uka87, MS97, GHHS10], that if the domain Ω is "nice[2]", then the Stokes operator $A_{\Omega,p}$ equipped with a suitable domain $\mathcal{D}(A_{\Omega,p})$ generates an analytic semigroup $\{T_{\Omega,p}(t)\}_{t\geq 0}$ on $L^p_\sigma(\Omega)$, $1 < p < \infty$, i.e. the solution to the abstract evolution equation (9) is given by $u(t) := T_{\Omega,p}(t)u_0$ for all initial values $u_0 \in L^p_\sigma(\Omega)$. The semigroup $\{T_{\Omega,p}(t)\}_{t\geq 0}$ is called *Stokes semigroup*. In Section 1.4 we introduce the Stokes operator in detail and we discuss properties of the Stokes semigroup.

Formally by applying the Helmholtz projection $\mathbb{P}_{\Omega,p}$ to (6)–(7) and by using the variation of constants formula the differential equation (6)–(7) can be rephrased as an integral equation:

$$u(t) = T_{\Omega,p}(t)u_0 - \int_0^t T_{\Omega,p}(t - s)\mathbb{P}_{\Omega,p}(u \cdot \nabla u)(s)\,\mathrm{d}s, \qquad t \in (0, T). \tag{10}$$

A function $u : (0, T) \to L^p_\sigma(\Omega)$ which satisfies the integral equation (10) is called *mild solution* to (6) – (7). By applying a suitable fixed point argument to (10) it is possible to construct local mild solutions to (6)–(7) or even global mild solutions for small data u_0. We refer to Fujita and Kato [FK64] for such an approach in the L^2-framework and to Kato [Kat84] and Giga [Gig86] for the L^p-case.

L^p-L^q-estimates for the Stokes semigroup

A crucial ingredient for the approaches of Kato and Giga to construct mild solutions is that the Stokes semigroup $\{T_{\Omega,p}(t)\}_{t\geq 0}$ on $L^p_\sigma(\Omega)$ is not only acting on $L^p_\sigma(\Omega)$ but also maps $L^p_\sigma(\Omega)$ into $L^q_\sigma(\Omega)$ for $q > p$. More precisely, for $1 < p \leq q < \infty$ one needs the following L^p-L^q-estimates for the Stokes semigroup on $L^p_\sigma(\Omega)$:

$$\|T_{\Omega,p}(t)f\|_{L^q(\Omega)} \leq Ct^{-\frac{d}{2}\left(\frac{1}{p}-\frac{1}{q}\right)}\|f\|_{L^p(\Omega)}, \qquad 0 < t < T, \quad f \in L^p_\sigma(\Omega). \tag{11}$$

For $T < \infty$ these estimates express a local behaviour of the Stokes semigroup at $t = 0$, and therefore they are referred to as local estimates. The validity of such local estimates basically follow from abstract properties of analytic semigroups and by using simple interpolation techniques. This allows the construction of local mild solutions to (6)–(7).

However, if one is interested in global mild solutions for small initial data, then such estimates need to hold for $T = \infty$ (in this case, we speak about global estimates). This means that one needs information about the asymptotic behaviour, i.e. the bahaviour as $t \to \infty$, of the Stokes semigroup. In particular for $p = q$ the estimate in (11) states that the Stokes semigroup $\{T_{\Omega,p}(t)\}_{t\geq 0}$ on $L^p_\sigma(\Omega)$ is uniformly bounded. For $p = 2$, the boundedness of the Stokes semigroup on arbitrary domains $\Omega \subset \mathbb{R}^d$ is rather "easy" to show by using abstract semigroup tools in the Hilbert space framework. We shall present the L^2-theory for the Stokes operator and the Stokes semigroup in Section 1.4.1 by making use of the form method which was mainly developed by Kato (see his monograph [Kat95, Chapter 6]). However, these techniques only work in the Hilbert space setting. For $p \neq 2$ the question whether the Stokes semigroup is uniformly bounded and whether the L^p-L^q-estimates (11) hold for $T = \infty$ is much more difficult and can be answered so far for special classes of domains only.

---

[2]Here one can think e.g. of $\Omega = \mathbb{R}^d$, $\Omega = \mathbb{R}^d_+$ or $\Omega \subset \mathbb{R}^d$ with a compact and smooth boundary.

In the following, fix $T = \infty$. For $\Omega = \mathbb{R}^d$ the estimates in (11) can be shown by using the explicit representation of the Stokes semigroup via the classical heat (also called Gaussian) kernel. Moreover, if $\Omega \subset \mathbb{R}^d$ is bounded, then one can show that the Stokes operator $A_{\Omega,p}$ is invertible which, by applying semigroup theoretic arguments, immediately yields that the estimates in (11) hold. However, for an exterior domain $\Omega \subset \mathbb{R}^d$, i.e. the exterior of a compact set $\mathcal{O} := \mathbb{R}^d \setminus \Omega$, the proof of (11) becomes much more involved. Such L^p-L^q-estimates for exterior domains were first proved by Iwashita [Iwa89] for $d \geq 3$, then later on by Maremonti and Solonnikov [MS97] for $d \geq 2$, and by Dan and Shibata [DS99a] for $d = 2$.

The works of Iwashita, Dan and Shibata are based on the so-called local energy decay approach. The basic idea of this approach is that, by the Laplace transform, the study of time decay of $\{T_{\Omega,p}(t)\}_{t \geq 0}$ is reduced to the study of solutions to the Stokes resolvent problem

$$\begin{cases} \lambda u - \Delta u + \nabla \mathrm{p} &= 0 \quad \text{in } \Omega, \\ \operatorname{div} u &= 0 \quad \text{in } \Omega, \\ u &= 0 \quad \text{on } \partial\Omega, \end{cases} \qquad (12)$$

for spectral parameters λ near the origin. This resolvent problem is then reduced via some localization (or cut-off) argument to a whole space problem and to a problem close to the boundary $\partial\Omega$. A survey on the local energy decay approach to the Stokes equations is given in [DKS98]. This approach was also used by Shibata and some of his collaborators to study related exterior domain problems, see e.g. [KS98, ES04, ES05, SS07b].

The basic strategy used in the proof of Maremonti and Solonnikov is completely different from the local energy decay approach. Instead of studying the resolvent problem (12), Maremonti and Solonnikov consider directly the non-stationary Stokes problem (8). Their starting point is the analysis of the Stokes problem in the L^2-framework. Then, by using localization, interpolation and duality arguments, they are able to prove, among other things, the estimates in (11) for all $1 < p \leq q < \infty$. Based on the L^2-theory introduced in Section 1.4.1 we also give a proof of the global estimates in (11), see Chapter 3. Although this is not a new result, we give an elementary and self-contained proof that nicely fits into the semigroup framework used in this thesis. Our basic approach follows the "philosophy" of Marmonti and Solonnikov to use L^2-theory as the starting point, but we also incorporate other ideas presented e.g. by Dan, Kobayashi, Shibata [DKS98] and Enomoto, Shibata [ES05]. We believe that our proof is easily accessible and that it highlights the basic ideas. However, note that the proof presented here is very much adapted to exterior domains and the basic approach does not directly carry over to general unbounded domains. In some sense, for our proof, but also for the proofs in [Iwa89, MS97, DS99a], it is crucial that the boundary of an exterior domain is compact. To our knowledge, the question whether the Stokes semigroup $\{T_{\Omega,p}(t)\}_{t \geq 0}$ is uniformly bounded on more general unbounded domains Ω (e.g. the domains considered in [GHHS10]), and for $p \neq 2$, is *still open*.

Viscous incompressible fluid flows around a rotating obstacle

The motion of compact obstacles or rigid bodies in a viscous fluid is a classical problem in fluid mechanics, and it is still in the focus of applied research, see the survey article by Galdi [Gal02]. In this context, as a mathematical prototype situation, it is interesting to consider the flow of viscous incompressible fluids around a *rotating* obstacle, where the rotation is *prescribed*. The rotation of the obstacle causes interesting mathematical problems and difficulties. Therefore, this problem has been attracting a lot of attention for the last 20 years, and a systematic study of this situation has been initiated. To describe the setting more precisely, let $\mathcal{O} \subset \mathbb{R}^3$ be a compact set, also referred to as the *obstacle*, having a smooth boundary. We shall assume that $\mathcal{O}$ is rotating with a constant angular velocity $\omega \in \mathbb{R}^3$. For notational simplicity let $M \in \mathbb{R}^{3 \times 3}$ be the matrix that represents the linear map $x \mapsto \omega \times x$. Then the rotation of $\mathcal{O}$ is governed by the semigroup e^{tM} and the exterior of the rotated obstacle at time $t \geq 0$ is given by

$$\Omega(t) := \{y(t) = e^{tM}x : x \in \Omega := \mathbb{R}^3 \setminus \mathcal{O}\}.$$

The motion of an homogeneous incompressible viscous fluid around the rotating obstacle $\mathcal{O}$ is described by the Navier-Stokes equations on the time-dependent domain $\Omega(t)$:

$$\begin{cases} v_t - \Delta v + v \cdot \nabla v + \nabla q &= 0 \quad \text{for } t \in (0,T), \, y \in \Omega(t), \\ \operatorname{div} v &= 0 \quad \text{for } t \in (0,T), \, y \in \Omega(t), \\ v &= My \quad \text{for } t \in (0,T), \, y \in \partial\Omega(t), \\ v|_{t=0} &= u_0 \quad \text{in } \Omega. \end{cases} \tag{13}$$

Here v and q denote the unknown velocity field and pressure of the fluid, respectively. The boundary condition $v|_{\partial\Omega} = Mx$ is the no-slip boundary condition, meaning that the velocity of the fluid at the boundary of the obstacle is equal to the velocity of the rotating obstacle.

To make the classical methods from the theory of the Navier-Stokes equations applicable to problem (13), the first step is to transform the system to the fixed reference domain $\Omega := \Omega(0)$. This is done by considering the equations in a rotating frame attached to the rotating obstacle $\mathcal{O}$. The transformed equations read as follows:

$$\begin{cases} u_t - \Delta u - Mx \cdot \nabla u + Mu + u \cdot \nabla u + \nabla p &= 0 \quad \text{in } (0,T) \times \Omega, \\ \operatorname{div} u &= 0 \quad \text{in } (0,T) \times \Omega, \\ u &= Mx \quad \text{on } (0,T) \times \partial\Omega, \\ u|_{t=0} &= u_0 \quad \text{in } \Omega. \end{cases} \tag{14}$$

Here u and p are the transformed velocity field and pressure of the fluid, respectively. The price to pay for the transformation is that the new equations contain two new linear terms: $Mx \cdot \nabla u$ and Mu. The coefficients of the drift term $Mx \cdot \nabla u$ grow linearly in the space variable x, and since Ω is unbounded, they are unbounded over ω. This is the particularity of the system (14), and this unbounded term leads to a number of technical difficulties.

The starting point in the analysis of (14) is the work of Borchers [Bor92], where weak solutions in $L^2_\sigma(\Omega)$ are constructed. In the case $\Omega = \mathbb{R}^d$ weak solutions and their decay properties were studied by Chen and Miyakawa [CM97] by using an explicit solution formula. Similarly to the classical Stokes operator, one can also consider a suitable linear operator which is relevant for the analysis of the modified Navier-Stokes equations (14). This operator on $L^p_\sigma(\Omega)$, $1 < p < \infty$, is given by

$$L_{\Omega,p}u := \mathbb{P}_{\Omega,p}\left(\Delta u + Mx \cdot \nabla u - Mu\right), \tag{15}$$

equipped with a suitable domain $\mathcal{D}(L_{\Omega,p})$. As mentioned above, the coefficients of the first order term $Mx \cdot \nabla u$ are unbounded. Therefore, this term cannot be considered as a "small" perturbation of the classical Stokes operator $A_{\Omega,p}$ and the generation of a semigroup by $L_{\Omega,p}$ does not follow from classical perturbation theory.

The operator $L_{\Omega,2}$ was first considered by Hishida [His99a] in $L_\sigma^2(\Omega)$, who showed that it generates a strongly continuous semigroup, which is however *not analytic*; see also [His99b, His00]. Nevertheless, he was able to show certain regularity properties of the semigroup which allowed him to prove local existence of a mild solution to (14) using a variant of the iteration scheme introduced by Fujita and Kato [FK64]. The fact that the semigroup is not analytic is caused by the unbounded coefficients. In fact, in [FN07] Farwig and Neustupa show that the essential spectrum of $L_{\Omega,2}$ consists of equally spaced half lines in the left half-plane parallel to the negative real-axis. This immediately yields that $L_{\Omega,2}$ cannot be the generator of an analytic semigroup. The results of Hishida were later extended to the L^p-setting by Hieber and Sawada [HS05] in the whole space case $\Omega = \mathbb{R}^d$ and by Geissert, Heck and Hieber [GHH06a] for exterior domains. They showed that $L_{\Omega,p}$ generates a strongly continuous semigroup on $L_\sigma^p(\Omega)$, $1 < p < \infty$, and based on a variant of Kato's iteration method (see [Kat84]) they also showed local existence of a mild solution in $L_\sigma^p(\Omega)$ for $p \geq 3$. Later, Hishida and Shibata [HS09] even derived global L^p-L^q-estimates for the semigroup generated by $L_{\Omega,p}$. Thus, they generalized the result of Iwashita for the Stokes operator $A_{\Omega,p}$ to the modified operator $L_{\Omega,p}$, also making use of the local energy decay approach. Based on these global estimates, Hishida and Shibata proved that (14) admits a mild solution in the L^3-setting which exists globally in time and converges to a steady state solution if the inital value u_0 and the angular velocity ω are sufficiently small.

Strong solutions to (14) were also obtained Galdi and Silvestre in [GS05]. They use a Faedo-Galerkin approximation and suitable a priori estimates to construct a local strong solution to (14) in the L^2-framework. They even show that the solution exists globally in time and converges to a steady state solution if the data are sufficiently small.

The stationary problem obtained from (14) by omitting the time-derivative u_t was studied by Galdi [Gal03] and Farwig and Hishida [FH07].

A different approach to study fluid flows around rotating obstacles which is based on a non-linear coordinate transformation was used e.g. in [DGH09].

Time-dependent angular and translational velocities and outflow conditions

Certainly, the situation described in (13) and (14) is rather simple in the sense that the obstacle $\mathcal{O}$ is "only" rotating with a constant angular velocity. It is reasonable to assume that $\mathcal{O}$ is in addition also translating with a prescribed constant velocity $-v_\infty$. Or equivalently, we supplement (13) by an outflow condition for the fluid, i.e. a prescribed velocity field for the fluid "far away" from the obstacle. Mathematically such a condition is given by

$$\lim_{|y| \to \infty} v(t,y) = v_\infty \neq 0 \qquad \text{in } (0,T), \qquad (16)$$

where $v_\infty \in \mathbb{R}^3$ is a given vector.

In both cases, suitable coordinate transformations yield the following new system:

$$\left\{ \begin{array}{rcll} u_t - \Delta u - Mx \cdot \nabla u + Mu & & & \\ \left. + \, \mathrm{e}^{-tM} v_\infty \cdot \nabla u + u \cdot \nabla u + \nabla p \right\} & = & 0 & \text{in } (0,T) \times \Omega, \\ \mathrm{div}\, u & = & 0 & \text{in } (0,T) \times \Omega, \\ u & = & Mx - \mathrm{e}^{-tM} v_\infty & \text{on } (0,T) \times \partial\Omega, \\ \lim_{|x| \to \infty} u(t,x) & = & 0 & \text{in } (0,T), \\ u|_{t=0} & = & u_0 & \text{in } \Omega. \end{array} \right. \tag{17}$$

Thus, a translation by $-v_\infty$, or equivalently, an outflow condition of the form (16), cause an additional linear term of the form $\mathrm{e}^{-tM} v_\infty \cdot \nabla u$. Note that the coefficients of this term depend in general on time t. The system of equations (17) was studied by Shibata, see [Shi08, Shi10], in the special case that v_∞ is parallel to the axis of rotation ω. This ensures that $\mathrm{e}^{-tM} v_\infty = v_\infty$. Thus, in this special case the coefficients are independent of time t and it is possible to base the analysis of (17) on the linear operator introduced in (15) perturbed by the lower order term $-v_\infty \cdot \nabla u$. In this sense, the results in [Shi08, Shi10] generalize the results from [HS09] where $v_\infty = 0$ was considered. A linear stationary problem related to (17) for $\Omega = \mathbb{R}^d$ was studied by Farwig [Far06] also assuming that v_∞ is parallel to ω, see also Galdi and Kyed [GK11].

It seems reasonable to remove the assumption that v_∞ is parallel to ω. As explained above, this involves treating the time-dependent term $\mathrm{e}^{-tM} v_\infty$. At the same time it is also physically reasonable to allow time-dependent angular velocities $\omega(t)$ and even a time-dependent translational velocity $-v_\infty(t)$, or equivalently a time-dependent outflow velocity $v_\infty(t)$. Denote by $\{Q(t)\}_{t \geq 0}$ the family of solution matrices of the linear ordinary differential equation

$$\left\{ \begin{array}{rcll} \partial_t Q(t)x & = & \omega(t) \times Q(t)x, & t \geq 0, \\ Q(0)x & = & x \in \mathbb{R}^d, \end{array} \right.$$

let $M(t) \in \mathbb{R}^{3 \times 3}$ be the matrix associated to the linear map $x \mapsto Q(t)^\top (\omega(t) \times x) Q(t)$ and set $c(t) := -Q(t)^\top v_\infty(t)$. Then the mathematically relevant equations on the reference domain $\Omega := \mathbb{R}^d \setminus \mathcal{O}$ are of the following form:

$$\left\{ \begin{array}{rcll} u_t - \Delta u - M(t)x \cdot \nabla u + M(t)u & & & \\ \left. - \, c(t) \cdot \nabla u + u \cdot \nabla u + \nabla p \right\} & = & 0 & \text{in } (0,T) \times \Omega, \\ \mathrm{div}\, u & = & 0 & \text{in } (0,T) \times \Omega, \\ u & = & M(t)x + c(t) & \text{on } (0,T) \times \partial\Omega, \\ u|_{t=0} & = & u_0 & \text{in } \Omega. \end{array} \right. \tag{18}$$

Such a *time-dependent* problem was first studied by Borchers [Bor92] in the framework of weak solutions. Local mild solutions to (18) in $L^2_\sigma(\Omega)$ were obtained by Hishida [His01]. However, Hishida assumes that $\omega(t) := (0,0,\omega_0(t))$ with $\omega_0(t) > 0$ and $v_\infty(t) \equiv 0$. In Hishida's setting the axis of rotation is fixed and the obstacle is not translating (or equivalently there is no outflow condition). So, Hishida does not consider the problem in full generality, and in fact, for the methods he used it is necessary to impose that the axis of rotation is fixed. Allowing a variable axis of rotation leads to additional mathematical difficulties. Time-periodic weak and strong solutions to (18), in the case that $M(t)$ and $c(t)$ are time-periodic with the same period, were obtained by Galdi and Silvestre [GS06] by using the Faedo-Galerkin method in the L^2-setting.

In Chapter 4 we shall consider problem (18) in full generality and we shall prove local existence of mild solutions for initial values $u_0 \in L^p_\sigma(\Omega)$, $p \geq 3$. The mathematical methods used here are not restricted to the case $d = 3$, but actually our approach works for arbitrary dimensions $d \geq 2$. Therefore, in Chapter 4 we actually consider (18) in $\Omega \subset \mathbb{R}^d$ for $d \geq 2$, where $M : (0, T) \to \mathbb{R}^{d \times d}$ and $c : (0, T) \to \mathbb{R}^d$ are prescribed.

A family of non-autonomous Ornstein-Uhlenbeck type operators

The approach to (18) used in Chapter 4 is based on the family of non-autonomous (time-dependent) operators

$$L_{\Omega,p}(t)u := \mathbb{P}_{\Omega,p}\left(\Delta u + M(t)x \cdot \nabla u + c(t) \cdot \nabla u - M(t)u\right), \qquad t \in (0, \infty), \quad x \in \Omega. \quad (19)$$

In unbounded domains, such operators are said to be of *Ornstein-Uhlenbeck type*, as they have linearly growing drift coefficients. We consider the following non-autonomous abstract linear evolution equation on $L^p_\sigma(\Omega)$:

$$\begin{cases} u'(t) - L_{\Omega,p}(t)u(t) &= 0, \quad t > s \geq 0, \\ u(s) &= u_0. \end{cases} \quad (20)$$

The difficulty in the analysis of (20) is again that the coefficients of the term $M(t)x \cdot \nabla u$ are unbounded over unbounded domains.

In Sections 4.1–4.3 we develop the L^p-theory for the family of non-autonomous operators introduced in (19) and we discuss well-posedness of (20) in the case $\Omega = \mathbb{R}^d$ and in the cases where $\Omega \subset \mathbb{R}^d$ is a bounded or exterior domain. Clearly, the (physically) interesting case is $\Omega \subset \mathbb{R}^d$ being an exterior domain, but mathematically $\Omega = \mathbb{R}^d$ is also interesting. This is due to the fact that in $\Omega = \mathbb{R}^d$ we already face the basic difficulties caused by the term $M(t)x \cdot \nabla u$. The case of bounded domains is "easy" and it is just included as we need bounded domain results in order to treat exterior domains.

The two main results, Theorem 4.6 and Theorem 4.19, state that for $\Omega = \mathbb{R}^d$ and for exterior domains $\Omega \subset \mathbb{R}^d$, the unique solution to (9) is given by $u(t) = S_\Omega(t, s)u_0$, where $\{S_{\Omega,p}(t, s)\}_{0 \leq s \leq t}$ is a strongly continuous evolution system on $L^p_\sigma(\Omega)$, $1 < p < \infty$. Moreover, it is shown that $\{S_{\Omega,p}(t, s)\}_{0 \leq s \leq t}$ enjoys certain regularity properties and we derive norm estimates for $\{S_{\Omega,p}(t, s)\}_{0 \leq s \leq t}$, see Theorem 4.9 for the case $\Omega = \mathbb{R}^d$ and Theorem 4.22 for exterior domains. In particular, we prove the following (local) L^p-L^q-estimates:

$$\|S_{\Omega,p}(t, s)f\|_{L^q(\Omega)} \leq C(t - s)^{-\frac{d}{2}\left(\frac{1}{p} - \frac{1}{q}\right)}\|f\|_{L^p(\Omega)}, \quad 0 < s < t < T < \infty, \quad f \in L^p_\sigma(\Omega). \quad (21)$$

In fact, for $\Omega = \mathbb{R}^d$ and under reasonable assumptions on $M : (0, \infty) \to \mathbb{R}^{d \times d}$ we can even show that (21) holds for $T = \infty$. Unfortunately, for exterior domains $\Omega \subset \mathbb{R}^d$ it is still unknown whether such estimates hold for $T = \infty$.

The starting point for our investigations in Chapter 4 is the whole space $\Omega = \mathbb{R}^d$, where we can use arguments from the Ornstein-Uhlenbeck theory, see e.g. [DPL07] and [GL08]. In particular it is possible to derive an explicit solution formula for the evolution system $\{S_{\mathbb{R}^d,p}(t,s)\}_{0 \leq s \leq t}$, see Section 4.1.1. Then by a suitable localization technique it is also possible to treat exterior domains in Section 4.3.

Based on the linear results we then construct a unique local mild solution to the non-linear problem (18) for exterior domains $\Omega \subset \mathbb{R}^d$ and arbitrary initial data $u_0 \in L^p_\sigma(\Omega)$, $p \geq d$, see Theorem 4.25. Since for exterior domains $\Omega \subset \mathbb{R}^d$ the estimates in (21) only hold for $T < \infty$, we cannot expect to obtain global mild solutions. However, for an arbitrarily fixed $T_0 < \infty$ we can show existence of a mild solution on $[0, T_0]$ if the given data are sufficiently small (depending on T_0), see Theorem 4.28. If we consider a problem of the form (18) in the whole space $\Omega = \mathbb{R}^d$, then we can even show existence and uniqueness of a global mild solution for sufficiently small initial data $u_0 \in L^d_\sigma(\mathbb{R}^d)$, see Theorem 4.29.

Note that the results in Chapter 4 generalize the results of Hieber and Sawada [HS05] and Geissert, Heck and Hieber [GHH06a] to the non-autonomous case. Some of the results in Chapter 4 have been already published in the papers [Han11], [GH11] jointly with M. Geissert and [HR10] jointly with A. Rhandi. For related results on scalar-valued Ornstein-Uhlenbeck equations see [HR11], where similar techniques are used.

Fluids with variable density in the exterior of a rotating obstacle

Another possible generalization of the situation described in (13) and (14) is to consider a more "complex" fluid model. As it was outlined above, there has been a great activity in the field of density-dependent Navier-Stokes equations. However, no results in this direction for fluids with a variable density in the exterior of a rotating obstacle seem to be known. In Chapter 5 we shall study this situation. Again let $\mathcal{O} \subset \mathbb{R}^3$ be a compact obstacle having a smooth boundary and we assume that $\mathcal{O}$ is rotating with a constant angular velocity $\omega \in \mathbb{R}^3$. Again $M \in \mathbb{R}^{3 \times 3}$ represents the linear map $x \mapsto \omega \times x$. As explained above, it is reasonable to consider the relevant equations for the fluid flow around a rotating obstacle $\mathcal{O}$ in a frame attached to the obstacle. In the case of incompressible fluids with a variable density the equations on the reference domain $\Omega := \mathbb{R}^d \setminus \mathcal{O}$ read as follows:

$$\begin{cases} \rho_t + u \cdot \nabla\rho - Mx \cdot \nabla\rho &= 0 & \text{in } (0,T) \times \Omega, \\ \rho u_t + \rho u \cdot \nabla u - \rho Mx \cdot \nabla u + \rho Mu + \nabla\mathrm{p} - \Delta u &= 0 & \text{in } (0,T) \times \Omega, \\ \operatorname{div} u &= 0 & \text{in } (0,T) \times \Omega, \\ u &= Mx & \text{on } (0,T) \times \partial\Omega, \\ (\rho, u)|_{t=0} &= (\rho_0, u_0) & \text{in } \Omega. \end{cases} \quad (22)$$

Here u and p are the unknown velocity field and pressure of the fluid, respectively. Moreover, ρ is the unknown density of the fluid. Furthermore, $u_0 : \Omega \to \mathbb{R}^3$ is the prescribed initial velocity field and $\rho_0 : \Omega \to \mathbb{R}$ is the initial density. The main result of Chapter 5, see Theorem 5.4, states that if the initial data (ρ_0, u_0) is taken from an appropriate function class, then (22) admits a unique local strong solution in the L^2-setting. Note that this result includes in particular the case $\rho_0 \equiv 1$ which was considered by Galdi and Silvestre in [GS05].

Recently, Fang, Hieber and Zhang [FHZ10] studied density-dependent incompressible viscous fluid flows in $\mathbb{R}^d$ subject to linearly growing initial data. Under a suitable transformation they obtain a new system of equations which is very similar to (22) in the sense that it also contains terms of the form $\rho Mx \cdot \nabla u$ and $Mx \cdot \nabla \rho$. After suitable assumptions on the data they prove local existence and uniqueness of a strong solution to these new equations and with that, they generalize the results of Danchin [Dan04] to linearly growing data. However, in order to prove their result, Fang, Hieber and Zhang use techniques from Fourier analysis. These techniques are only applicable in the whole space situation. Therefore, to treat the case of exterior domains $\Omega \subset \mathbb{R}^d$ we need completely different methods. Inspired by the works of Galdi and Silvestre on the homogeneous Navier-Stokes flow in the exterior of a rotating obstacle, see [GS02, GS05], and by work on inhomogeneous fluids in bounded domains, see [Kim87, BRMFC03, CK03, CK04], we shall use an approach based on the Faedo-Galerkin method in the L^2-framework. The crucial point of this approach is to obtain *a priori* estimates for (ρ, u), or more precisely for the constructed sequence of approximating solutions, in suitable (higher order) norms. Here the terms $\rho Mx \cdot \nabla u$ and $Mx \cdot \nabla \rho$ cause several technical difficulties. In order to treat these "bad" terms we use some of the arguments of Galdi and Silvestre in [GS05], but it is also necessary to develop new arguments since the presence of the additional equation for the density causes additional difficulties which were not faced in [GS05]. In particular, in order to control the gradient of the density ρ in the L^2-norm it is necessary to control the velocity field u in $W^{2,6}(\Omega)^3$. Thus, it is not possible to work in the L^2-context only, but we also need to derive estimates in the L^p-framework. The main tool to obtain the estimate in $W^{2,6}(\Omega)^3$ is to use elliptic L^p-theory of the modified Stokes system with rotating effect (see below).

The proof of Theorem 5.4 involves a fair amount of technicalities and therefore in Section 5.2 we first give an overview of the basic steps and ideas without going too much into technical details. All technicalities and the rigorous analysis are then provided in Sections 5.3–5.6. The "heart" of Chapter 5 is Section 5.5, where all the necessary a priori estimates are derived.

The modified stationary Stokes system with rotating effect

As explained above, in Chapter 5 we need elliptic L^p-estimates for the modified Stokes system with rotating effect. More precisely, the relevant system for the purposes of Chapter 5 is of the form

$$\begin{cases} -\Delta u - Mx \cdot \nabla u + Mu + \nabla \mathrm{p} &= g \quad \text{in } B_R, \\ \operatorname{div} u &= 0 \quad \text{in } B_R, \\ u &= 0 \quad \text{on } \partial B_R, \end{cases} \qquad (23)$$

where $B_R \subset \mathbb{R}^d$, $d \geq 2$, denotes the ball[3] around the origin with radius $R > 1$, $M \in \mathbb{R}^{d \times d}$ is a skew-symmetric matrix and $g \in L^p(B_R)^d$ is a given vector field. We are then interested in elliptic estimates of the form

$$\|\nabla^2 u\|_{L^p(B_R)} + \|Mx \cdot \nabla u - Mu\|_{L^p(B_R)} + \|\nabla \mathrm{p}\|_{L^p(B_R)} \leq C\|g\|_{L^p(B_R)}. \qquad (24)$$

---

[3]For the purposes of Chapter 5 it is relevant that we consider the system (23) on a ball B_R. This is due to the fact the exterior domain Ω is approximated by a sequence of bounded domains $\Omega_R := \Omega \cap B_R$. Therefore it is necessary to have estimates on balls B_R that do not depend on the actual value of R.

Since we consider a bounded domain B_R, such estimates actually follow directly from the classical elliptic Stokes theory on bounded domains. However, for our purposes the crucial point is to show that the constant $C > 0$ is *independent* of the actual value of $R > 1$. This is by far nontrivial as (23) contains the term $Mx \cdot \nabla u$, where the size of the coefficients actually grows with R. Therefore, it is not possible to use classical elliptic Stokes theory on bounded domains and we need new arguments.

A stationary problem of the form (23) in the whole space case $\mathbb{R}^3$ was first studied by Farwig, Hishida and Müller [FHM04] who proved existence of a solution (u, p) and derived elliptic L^p-estimates for the solution. Their proof makes heavily use of methods from Fourier analysis. However, their result and proof cannot be directly modified to hold in a ball B_R with a constant independent of $R > 1$. Nevertheless, the validity of such elliptic estimates in the whole space gives hope that these estimates should also hold in a ball B_R with a constant independent of $R > 1$.

A simpler proof of the result of Farwig, Hishida and Müller was recently given by Galdi and Kyed [GK11]. The basic idea of their proof is to transform the stationary problem (23) in $\mathbb{R}^3$ to a classical non-stationary Stokes problem and then use maximal regularity results for the Stokes system in $\mathbb{R}^3$. Somehow, the crucial point in their proof is that the orthogonal transformation e^{tM} leaves the domain $\mathbb{R}^3$ invariant. The proof of Galdi and Kyed is more elementary than the proof in [FHM04], and in fact, it turns out that the basic ideas of this proof can also be used to consider the system (23) in a ball B_R. This is due to the fact that the orthogonal transformation e^{tM} leaves B_R invariant, too.

The elliptic estimates in (24) with a constant independent of $R > 1$ are stated in Theorem 2.12. This result is a major tool in Chapter 5 and it is also interesting in its own right as it can be considered as a nontrivial variant of the results in [FHM04] and [GK11].

1 Preliminaries

In this chapter the basic tools and concepts that will be used in the remainder of this thesis are introduced. This includes the basic notations, the basic function spaces, important results and methods from the theory of abstract Cauchy problems and specific concepts in the theory of the Navier-Stokes equations. Most of the material presented here is standard. It is included for reference, self-containedness and for providing the right framework to mathematically formulate the problems of this thesis.

1.1 Notations, function spaces and basic inequalities

In this section we introduce the basic notations and we describe the basic function spaces, in particular Sobolev spaces. More specialized notations and function spaces are introduced later within the text. In particular, the relevant function spaces for hydrodynamics, function spaces of solenoidal vector fields, are introduced separately in Section 1.3.

General conventions and basic notations

Throughout this thesis we always use generic constants $C > 0$ which may change from line to line. Their dependence on parameters is expressed only if necessary. In these cases we write $C := C(\alpha, \beta, ...)$ if the constant depends on the quantities $\alpha, \beta, ...$

In principle all function spaces considered in this thesis are real-valued[1]. Therefore we shall introduce all spaces in the real-valued case only. However, for theoretical investigations, such as spectral methods, form methods, the Fourier transform and complex interpolation, we actually need complex-valued spaces. In these cases we use the usual complexification of the real-valued function spaces and we keep the same notations. As it is standard to change from real-valued function spaces to complex-valued function spaces, and vice versa, no distinction between real-valued and complex-valued spaces is made, unless it is important.

The natural numbers are denoted as usual by $\mathbb{N} := \{1, 2, ...\}$, whereas $\mathbb{N}_0 := \mathbb{N} \cup \{0\}$. By $\mathbb{Z} := \{..., -1, 0, 1, ...\}$ we denote the integers. The real and complex numbers are denoted by $\mathbb{R}$ and $\mathbb{C}$, respectively, and we write $\operatorname{Re} z$ and $\operatorname{Im} z$ for the real and imaginary part of $z \in \mathbb{C}$, respectively. By $\overline{z}$ we denote the complex conjugate of $z \in \mathbb{C}$.

---

[1]This is also reasonable, as we use the function spaces to find solutions to the Navier-Stokes equations, i.e. to find the velocity field and the pressure of a fluid.

The d-dimensional *Euclidean* space is denoted by $\mathbb{R}^d$. The scalar product on $\mathbb{R}^d$ is given by

$$x \cdot y := \sum_{i=1}^{d} x_i y_i, \qquad x = (x_1, \ldots, x_d), y = (y_1, \ldots, y_d) \in \mathbb{R}^d.$$

The Euclidian norm of a point $x \in \mathbb{R}^d$ is defined as $|x| := \sqrt{x \cdot x}$. For every $x_0 \in \mathbb{R}^d$ the open ball with radius $R > 0$ centered at x_0 is denoted by $B_R(x_0) := \{x \in \mathbb{R}^d : |x - x_0| < R\}$. To make notation shorter we simply write $B_R := B_R(0)$. The trace of a matrix $M \in \mathbb{R}^{d \times d}$ is denoted by $\operatorname{tr} M$ and the transpose of the matrix M is denoted by $M^\top$. Analogously, $x^\top$ denotes the transpose of a vector $x \in \mathbb{R}^d$. If we write MN and Mx we mean the usual product between two matrices and between a matrix and a vector, respectively. For $M, N \in \mathbb{R}^{d \times d}$ we define $M : N := \operatorname{tr}(MN^\top)$ and $|M| := \sqrt{M : M}$. If $x \in \mathbb{R}^d$ and $M \in \mathbb{R}^{d \times d}$, then by $x \cdot M$ we mean the vector $x \cdot M := (x \cdot M_{\cdot j})_{j=1}^{d}$, where $M_{\cdot j}$ denotes the j-th column of M.

Let $\emptyset \neq \Omega \subset \mathbb{R}^d$, $d \in \mathbb{N}$, be open. We consider real-valued functions $u : \Omega \to \mathbb{R}$ or real-valued vector fields $u : \Omega \to \mathbb{R}^d$, $d \geq 2$. By $\operatorname{supp} u$ we denote the support of u, that is

$$\operatorname{supp} u := \overline{\{x \in \Omega : u(x) \neq 0\}}.$$

We use the abbreviations

$$\partial_i u := \frac{\partial u}{\partial x_i}, \qquad \partial_{ij} u := \frac{\partial^2 u}{\partial x_i x_j}, \qquad \partial^\alpha u := \partial_1^{\alpha_1} \partial_2^{\alpha_2} \ldots \partial_d^{\alpha_d} u,$$

where $\alpha = (\alpha_1, \ldots, \alpha_d) \in \mathbb{N}_0^d$ is a given multi-index of length $|\alpha| := \sum_{i=1}^{d} \alpha_i$. This notation is used in the classical as well as in the distributional sense. For a function $u : \Omega \to \mathbb{R}$ the *gradient* ∇u is given by $\nabla u := (\partial_1 u, \ldots, \partial_d u)^\top$ and $\nabla^2 u$ stands for the matrix of all second order partial derivatives. If $u : \Omega \to \mathbb{R}^d$, $d \geq 2$, is vector field, then we set

$$\operatorname{div} u := \sum_{i=1}^{d} \partial_i u_i, \qquad \nabla u := (\partial_i u_j)_{i,j=1}^{d}, \qquad \nabla^2 u := (\partial_i \partial_j u_k)_{i,j,k=1}^{d}.$$

Moreover, the *Laplacian* of $u : \Omega \to \mathbb{R}$ and $u : \Omega \to \mathbb{R}^d$ is defined by

$$\Delta u := \sum_{i=1}^{d} \partial_{ii} u.$$

If u is defined on I or $I \times \Omega$, where $I \subset \mathbb{R}_+$ is an interval denoting time, then the time derivative of u is denoted by u', u_t, $\partial_t u$ or $\frac{\mathrm{d}}{\mathrm{d}t} u$.

Banach spaces and linear operators

Let $(X, \|\cdot\|_X)$ and $(Y, \|\cdot\|_Y)$ be two Banach spaces. By $Y \hookrightarrow X$ we mean that Y is continuously embedded into X. The dual pairing between X and its dual space X' is denoted by $\langle \cdot, \cdot \rangle_{X,X'}$. The space of continuous linear mappings from X to Y is denoted by $\mathscr{L}(X,Y)$ and we set $\mathscr{L}(X) := \mathscr{L}(X, X)$. The operator norm of some linear operator $T \in \mathscr{L}(X, Y)$ is denoted by $\|T\|_{\mathscr{L}(X,Y)}$. For a closed linear operator A on a Banach space X we write $\mathcal{D}(A)$ to denote its domain and $\mathcal{R}(A)$ for its range. The spectrum and the resolvent set of A are denoted by $\sigma(A)$ and $\rho(A)$, respectively. For $\lambda \in \rho(A)$, $R(\lambda, A) := (\lambda - A)^{-1}$ denotes the resolvent of the operator A at λ. A^* denotes the adjoint of A.

Spaces of continuous functions

Let $k \in \mathbb{N}_0 \cup \{\infty\}$ and $\emptyset \neq \Omega \subset \mathbb{R}^d$, $d \in \mathbb{N}$, be open. The vector space of all continuous functions on Ω with values in a Banach space X is denoted by $C(\Omega; X)$ and by $C^k(\Omega; X)$ we denote the space of all k-times continuously differentiable functions on Ω with values in X. The space of all bounded continuous functions is denoted by $C_b(\Omega; X)$. The space of α-Hölder continuous functions $C^{0,\alpha}(\Omega; X)$, where $\alpha \in (0,1]$, is defined by all $u \in C_b(\Omega; X)$ such that

$$\sup_{x,y \in \Omega, x \neq y} \frac{\|u(x) - u(y)\|_X}{|x - y|^\alpha} < \infty.$$

If $\alpha = 1$ we say that u is Lipschitz-continuous. To simplify notation, we just write $C^k(\Omega)$ instead of $C^k(\Omega; \mathbb{R})$. Moreover, since a vector field $u : \Omega \to \mathbb{R}^d$ is continuous and differentiable if and only if each of its component functions is so, we also write $C^k(\Omega)^d := C^k(\Omega; \mathbb{R}^d)$. This notational convention also applies to all function spaces defined in the following: By $C^k_c(\Omega)$ we denote the subspace of functions in $C^k(\Omega)$ that are compactly supported. Now let $k \in \mathbb{N}_0$. For $\alpha \in (0,1]$ we denote by $C^{k,\alpha}(\Omega)$ the space of all k-times continuously differentiable functions $u : \Omega \to \mathbb{R}$ such that the k-th derivatives are α-Hölder continuous. By $C^k_b(\Omega)$ we denote the space of functions having bounded, continuous derivatives up to k-th order. By $C^k(\overline{\Omega})$ we denote the subspace of $C^k_b(\Omega)$ having bounded, uniformly continuous derivatives up to k-th order on Ω. This is a convenient notation that is frequently used in the literature. However, this notation leads to ambiguities if Ω is unbounded, e.g. $C(\overline{\mathbb{R}^d}) \neq C(\mathbb{R}^d)$ although $\overline{\mathbb{R}^d} = \mathbb{R}^d$. By $C^{k,\alpha}(\overline{\Omega})$ we denote the subspace of $C^k(\overline{\Omega})$ consisting of functions whose derivatives up to order k are α-Hölder continuous. Similarly, the spaces $C^k(\overline{I}; X)$ and $C^{\alpha,k}(\overline{I}; X)$ are defined for $I \subset \mathbb{R}$ finite and X a Banach space.

Classes of domains

A *domain* $\Omega \subset \mathbb{R}^d$ is an open and connected subset. By definition a domain is always non-empty. Let Ω be a domain with a compact boundary, i.e. Ω is either a bounded domain or Ω is a domain complement in $\mathbb{R}^d$ of a compact (not necessarily connected) set. The latter is called an *exterior* domain. We say Ω is of *class C^k* if for every $x_0 \in \partial\Omega$ there exist a ball $B_R(x_0)$ for some radius $R > 0$ and a C^k-function φ defined on a domain $D \subset \mathbb{R}^{d-1}$ such that – upon relabeling and reorienting the axes of the coordinate system if necessary – the set $\partial\Omega \cap B_R(x_0)$ can be represented by an equation of the type $x_d = \varphi(x_1, \ldots, x_{d-1})$ and each $x \in \Omega \cap B_R(x_0)$ satisfies $x_d < \varphi(x_1, \ldots, x_{d-1})$. We say that Ω is *locally Lipschitz* if for every $x_0 \in \partial\Omega$ we may choose a Lipschitz continuous function φ in the foregoing construction. Note that domains of class C^k or domains that are locally Lipschitz have by our definition compact boundaries. However, to emphasise this we sometimes still explicitly state it although it is redundant. For a domain of class C^1 we denote by $\nu := (\nu_1, \ldots, \nu_d)$ the outward pointing *unit normal vector field* to $\partial\Omega$.

Lebesgue and Sobolev spaces

Let $\emptyset \neq \Omega \subset \mathbb{R}^d$, $d \in \mathbb{N}$, be open. For $1 \leq p \leq \infty$, $L^p(\Omega; X)$ denotes the *Lebesgue-Bochner space* of functions which take values in a Banach space X. We simply write $L^p(\Omega) := L^p(\Omega, \mathbb{R})$ and $L^p(\Omega)^d := L^p(\Omega, \mathbb{R}^d)$. The norms of $L^p(\Omega)$ and $L^p(\Omega)^d$ are both denoted by $\| \cdot \|_{L^p(\Omega)}$ as it should always be clear from the context if we consider $u : \Omega \to \mathbb{R}$ or $u : \Omega \to \mathbb{R}^d$.

3

We only give definitions for scalar-valued functions $u : \Omega \to \mathbb{R}$, as all definitions carry over vector fields $u : \Omega \to \mathbb{R}^d$, $d \geq 2$, by considering the vector fields component-wise. To denote function spaces of vector fields, the same convention as for L^p-functions holds. In particular, for denoting norms we do not distinguish between scalar-valued function spaces and function spaces of vector fields.

The space $L^2(\Omega)$ is a Hilbert space with respect to the scalar product

$$\langle u, v \rangle_{L^2(\Omega)} := \int_\Omega u \cdot v \, \mathrm{d}x, \qquad u, v \in L^2(\Omega).$$

If it is clear from the context we denote the L^2-scalar product simply by $\langle \cdot, \cdot \rangle_\Omega$. Similarly, the dual pairing between $L^p(\Omega)$ and $L^{p'}(\Omega)$, where $\frac{1}{p} + \frac{1}{p'} = 1$, is denoted by

$$\langle u, v \rangle_{L^p(\Omega), L^{p'}(\Omega)} := \int_\Omega u \cdot v \, \mathrm{d}x, \qquad u \in L^p(\Omega), \, v \in L^{p'}(\Omega).$$

If there is no danger of misinterpretation we simply write $\langle \cdot, \cdot \rangle_\Omega$. By $L^p_{\mathrm{loc}}(\Omega)$ we denote the space of all functions u such that $u \in L^p(\Omega')$ for all bounded Ω' with $\overline{\Omega'} \subset \Omega$. For $f \in L^p(\mathbb{R}^d)$ and $g \in L^q(\mathbb{R}^d)$ we define the convolution $f * g$ by

$$(f * g)(x) := \int_{\mathbb{R}^d} f(x - y) f(y) \, \mathrm{d}y,$$

whenever the integral above is well-defined. By $\mathcal{S}(\mathbb{R}^d)$ we mean the Schwartz space of rapidly decreasing functions. The Fourier transform on $\mathcal{S}(\mathbb{R}^d)$ is denoted by

$$\mathcal{F}u(\xi) := \widehat{u}(\xi) := \frac{1}{(2\pi)^{d/2}} \int_{\mathbb{R}^d} \mathrm{e}^{-ix \cdot \xi} u(x) \, \mathrm{d}x, \qquad u \in \mathcal{S}(\mathbb{R}^d),$$

whereas on the dual space $\mathcal{S}'(\mathbb{R}^d)$ the Fourier transform is defined by the relation

$$\langle \mathcal{F}u, \varphi \rangle_{\mathcal{S}'(\mathbb{R}^d), \mathcal{S}(\mathbb{R}^d)} := \langle u, \mathcal{F}\varphi \rangle_{\mathcal{S}'(\mathbb{R}^d), \mathcal{S}(\mathbb{R}^d)} \qquad u \in \mathcal{S}'(\mathbb{R}^d), \, \varphi \in \mathcal{S}(\mathbb{R}^d).$$

Next we come to the notion of *Sobolev spaces* which plays a central role in this thesis.

Definition 1.1 (Sobolev spaces). Let $\emptyset \neq \Omega \subset \mathbb{R}^d$, $d \in \mathbb{N}$, be open. Let $1 \leq p \leq \infty$ and $m \in \mathbb{N}_0$. Then the *Sobolev space of order m* is defined by

$$W^{m,p}(\Omega) := W^{m,p}(\Omega; \mathbb{R}) := \{u \in L^p(\Omega) : \partial^\alpha u \in L^p(\Omega) \text{ for all } |\alpha| \leq m\},$$

where $\partial^\alpha u$ has to be understood in the sense of distributions. The norm $\| \cdot \|_{W^{m,p}(\Omega)}$ is defined by

$$\|u\|_{W^{m,p}(\Omega)}^p := \sum_{|\alpha| \leq m} \|\partial^\alpha u\|_{L^p(\Omega)}^p \qquad \text{for } 1 \leq p < \infty$$

and

$$\|u\|_{W^{m,\infty}(\Omega)} := \sum_{|\alpha| \leq m} \|\partial^\alpha u\|_{L^\infty(\Omega)}.$$

Remark 1.2. Similarly, one defines the space $W^{m,p}(I; X)$, where $I \subset \mathbb{R}$ and X is a Banach space.

The closure of $C_c^\infty(\Omega)$ with respect to the norm of $W^{m,p}(\Omega)$ is denoted by $W_0^{m,p}(\Omega)$. In the case $p = 2$ we also write $H^m(\Omega) := W^{m,2}(\Omega)$ and $H_0^m(\Omega) := W_0^{m,2}(\Omega)$. The space $H^m(\Omega)$ is a Hilbert space with respect to the scalar product

$$\langle u, v \rangle_{H^m(\Omega)} := \sum_{|\alpha| \leq m} \int_\Omega \partial^\alpha u \cdot \partial^\alpha v \, dx.$$

For $m = 0$ the Sobolev spaces $W^{m,p}(\Omega)$ and $W_0^{m,p}(\Omega)$ are just the usual Lebesgue spaces $L^p(\Omega)$. By $W_{\text{loc}}^{m,p}(\Omega)$ we denote the space of all u such that $\partial^\alpha u \in L_{\text{loc}}^p(\Omega)$ for all $|\alpha| \leq m$.

Next we introduce *homogeneous Sobolev spaces* which are crucial for unbounded domains.

Definition 1.3 (Homogeneous Sobolev spaces). Let $\Omega \subset \mathbb{R}^d$, $d \in \mathbb{N}$, be a domain. For $1 \leq p < \infty$ and $m \in \mathbb{N}$ define

$$D^{m,p}(\Omega) := \{u \in L_{\text{loc}}^1(\Omega) : \partial^\alpha u \in L^p(\Omega) \text{ for } |\alpha| = m\}.$$

On $D^{m,p}(\Omega)$ introduce the semi-norm $\|\cdot\|_{\widehat{W}^{m,p}(\Omega)}$ defined by

$$\|u\|_{\widehat{W}^{m,p}(\Omega)}^p := \sum_{|\alpha|=m} \|\partial^\alpha u\|_{L^p(\Omega)}^p.$$

Then the *homogeneous Sobolev space* $\widehat{W}^{m,p}(\Omega)$ is defined as

$$\widehat{W}^{m,p}(\Omega) := D^{m,p}(\Omega)/\mathcal{P}_{m-1},$$

where $\mathcal{P}_{m-1}$ denotes the space of all polynomials of order at most $m - 1$. $\widehat{W}^{m,p}(\Omega)$ is a Banach spaces with respect to the norm $\|\cdot\|_{\widehat{W}^{m,p}(\Omega)}$.

Remarks 1.4. 1. When it is more convenient to work with functions instead equivalence classes we shall use the spaces $D^{m,p}(\Omega)$, however we have to keep in mind that these spaces arc not Banach spaces.

2. If $u \in D^{m,p}(\Omega)$, then it holds that $u \in W_{\text{loc}}^{m,p}(\Omega)$, see [Gal94a, Remark 5.1].

Next we extend Definition 1.1 to the negative scale. This is done by duality.

Definition 1.5 (Sobolev spaces of negative order). Let $\Omega \subset \mathbb{R}^d$, $d \in \mathbb{N}$, be a domain, $1 \leq p < \infty$ and $m \in \mathbb{N}$. Moreover, let p' be the dual exponent of p, i.e. $\frac{1}{p} + \frac{1}{p'} = 1$. The *Sobolev space of negative order* $-m$ is defined by

$$W^{-m,p}(\Omega) := \left(W_0^{m,p'}(\Omega) \right)'.$$

Moreover, define

$$W_0^{-m,p}(\Omega) := \left(W^{m,p'}(\Omega) \right)'.$$

We shall also make use of intermediate spaces which are defined via interpolation theory. For a detailed presentation of interpolation theory we refer to the monograph by Triebel

[Tri95]. In the following let X, Y be Banach spaces that form an interpolation couple. Then the *complex interpolation space* between X and Y is denoted by $[X, Y]_\theta$, where $\theta \in (0, 1)$.

Definition 1.6 (Bessel potential spaces, [AF03, 7.57]). Let $m \in \mathbb{N}$ and $\Omega \subset \mathbb{R}^d$ be a domain of class C^m. Moreover, let $1 < p < \infty$ and $0 < s \leq m$. The *Bessel potential space* $H^{s,p}(\Omega)$ is defined by

$$H^{s,p}(\Omega) := [L^p(\Omega), W^{m,p}(\Omega)]_{\frac{s}{m}}.$$

Bessel potential spaces of integer order coincide with the Sobolev spaces defined in Definition 1.1, i.e. $H^{s,p}(\Omega) = W^{s,p}(\Omega)$ for $s \in \mathbb{N}$.

The *real interpolation space* between X and Y is denoted by $(X, Y)_{\theta,q}$, where $\theta \in (0, 1)$ and $1 \leq q \leq \infty$.

Definition 1.7 (Besov spaces, [AF03, 7.32]). Let $m \in \mathbb{N}$ and $\Omega \subset \mathbb{R}^d$ be a domain of class C^m. Moreover, let $1 \leq p < \infty$, $1 \leq q \leq \infty$ and $0 < s < m$. The *Besov space* $B^s_{p,q}(\Omega)$ is defined by

$$B^s_{p,q}(\Omega) := (L^p(\Omega), W^{m,p}(\Omega))_{\frac{s}{m},q}.$$

For s being an integer these intermediate spaces may not coincide with the Sobolev spaces defined in Definition 1.1. Even for $\Omega = \mathbb{R}^d$ it is not true that $B^s_{p,q}(\Omega) = W^{s,p}(\Omega)$ holds, unless $p = q = 2$, see [AF03, 7.33].

Remark 1.8. Set $\theta := s/m$. Then there exists a constant $C := C(\theta, q) > 0$ such that

$$\|u\|_{B^s_{p,q}(\Omega)} \leq C \|u\|^{1-\theta}_{L^p(\Omega)} \|u\|^{\theta}_{W^{m,p}(\Omega)}.$$

See [Tri95, Section 1.3.3].

Definition 1.9 (Sobolev-Slobodeckiĭ spaces, [Tri95, 4.2.1]). Let $m \in \mathbb{N}$ and $\Omega \subset \mathbb{R}^d$ be a domain of class C^m. Moreover, let $1 \leq p < \infty$ and $0 < s < m$. The *Sobolev-Slobodeckiĭ* space $W^s_p(\Omega)$ is defined by

$$W^s_p(\Omega) := \begin{cases} W^{s,p}(\Omega) & \text{if } s \in \mathbb{N}, \\ B^s_{p,p}(\Omega) & \text{if } s \notin \mathbb{N}. \end{cases}$$

If $\Omega \subset \mathbb{R}^d$ is a bounded domain of class C^m, then by Theorem 4.3.1.2 and by Remark 4.4.1.2 in [Tri95] the norm of $W^s_p(\Omega)$ with $s \notin \mathbb{N}$ is equivalent to the norm defined by

$$\|u\|_{L^p(\Omega)} + \sum_{|\alpha|=[s]} \int_{\Omega \times \Omega} \frac{|\partial^\alpha u(x) - \partial^\alpha(y)|^p}{|x - y|^{d+(s-[s])p}} \, dx \, dy. \tag{1.1}$$

Here $[\cdot]$ denote the Gaussian brackets. The norm in (1.1) is used to introduce the spaces $W^s_p(\Omega)$ in the monograph by Grisvard [Gri85].

Trace inequalities

Having Sobolev-Slobodeckiĭ spaces at hand we can now say a few words about boundary values of functions in Sobolev spaces. In the following let $\Omega \subset \mathbb{R}^d$ be a domain of class C^1. For functions $u \in C(\overline{\Omega})$ we define a linear operator tr by setting $\mathrm{tr}(u) = u|_{\partial\Omega}$. A classical trace theorem, see e.g. [AF03, Theorem 5.36], states that the mapping tr has a unique extension as a bounded linear operator $\mathrm{tr} : W^{1,p}(\Omega) \to L^p(\partial\Omega)$. The operator tr is called *trace operator* and $\mathrm{tr}(u)$ is called *trace* of u. For the purposes of this thesis this classical trace theorem is not sufficient, but we shall make use of the following result.

Proposition 1.10 ([Gri85, Theorem 1.5.1.2]). *Let $1 < p < \infty$ and let $\Omega \subset \mathbb{R}^d$ be a domain with compact boundary that is of class C^1. Then the mapping tr has a unique extension as a bounded linear operator*

$$\mathrm{tr} : W_p^{1/p+\delta}(\Omega) \to L^p(\partial\Omega)$$

for all $0 < \delta \leq 1 - \frac{1}{p}$.

Note that in [Gri85] Proposition 1.10 is actually only stated for bounded domains Ω, however the result directly carries over to exterior domains. In Chapter 5 we frequently apply Proposition 1.10 in the following form.

Corollary 1.11. *Let $1 < p < \infty$, $\Omega \subset \mathbb{R}^d$ be a domain with compact boundary that is of class C^1 and let $u \in W^{1,p}(\Omega)$. Then for every $\varepsilon > 0$ there exists a constant $C := C(\varepsilon) > 0$ such that*

$$\|\mathrm{tr}(u)\|_{L^p(\partial\Omega)} \leq C\|u\|_{L^p(\Omega)} + \varepsilon\|\nabla u\|_{L^p(\Omega)}.$$

Proof. Let $0 < \delta < 1 - 1/p$. From Proposition 1.10 and the definition of $W_p^{1/p+\delta}(\Omega)$ as a real interpolation space, see Remark 1.8, it follows that

$$\|\mathrm{tr}(u)\|_{L^p(\partial\Omega)} \leq C\left(\|u\|_{L^p(\Omega)} + \|u\|_{L^p(\Omega)}^{1-\frac{1}{p}-\delta}\|\nabla u\|_{L^p(\Omega)}^{\frac{1}{p}+\delta}\right).$$

Applying Young's inequality, see Lemma 1.21 below, yields the assertion. $\qquad\square$

Remark 1.12. In this thesis we often write $u|_{\partial\Omega}$ to denote the trace as this notation seems more intuitive. We keep in mind that this has to be interpreted via the trace operator tr.

Sobolev embeddings and interpolation inequalities

Now let us state results on Sobolev embeddings and collect interpolation inequalities.

Proposition 1.13 (Extension operator, [Ste70, Chapter VI]). *Let $\Omega \subset \mathbb{R}^d$ be a domain with compact boundary that is locally Lipschitz. There exists an operator E that maps functions on Ω to functions on $\mathbb{R}^d$ such that*

(i) $Ef|_\Omega = f$,

(ii) $E : W^{m,p}(\Omega) \to W^{m,p}(\mathbb{R}^d)$ *is continuous for all $m \in \mathbb{N}_0$ and all $1 \leq p \leq \infty$.*

The operator E is called *(total) extension operator*.

Proposition 1.14 (Sobolev embedding, [AF03, 4.12]). *Let $\Omega \subset \mathbb{R}^d$ be the whole space $\mathbb{R}^d$ or a domain with compact boundary that is locally Lipschitz. Moreover, let $1 \leq p < \infty$, $j \in \mathbb{N}_0$ and $m \in \mathbb{N}$.*

   (a) If $mp > d$, then $W^{m,p}(\Omega) \hookrightarrow L^q(\Omega)$ for $p \leq q \leq \infty$.

   (b) If $mp = d$, then $W^{m,p}(\Omega) \hookrightarrow L^q(\Omega)$ for $p \leq q < \infty$.

   (c) If $mp < d$, then $W^{m,p}(\Omega) \hookrightarrow L^q(\Omega)$ for $p \leq q \leq p^* = dp/(d - mp)$.

   (d) If $mp > d > (m-1)p$, then $W^{j+m,p}(\Omega) \hookrightarrow C^{j,\lambda}(\overline{\Omega})$ for $0 < \lambda \leq m - \frac{d}{p}$.

   (e) If $d = (m-1)p$, then $W^{j+m,p}(\Omega) \hookrightarrow C^{j,\lambda}(\overline{\Omega})$ for $0 < \lambda < 1$.

The embedding constants for the emdeddings above depend only on d, m, p, q, j and the regularity of the boundary, but they are independent of the actual size of the domain.

Later in Chapter 5 it is very important that the embedding constants are indeed independent of the actual size of the domain.

Proposition 1.15 (Sobolev inequality, [AF03, Theorem 4.31]). *Let $\Omega \subset \mathbb{R}^d$ be a domain. Then for $1 \leq p < d$ there exists a constant $C := C(d,p) > 0$ such that*

$$\|u\|_{L^{p^*}(\Omega)} \leq C \|\nabla u\|_{L^p(\Omega)} \qquad \text{for all } u \in W_0^{1,p}(\Omega),$$

where $p^* = dp/(d - p)$.

If the domain Ω is bounded, we have the following compact embeddings.

Proposition 1.16 (Rellich, [AF03, 6.3]). *Let $\Omega \subset \mathbb{R}^d$ be a bounded domain that is locally Lipschitz. Moreover, let $1 \leq p < \infty$, $m \in \mathbb{N}$. Then the embedding $W^{m,p}(\Omega) \hookrightarrow W^{m-1,p}(\Omega)$ is compact.*

An important tool in this thesis is the Gagliardo-Nirenberg inequality which in some sense can be considered as a generalization of the inequality in Proposition 1.15. For more information we refer to [Gag59], [Nir59], [CM04] and the references therein.

Proposition 1.17 (Gagliardo-Nirenberg inequality). *Let $1 \leq r, q \leq \infty$ and let $\Omega \subset \mathbb{R}^d$ be the whole space $\mathbb{R}^d$ or a domain with compact boundary that is locally Lipschitz. Furthermore, let $j, m \in \mathbb{N}_0$ with $0 \leq j < m$. If $1 \leq p \leq \infty$ and $\frac{j}{m} \leq \theta < 1$ such that*

$$\frac{1}{p} = \frac{j}{d} + \theta \left(\frac{1}{r} - \frac{m}{d} \right) + (1 - \theta) \frac{1}{q},$$

then

$$\|\nabla^j u\|_{L^p(\Omega)} \leq C_1 \|\nabla^m u\|_{L^r(\Omega)}^{\theta} \|u\|_{L^q(\Omega)}^{1-\theta} + C_2 \|u\|_{L^q(\Omega)} \tag{1.2}$$

holds for all $u \in L^q(\Omega) \cap D^{m,r}(\Omega)$. Here the constants $C_1, C_2 > 0$ depend only on Ω, j, m, q, r. If $m - j - \frac{d}{r} \notin \mathbb{N}_0$, then (1.2) even holds true for all $\frac{j}{m} \leq \theta \leq 1$.
Moreover, (1.2) holds also with $C_2 = 0$ if $\Omega = \mathbb{R}^d$, Ω is an exterior domain, or Ω is bounded and $u \in W_0^{m,p}(\Omega)$.

Proposition 1.18 (Riesz-Thorin, [Lun09, Chapter 2]). *Let $\Omega \subset \mathbb{R}^d$ be a domain and let $1 \leq p_1, p_2, q_1, q_2 \leq \infty$. Moreover, let*

$$T : L^{p_0}(\Omega) + L^{p_1}(\Omega) \to L^{q_0}(\Omega) + L^{q_1}(\Omega)$$

be a linear operator such that

$$T \in \mathscr{L}(L^{p_0}(\Omega), L^{q_0}(\Omega)) \cap \mathscr{L}(L^{p_1}(\Omega), L^{q_1}(\Omega)).$$

Then $T \in \mathscr{L}(L^{p_\theta}(\Omega), L^{q_\theta}(\Omega))$, $0 < \theta < 1$, with

$$\frac{1}{p_\theta} = \frac{1-\theta}{p_0} + \frac{\theta}{p_1}, \qquad \frac{1}{q_\theta} = \frac{1-\theta}{q_0} + \frac{\theta}{q_1}$$

and

$$\|T\|_{\mathscr{L}(L^{p_\theta}(\Omega), L^{q_\theta}(\Omega))} \leq \|T\|_{\mathscr{L}(L^{p_0}(\Omega), L^{q_0}(\Omega))}^{1-\theta} \|T\|_{\mathscr{L}(L^{p_1}(\Omega), L^{q_1}(\Omega))}^{\theta}.$$

Another very useful estimate is given by the Poincaré inequality. The following formulation of the Poincaré inequality is classical.

Proposition 1.19 (Poincaré inequality). *Let $\Omega \subset \mathbb{R}^d$ be a bounded domain that is locally Lipschitz and let $1 < p < \infty$. There exists a constant $C := C(\Omega) > 0$ such that*

$$\|u\|_{L^p(\Omega)} \leq C \left(\|\nabla u\|_{L^p(\Omega)} + \left| \int_\Omega u \, dx \right| \right)$$

for all $u \in W^{1,p}(\Omega)$ and

$$\|u\|_{L^p(\Omega)} \leq C \|\nabla u\|_{L^p(\Omega)}$$

for all $u \in W_0^{1,p}(\Omega)$.

In this thesis we frequently apply the Poincaré inequality in the following special situation which does not directly fit into the setting of Proposition 1.19.

Lemma 1.20. *Let $\Omega \subset \mathbb{R}^d$ be an exterior domain of class C^1 and let $R > 0$ be sufficiently large such that $\mathcal{O} := \mathbb{R}^d \setminus \Omega \subset B_R$. Moreover, let $1 < p < \infty$ and let $u \in W_0^{1,p}(\Omega)$. Then there exists a constant $C := C(\Omega, R) > 0$ such that*

$$\|u\|_{L^p(\Omega \cap B_R)} \leq C \|\nabla u\|_{L^p(\Omega \cap B_R)}.$$

Lemma 1.20 is a special case of [HDR10, Theorem 3.5] which essentially follows from [Zie89, Theorem 4.8.1].

Other useful inequalities

To conclude this section we collect some well-know inequalities which are included here for completeness as they are frequently used, especially in Chapter 5.

Lemma 1.21 (Young's inequality, general version). *Let $N \in \mathbb{N}$, let $p_i \in [0, \infty)$, $i = 1, \ldots, N$ be such that $\sum_{i=1}^{N} \frac{1}{p_i} = 1$ and let $c_i > 0$ be such that $\prod_{i=1}^{N} c_i = 1$. Then*

$$\prod_{i=1}^{N} a_i \leq \sum_{i=1}^{N} \frac{c_i^{p_i}}{p_i} a_i^{p_i}$$

holds for all non-negative real numbers a_i, $i = 1, \ldots, N$.

Lemma 1.22 (Young's inequality for convolutions). *Let $\emptyset \neq \Omega \subset \mathbb{R}^d$ be open. If $f \in L^p(\Omega)$ and $g \in L^q(\Omega)$, then $f * g \in L^r(\Omega)$, and*

$$\|f * g\|_{L^r(\Omega)} \leq \|f\|_{L^p(\Omega)} \|g\|_{L^q(\Omega)},$$

where $\frac{1}{p} + \frac{1}{q} = 1 + \frac{1}{r}$ and $p, q, r \geq 1$.

Lemma 1.23 (Interpolation inequality for L^p-spaces). *Let $1 < p < q < r < \infty$ and let $\emptyset \neq \Omega \subset \mathbb{R}^d$ be open. Then $L^p(\Omega) \cap L^r(\Omega) \subset L^q(\Omega)$ and*

$$\|f\|_{L^q(\Omega)} \leq \|f\|_{L^p(\Omega)}^{\theta} \|f\|_{L^r(\Omega)}^{1-\theta}, \qquad f \in L^p(\Omega) \cap L^r(\Omega)$$

holds, where $q^{-1} = \theta p^{-1} + (1 - \theta) r^{-1}$, $\theta \in [0, 1]$.

Lemma 1.24 (Gronwall's inequality, [MPF91, Chapter XII]). *Let $u(\cdot), k(\cdot)$ be real-valued non-negative functions on $[0, T]$. Assume that $u(\cdot)$ is continuous on $[0, T]$ and $k(\cdot)$ is integrable on $[0, T]$.*

(a) **Linear version:** *If $a \geq 0$ is a constant and $u(\cdot)$ satisfies the integral inequality*

$$u(t) \leq a + \int_0^t k(s) u(s) \, \mathrm{d}s$$

for $t \in [0, T]$, then

$$u(t) \leq a \exp\left(\int_0^t k(s) \, \mathrm{d}s\right)$$

for $t \in [0, T]$.

(b) **Non-linear version:** *Let $a, b \geq 0$ be constants. Moreover, let $g(\cdot)$ be a positive nondecreasing function on $[0, \infty)$. If $u(\cdot)$ satisfies the integral inequality*

$$u(t) \leq a + b \int_0^t k(s) g(u(s)) \, \mathrm{d}s$$

for $t \in [0, T]$, then there exists a $0 < T_0 \leq T$ such that

$$u(t) \leq G^{-1}\left(G(a) + b \int_0^t k(s) \, \mathrm{d}s\right)$$

for $t \in [0, T_0]$, where $G(t) := \int_0^t (g(s))^{-1} \, \mathrm{d}s$, $t > 0$. Here T_0 has to be chosen in such a way that for $t \in [0, T_0]$ the function $G(a) + b \int_0^t k(s) \, \mathrm{d}s$ belongs to the domain of G^{-1}.

1.2 Abstract Cauchy problems

In this section abstract linear evolution equations are considered. Our presentation is divided into two parts: First we study *autonomous* equations, the case where the underlying system does not depend on time. The study of such systems leads to the concept of strongly continuous operator semigroups, also called C_0-semigroups. This theory is presented in great detail in the monographs [Paz83], [EN00] and [ABHN01], the latter two serving as our main references. A special emphasis in our presentation is placed on Hilbert space methods. In the second part we consider *non-autonomous* equations, here the underlying system does depend on time. The theory of non-autonomous Cauchy problems is by far not as well-developed as in the autonomous case. We refer to [Paz83, Chapter 5], [Lun95, Chapter 6] or the survey article [Sch02].

1.2.1 Autonomous abstract Cauchy problems and C_0-semigroups

In a Banach space X we consider the *autonomous abstract Cauchy problem*

$$\begin{cases} u'(t) &=& Au(t), \quad t > 0, \\ u(0) &=& x, \end{cases} \tag{1.3}$$

where A is a closed linear operator on X with domain $\mathcal{D}(A)$ and $x \in X$ is a given initial value. By a *classical* solution to (1.3) we mean a function $u(\cdot; x) \in C([0, \infty); X) \cap C^1((0, \infty); X)$ such that $u(t; x) \in \mathcal{D}(A)$ for all $t > 0$ and (1.3) holds.

Definition 1.25. For a linear operator $(A, \mathcal{D}(A))$ on a Banach space X the abstract Cauchy problem (1.3) is called *well-posed* if $\mathcal{D}(A)$ is dense in X and the following statements are true:

1. **Existence and uniqueness:** For every $x \in \mathcal{D}(A)$ there exists a unique classical solution $t \mapsto u(t; x) \in \mathcal{D}(A)$ to (1.3).

2. **Continuous dependence:** The solution depends continuously on the data, i.e. for $\mathcal{D}(A) \ni x_n \to 0$ one has $u(t; x_n) \to 0$, uniformly for t in compact subsets of $[0, \infty)$.

If we start from the well-posed problem (1.3) with a unique solution $u(\cdot; x)$ for initial value $x \in \mathcal{D}(A)$, then we can define a family of linear operators $\{T(t)\}_{t \geq 0} \subset \mathscr{L}(X)$ by setting

$$T(t)x := u(t; x), \qquad t \geq 0, \ x \in \mathcal{D}(A). \tag{1.4}$$

The uniqueness of the solution $u(\cdot; x)$ implies that $T(t+s)x = T(t)T(s)x$ holds for all $t, s \geq 0$ and $x \in \mathcal{D}(A)$. This observation motivates the following definition.

Definition 1.26. A one-parameter family of bounded linear operators $\{T(t)\}_{t \geq 0}$ on a Banach space X is called a *strongly continuous semigroup*, or *C_0-semigroup*, if the following holds:

(i) $T(0) = \text{Id}$, where Id is the identity on the Banach space X.

(ii) $T(t + s) = T(t)T(s)$ for all $t, s \geq 0$.

(iii) The map $t \mapsto T(t)x$ is continuous from $\mathbb{R}_+$ into X for all $x \in X$.

11

If $\{T(t)\}_{t\geq0}$ is a C_0-semigroup, then there exist constants $M \geq 1$ and $\omega \in \mathbb{R}$ such that

$$\|T(t)\|_{\mathscr{L}(X)} \leq Me^{\omega t}, \qquad t \geq 0. \tag{1.5}$$

The infimum over all such possible ω is called *growth-bound* of $\{T(t)\}_{t\geq0}$ and is denoted by $\omega_0(T)$. We say that the semigroup $\{T(t)\}_{t\geq0}$ is

- *quasi-contractive* if we can chose $M = 1$,

- *contractice* if $\|T(t)\|_{\mathscr{L}(X)} \leq 1$ for all $t \geq 0$, and

- *(uniformly) bounded* if we can chose $\omega = 0$.

Above we have seen that by (1.4) the solution to a well-posed autonomous Cauchy problem is given via a C_0-semigroup. Now the question arises if a given C_0-semigroup is always related to some autonomous Cauchy problem. For that we first need to connect a given C_0-semigroup $\{T(t)\}_{t\geq0}$ to some linear operator A.

Definition 1.27. Let X be a Banach space and $\{T(t)\}_{t\geq0}$ be a C_0-semigroup on X. The operator $A : \mathcal{D}(A) \to X$ defined by

$$Ax := \lim_{h \to 0} \frac{1}{h}\left(T(h)x - x\right)$$

with domain

$$\mathcal{D}(A) := \left\{x \in X : \lim_{h \to 0} \frac{1}{h}\left(T(h)x - x\right) \text{ exists}\right\}$$

is called the *generator* of $\{T(t)\}_{t\geq0}$.

Remarks 1.28.   1. The generator of a C_0-semigroup is always closed, densely defined and unique, see [EN00, Chapter II, Theorem 1.4].

2. The generator A of a C_0-semigroup $\{T(t)\}_{t\geq0}$ always commutes with the semigroup, i.e. $AT(t) = T(t)A$ for all $t \geq 0$.

Property (ii) in Definition 1.26 is called *semigroup property* of $\{T(t)\}_{t\geq0}$. Note that this algebraic property ensures that the differentiability of the map $t \mapsto T(t)x \in X$ is already implied by right differentiability at $t = 0$, see [EN00, Chapter II, Lemma 1.1]. Thus we obtain the following connection between C_0-semigroups and abstract Cauchy problems.

Proposition 1.29 ([EN00, Chapter II, Theorem 6.7])**.** *The abstract Cauchy problem* (1.3) *is well-posed if and only if the operator* $(A, \mathcal{D}(A))$ *is the generator of a C_0-semigroup $\{T(t)\}_{t\geq0}$. In this case the solution to* (1.3) *is given by* $u(t;x) = T(t)x$.

By Proposition 1.29 the solvability of (1.3) is reduced to the question whether a given linear operator A is the generator of a C_0-semigroup. In the literature there are several well-known generation results which state necessary and sufficient conditions for an operator A to be the generator of a C_0-semigroup in terms of (spectral) properties of A. We refer to [ABHN01, Chapter 3] or [EN00, Chapter II] for a detailed overview. Here we just give one classical result in the case of contraction semigroups.

Proposition 1.30 (Hille-Yosida, [EN00, Theorem II.3.5]). *Let A be a linear operator on a Banach space X. The following statements are equivalent:*

(i) *A generates a contractive C_0-semigroup on X.*

(ii) *The domain $\mathcal{D}(A)$ is dense in X, and for every $\lambda > 0$ one has $\lambda \in \rho(A)$ and*

$$\|\lambda R(\lambda, A)\| \leq 1.$$

Let us return to our starting point, the Cauchy problem (1.3). It is clear that in general we can only expect to find classical solutions to (1.3) if $x \in \mathcal{D}(A)$. However, in the following we want to consider a special situation in which we can even find solutions to (1.3) for all initial value $x \in X$. This brings us to the concept of *analytic* semigroups, also called *holomorphic* semigroups.

Definition 1.31. Let $\theta \in (0, \frac{\pi}{2}]$. A semigroup $\{T(t)\}_{t \geq 0}$ on X is called *analytic* (or *holomorphic*) of angle θ if it has an analytic extension to the sector

$$\Sigma_\theta := \{z \in \mathbb{C} \setminus \{0\} : |\arg z| < \theta\}$$

which is bounded on $\Sigma_{\theta'} \cap \{z \in \mathbb{C} : |z| \leq 1\}$ for all $\theta' \in (0, \theta)$. The extension of $\{T(t)\}_{t \geq 0}$ to Σ_θ is denoted by $\{T(z)\}_{z \in \Sigma_\theta}$.

Definition 1.32. Let $\theta \in (0, \frac{\pi}{2}]$. A semigroup $\{T(t)\}_{t \geq 0}$ on X is called *bounded analytic* of angle θ if it has a bounded analytic extension to the sector $\Sigma_{\theta'}$ for each $\theta' \in (0, \theta)$.

Remarks 1.33. 1. An operator A generates an analytic semigroup if and only if there exists a constant $\omega \geq 0$ such that $A - \omega$ is the generator of a bounded analytic semigroup, see [ABHN01, Proposition 3.7.4].

2. For the generator A of a semigroup $\{T(t)\}_{t \geq 0}$ the *spectral bound* $s(A)$ is defined as $s(A) := \sup\{\operatorname{Re}\lambda : \lambda \in \sigma(A)\}$. It always holds that $s(A) \leq \omega_0(T)$. If A is the generator of an analytic semigroup, then $s(A) = \omega_0(T)$ holds, see [ABHN01, Theorem 5.1.12].

The following result gives a characterization of generators of bounded analytic semigroups in terms of resolvent estimates.

Proposition 1.34 ([ABHN01, Theorem 3.7.11 and Corollary 3.7.12]). *For a linear operator $(A, \mathcal{D}(A))$ on a Banach space X the following statements are equivalent:*

(i) *A generates a bounded analytic semigroup on X.*

(ii) *$\{z \in \mathbb{C} : \operatorname{Re} z > 0\} \subset \rho(A)$ and*

$$\sup_{\operatorname{Re}\lambda > 0} \|\lambda R(\lambda, A)\| < \infty.$$

Bounded analytic semigroups can also be characterized as follows.

13

Proposition 1.35 (Real characterization, [ABHN01, Theorem 3.7.19].). *Let $\{T(t)\}_{t \geq 0}$ be a bounded semigroup on X with generator A. Then $\{T(t)\}_{t \geq 0}$ is bounded analytic if and only if $T(t)x \in \mathcal{D}(A)$ for all $t > 0$, $x \in X$, and*

$$\sup_{t>0} \|tAT(t)\| < \infty.$$

Remarks 1.36. 1. Let A be the generator of an analytic semigroup $\{T(t)\}_{t \geq 0}$ on X. Then $T(t)x \in \mathcal{D}(A^n)$ for all $t > 0$, $x \in X$, and $n \in \mathbb{N}$.

2. If A generates an analytic semigroup $\{T(t)\}_{t \geq 0}$ on X, then for every fixed $T_0 > 0$ there exists a constant $C := C(T_0) > 0$ such that

$$\|AT(t)x\| \leq C\frac{1}{t}\|x\|, \qquad t \in (0, T_0], \ x \in X. \tag{1.6}$$

As a consequence of Proposition 1.35 and Remarks 1.36 we see that if A generates an analytic semigroup, then for all $x \in X$ we obtain a solution u to (1.3) such that

$$u \in C^\infty((0, \infty); X) \cap C([0, \infty); X) \cap C((0, \infty); \mathcal{D}(A)).$$

This observation is often referred to as the *smoothing effect* of analytic semigroups.

Generators of analytic semigroups are stable under small perturbations.

Proposition 1.37 ([ABHN01, Theorem 3.7.23]). *Let A be the generator of an analytic semigroup on X. Let $B : \mathcal{D}(A) \to X$ be an operator such that for every $\varepsilon > 0$ there exists a constant $b \geq 0$ sucht hat*

$$\|Bx\| \leq \varepsilon\|Ax\| + b\|x\|$$

for all $x \in \mathcal{D}(A)$. Then $A + B$ generates an analytic semigroup on X.

Until now we collected the basic results for C_0-semigroups and analytic semigroups and we gave some characterizations for generators of semigroups. Basically, we saw that the generation of a semigroup by some operator A is closely related to certain spectral properties of A. In a lot of concrete situations the resolvent estimates stated in Proposition 1.30 and Proposition 1.34 are quite difficult to check. However, if one works on Hilbert spaces things are getting much easier, as in this setting there are very efficient tools for solving evolutionary problems.

From now on let H, V be Hilbert spaces such that $V \hookrightarrow H$. The scalar product on H is denoted by $\langle \cdot, \cdot \rangle_H$.

Definition 1.38. An operator A on H is called *accretive* if

$$\mathrm{Re}\,\langle Ax, x \rangle_H \geq 0, \qquad x \in \mathcal{D}(A).$$

An operator A is called *m-accretive* if it is accretive and $A + \mathrm{Id}$ is surjective.

Based on this notion we obtain a very convenient characterization for generators of contraction semigroups on Hilbert spaces, see [Haa06, Proposition C.7.3] or [ABHN01, Theorem 3.4.5][2].

Proposition 1.39 (Lumer-Phillips). *Let A be a linear operator on a Hilbert space H. The following statements are equivalent:*

(i) $-A$ *generates a contraction semigroup on H.*

(ii) A *is m-accretive.*

Due to reflexivity of Hilbert spaces, each m-accretive operator A is densely defined, see [Kat95, Section V.10]. Accretivity of A can be reformulated by the condition

$$\|(\lambda + A)x\| \geq \lambda \|x\|, \qquad \lambda > 0, \; x \in \mathcal{D}(A).$$

If A is m-accretive, then it is an immediate consequence that $\lambda + A$ is invertible for all $\operatorname{Re} \lambda > 0$ and

$$\|(\lambda + A)^{-1}\| \leq \frac{1}{\operatorname{Re} \lambda}, \qquad \operatorname{Re} \lambda > 0.$$

This shows that Proposition 1.39 is basically a consequence of Proposition 1.30 .

Given a linear operator A on the Hilbert space H we call

$$W(A) := \{\langle Ax, x \rangle_H : x \in \mathcal{D}(A), \|x\|_H = 1\} \subset \mathbb{C}$$

the *numerical range* of A. It is easily seen that for an accretive operator A be have

$$W(A) \subset \{z \in \mathbb{C} : \operatorname{Re} z \geq 0\}.$$

The next result reformulates Proposition 1.39 for generators of semigroups that are contractive on a sector, see [Haa06, Proposition 7.1.1] or [Kat95, Theorem IX.1.24].

Proposition 1.40. *Let A be a linear operator on an Hilbert space H. The following statements are equivalent:*

(i) *The operator A is m-accretive and the numerical range $W(A)$ is contained in the sector Σ_ϕ for some $\phi \in [0, \frac{\pi}{2})$.*

(ii) $-A$ *generates an analytic C_0-semigroup $\{T(z)\}_{z \in \Sigma_\theta}$ of angle $\theta := \frac{\pi}{2} - \phi$ such that*

$$\|T(z)\|_{\mathscr{L}(H)} \leq 1, \qquad z \in \Sigma_\theta.$$

A very efficient way to define generators of holomorphic semigroups on Hilbert spaces is the *form method* which was mainly developed by Kato, see his monograph [Kat95, Chapter VI]. See also the survey article by Arendt [Are04, Chapter 5].

---

[2]Note that in [ABHN01] different terminology is used. There an operator A is called dissipative if $-A$ is accretive.

Definition 1.41. A closed form on H is a sesquilinear form $a : V \times V \to \mathbb{C}$ which is continuous, i.e.

$$|a(u,v)| \leq M\|u\|_V\|v\|_V, \qquad u, v \in V$$

for some $M \geq 0$, and elliptic, i.e.

$$\operatorname{Re} a(u,u) + \delta\|u\|_H^2 \geq \alpha\|u\|_V^2, \qquad u \in V$$

for some $\delta \in \mathbb{R}$ and some $\alpha > 0$. The space V is called the domain of the form a.

Let a be a closed form on H with dense domain V. Then we associate an operator A with the closed form a by setting

$$
\begin{aligned}
\mathcal{D}(A) &:= \{u \in V : \text{there exists } v \in H \text{ such that } a(u,\varphi) = \langle v, \varphi\rangle_H \text{ for all } \varphi \in V\}, \\
Au &:= v.
\end{aligned}
$$

Note that v is well-defined since V is dense in H. We call A the *operator induced by the form* $a : V \times V \to \mathbb{C}$.

If an operator A is induced by a form $a : V \times V \to \mathbb{C}$, then the generation of a semigroup is automatic and a lot of properties of the operator and the semigroup follow from properties of the form. In fact, the shifted operator $A + \delta$, where δ is the constant from Definition 1.41, is automatically m-accretive. More precisely, for every $x \in \mathcal{D}(A)$ we have

$$\langle (A+\delta)x, x\rangle_H = a(x,x) + \delta\|x\|_H^2 \geq \alpha\|x\|_V^2 \geq 0,$$

and the fact that $A + \delta + \operatorname{Id}$ is surjective is a direct consequence of the classical Lax-Milgram lemma. Moreover, by using the continuity and the ellipticity of $a : V \times V \to \mathbb{C}$, it is not difficult to show that the numerical range $W(A)$ is contained in some sector $\Sigma_\phi + \delta$. Thus, by using standard rescaling arguments for semigroup generators, see [EN00, Section II.2], we obtain the following result as a consequence of Proposition 1.40.

Proposition 1.42. *Let A be an operator induced by some form $a : V \times V \to \mathbb{C}$. Then the operator $-A$ generates an analytic semigroup $\{T(z)\}_{z \in \Sigma_\theta}$ for some $\theta \in (0, \frac{\pi}{2}]$ such that*

$$\|T(z)\| \leq e^{\omega|z|}, \qquad z \in \Sigma_\theta$$

for some $\omega \in \mathbb{R}$.

Remarks 1.43.   1. If $a(u,u) \geq 0$ for all $u \in V$, then $W(A) \subset [0,\infty)$ and Proposition 1.42 holds with $\omega = 0$ and $\theta = \frac{\pi}{2}$.

2. A form $a : V \times V \to \mathbb{C}$ is called *symmetric* if $a(u,v) = \overline{a(v,u)}$ for all $u, v \in V$. An operator A induced by a symmetric and densely defined form is self-adjoint.

3. A form $a : V \times V \to \mathbb{C}$ is symmetric if and only if $W(A) \subset \mathbb{R}$.

4. An operator A induced by some form $a : V \times V \to \mathbb{C}$ is injective if and only if $\{u \in V : a(u,u) = 0\} = \{0\}$.

To conclude this section on autonomous Cauchy problems let us say a few words about inhomogeneous problems of the form

$$\begin{cases} u'(t) - Au(t) & = \ f(t), \quad t > 0, \\ u(0) & = \ x, \end{cases} \tag{1.7}$$

where $x \in X$ and $f \in L^1_{loc}(\mathbb{R}_+; X)$. If $f \in C([0, \infty); X)$, then as in the homogeneous case (1.3), a *classical* solution to (1.7) is a function $u(\cdot; x) \in C([0, \infty); X) \cap C^1((0, \infty); X)$ such that $u(t) \in \mathcal{D}(A)$ for all $t > 0$ and u satisfies (1.7). If the operator $(A, \mathcal{D}(A))$ generates a C_0-semigroup $\{T(t)\}_{t \geq 0}$, then the natural candidate for a solution to (1.7) is given by the *variation of constants formula*

$$u(t) := T(t)x + \int_0^t T(t-s)f(s)\,\mathrm{d}s. \tag{1.8}$$

Therefore we say that u is a *mild solution* to (1.7) if $(A, \mathcal{D}(A))$ generates a C_0-semigroup $\{T(t)\}_{t \geq 0}$ and (1.8) holds. From [ABHN01, Proposition 3.1.16] it follows that (1.7) has a unique mild solution for all $x \in X$ if $(A, \mathcal{D}(A))$ is a generator of a C_0-semigroup. Moreover, if u is a classical solution to (1.7), then u is also a mild solution.

Definition 1.44. Let $J = (0, T_0)$ for some $T_0 > 0$ and let $1 < q < \infty$.

 (a) Let $f \in L^q(J; X)$. A function u is called *strong solution* to (1.7) on the interval J if $u \in W^{1,q}(J; X) \cap L^q(J; \mathcal{D}(A))$, $u(0) = x$ (in the sense of traces) and u satisfies (1.7) almost everywhere in J.

 (b) An operator $(A, \mathcal{D}(A))$ is said to admit *maximal L^q-regularity* on J in the Banach space X if for every $f \in L^q(J; X)$ there exists a unique strong solution u to (1.7) with initial condition $u(0) = 0$.

Remark 1.45. Let u be a strong solution to (1.7). By the Sobolev embedding $W^{1,q}(J; X) \hookrightarrow C(\overline{J}; X)$ is makes sense to write $u(0)$.

For a survey on maximal L^q-regularity we refer to [Dor93].

Remarks 1.46.   1. If $(A, \mathcal{D}(A))$ admits maximal L^q-regularity for some $1 < q < \infty$, then the operator has maximal L^q-regularity for all $1 < q < \infty$.

 2. If $(A, \mathcal{D}(A))$ admits maximal L^q-regularity, then in particular A generates an analytic semigroup. The converse is in general not true.

In [Ama95, Section III 4.10] it is shown that if A admits maximal L^q-regularity, then the trace space $\{u(0) : u \in W^{1,q}(J; X) \cap L^q(J; \mathcal{D}(A))\}$ can be characterized as the real interpolation space $(X, \mathcal{D}(A))_{1-1/q,q}$.

Proposition 1.47 ([Ama95, Theorem 4.10.7 and Remark 4.10.9]). *Suppose $(A, \mathcal{D}(A))$ admits maximal L^q-regularity on $J = (0, T_0)$ for some $T_0 > 0$. Then for every $f \in L^q(J; X)$ and every $x \in (X, \mathcal{D}(A))_{1-1/q,q}$ there exists a unique strong solution u to (1.7). In addition, there exists a constant $C > 0$ independent of u, f and x such that*

$$\|u'\|_{L^q(J;X)} + \|Au\|_{L^q(J;X)} \leq C\Big(\|f\|_{L^q(J;X)} + \|x\|_{(X,\mathcal{D}(A))_{1-1/q,q}}\Big).$$

1.2.2 Non-autonomous abstract Cauchy problems and evolution systems

In this section we consider a generalization of the situation in Section 1.2.1. We replace the fixed linear operator $(A, \mathcal{D}(A))$ in the abstract Cauchy problem (1.3) by a family of linear operators $\{A(t), \mathcal{D}(A(t))\}_{t \geq 0}$ depending on time $t \geq 0$. This means we consider the *non-autonomous abstract Cauchy problem*

$$\begin{cases} u'(t) & = & A(t)u(t), \quad t > s \geq 0, \\ u(s) & = & x, \end{cases} \tag{1.9}$$

on a Banach space X for a given initial value $x \in X$. By a *classical* solution to (1.9) we mean a function $u(\cdot; s, x) \in C([s, \infty), X) \cap C^1((s, \infty), X)$ such that $u(t; s, x) \in \mathcal{D}(A(t))$ for all $t > s$ and (1.9) holds. Analogously to Definition 1.25, by well-posedness of (1.9) we mean the following.

Definition 1.48. For a family $\{A(t), \mathcal{D}(A(t))\}_{t \geq 0}$ of linear operators on a Banach space X, the non-autonomous Cauchy problem (1.9) is called *well-posed* (on regularity spaces $\{Y_s\}_{s \geq 0}$) if the following statements are true:

1. **Existence and uniqueness:** There are dense subspaces $Y_s \subset \mathcal{D}(A(s))$ of X such that for $x \in Y_s$ there is a unique classical solution $t \mapsto u(t; s, x) \in Y_t$ to (1.9).

2. **Continuous dependence:** The solution depends continuously on the data, i.e. for $s_n \to s$ and $Y_{s_n} \ni x_n \to x \in Y_s$ one has $\tilde{u}(t; s_n, x_n) \to \tilde{u}(t; s, x)$ uniformly for t in compact subsets of $[0, \infty)$, where $\tilde{u}(t; s, x) := u(t; s, x)$ for $t \geq s$ and $\tilde{u}(t; s, x) := x$ for $t < s$.

The following notation is very useful in this section: Set

$$\Lambda := \{(t, s) \in \mathbb{R}^2 : 0 \leq s \leq t\}.$$

Moreover, set $\partial_t := \frac{\partial}{\partial t}$ and $\partial_s := \frac{\partial}{\partial s}$.

As in the autonomous case, solutions to (1.9) are given by a family of solution operators that obey certain algebraic properties.

Definition 1.49. A two-parameter family of bounded linear operators $\{U(t, s)\}_{(t,s) \in \Lambda}$ on a Banach space X is called a *(strongly continuous) evolution system* if the following conditions hold:

(i) $U(s, s) = \text{Id}$ for $s \geq 0$, where Id is the identity on the Banach space X.

(ii) $U(t, s) = U(t, r)U(r, s)$ for $0 \leq s \leq r \leq t < \infty$.

(iii) The map $(t, s) \mapsto U(t, s)x$ is continuous from Λ in X for all $x \in X$.

Remark 1.50. If $\{T(t)\}_{t \geq 0}$ is a C_0-semigroup on X, then $U(t, s) := T(t - s)$ for $(t, s) \in \Lambda$ defines an evolution system on X.

Condition (ii) in Definition 1.49 is a generalization of the semigroup property in Definition 1.26. However, in general and in contrast to the autonomous case, this algebraic property of an evolution system does not imply any differentiability on a dense subspace. Therefore we need to make the following definition.

Definition 1.51. We say the evolution system $\{U(t,s)\}_{(t,s)\in\Lambda}$ *solves the Cauchy problem* (1.9) (on regularity spaces $\{Y_s\}_{s\geq 0}$) if there are dense subspaces $Y_s \subset X$, $s \in [0,\infty)$, such that $U(t,s)Y_s \subset Y_t \subset \mathcal{D}(A(t))$ for $t \geq s$ and the function $t \mapsto U(t,s)x$ is a solution to (1.9) for $s \in [0,\infty)$ and $x \in Y_s$.

Remark 1.52. In analogy to semigroups we also say that the evolution system $\{U(t,s)\}_{(t,s)\in\Lambda}$ is *generated* by $\{A(t), \mathcal{D}(A(t))\}_{t\geq 0}$ if $\{U(t,s)\}_{(t,s)\in\Lambda}$ solves the Cauchy problem (1.9).

Analogously to Proposition 1.29 we have the following result. A proof can be found e.g. in the Ph.D. thesis of Nickel [Nic96].

Proposition 1.53. *The Cauchy problem (1.9) is well-posed on $\{Y_s\}_{s\geq 0}$ if and only if there is an evolution system solving (1.9) on $\{Y_s\}_{s\geq 0}$.*

In contrast to the autonomous case, there is no characterization of well-posedness of (1.9) in terms of (spectral) properties of $\{A(t), \mathcal{D}(A(t))\}_{t\geq 0}$. However, there are several sufficient conditions that guarantee well-posedness of (1.9). These conditions are generally divided into two cases: A *hyperbolic* and a *parabolic* case. Roughly speaking, the parabolic case corresponds to the case of analytic semigroups, whereas in the hyberbolic case the stability of certain products is assumed. In both cases certain continuity assumptions for the map $t \mapsto A(t)$ are required. There is great variety of results in this context with different kinds of assumptions. Here we restrict ourselves to rather simple versions for each type.

Assumption 1.54 (Hyperbolic case). Let $\{A(t), \mathcal{D}(A(t))\}_{t\geq 0}$ be a family of closed linear operators on a Banach space X. We suppose that that the following conditions hold:

(H1) The family $\{A(t), \mathcal{D}(A(t))\}_{t\geq 0}$ is *stable*, i.e., all the operators $A(t)$, $t \geq 0$, are generators of C_0-semigroups, and for every $T > 0$ there exist constants $M := M(T) \geq 1$ and $\omega := \omega(T) \in \mathbb{R}$ such that

$$(\omega, \infty) \subset \rho(A(t)) \quad \text{for all } t \in [0,T]$$

and

$$\left\| \prod_{j=1}^{k} R(\lambda, A(t_j)) \right\| \leq M(\lambda - \omega)^{-k}$$

for all $\lambda > \omega$ and every finite sequence $0 \leq t_1 \leq t_2 \leq \ldots \leq t_k \leq T$, $k \in \mathbb{N}$.

(H2) There exists a densely embedded subspace $Y \hookrightarrow X$ which is a core for every $A(t)$ such that the family of the parts $\{A_{|Y}(t), \mathcal{D}(A) \cap Y\}_{t\geq 0}$ is a stable family on the space Y.

(H3) For every $T > 0$ the mapping $[0,T] \ni t \mapsto A(t)$ is continuous in the $\mathscr{L}(Y,X)$-norm.

Remark 1.55. If $\{A(t), \mathcal{D}(A(t))\}_{t\geq 0}$ is a family of generators of contraction semigroups, then (H1) in Assumption 1.54 automatically holds.

Assumption 1.56 (Parabolic case). Let $\{A(t), \mathcal{D}(A(t))\}_{t \geq 0}$ be a family of closed linear operators on a Banach space X. We suppose that that the following conditions hold:

(P1) There exists a subspace Y of X such that $Y \hookrightarrow X$ and for every $t \geq 0$ the operator $(A(t), \mathcal{D}(A(t))$ is the generator of an analytic semigroup on X with $\mathcal{D}(A(t)) \simeq Y$ (here we mean $\mathcal{D}(A(t)) = Y$ with equivalent norms).

(P2) For every $T > 0$, the mapping $[0, T] \ni t \mapsto A(t)$ is α-Hölder continuous in the $\mathscr{L}(Y, X)$-norm for some $\alpha \in (0, 1)$.

Lemma 1.57 ([Lun95, Lemma 6.1.1]). *Let $T > 0$ and let $\{A(t), \mathcal{D}(A(t))\}_{t \geq 0}$ satisfy Assumption 1.56. Then there exists a constant $\gamma := \gamma(T) > 1$ such that for all $y \in Y$ it holds that*

$$\gamma^{-1} \|y\|_Y \leq \|y\| + \|A(t)x\| \leq \gamma \|y\|_Y.$$

Proposition 1.58 (Hyperbolic case, [Paz83, Theorem 3.1]). *Assume that Assumption 1.54 holds. There exists a unique evolution system $\{U(t, s)\}_{(t,s) \in \Lambda}$ on X with the following properties:*

(a) *For every $x \in Y$ and $s \geq 0$, the map $t \mapsto U(t, s)x$ is differentiable in (s, ∞) and*

$$\partial_t U(t, s)x = A(t)U(t, s)x.$$

(b) *For every $x \in Y$ and $t > 0$, the map $s \mapsto U(t, s)x$ is differentiable in $[0, t)$ and*

$$\partial_s U(t, s)x = -U(t, s)A(s)x.$$

(c) *For every $T > 0$ there exists a constant $C := C(T) > 0$ such that*

$$\|U(t, s)x\| \leq C\|x\|, \qquad x \in X, \ 0 \leq s \leq t \leq T.$$

Proposition 1.59 (Parabolic case, [Lun95, Chapter 6]). *Assume that Assumption 1.56 holds. There exists a unique evolution system $\{U(t, s)\}_{(t,s) \in \Lambda}$ on X with the following properties:*

(a) *For $(t, s) \in \Lambda$ with $t \neq s$, the operator $U(t, s)$ maps X into Y.*

(b) *The map $t \mapsto U(t, s)$ is differentiable in (s, ∞) with values in $\mathscr{L}(X)$ and*

$$\partial_t U(t, s) = A(t)U(t, s).$$

(c) *For every $x \in Y$ and $t > 0$, the map $s \mapsto U(t, s)x$ is differentiable in $[0, t)$ and*

$$\partial_s U(t, s)x = -U(t, s)A(s)x.$$

(d) *For every $T > 0$ there exists a constant $C := C(T) > 0$ such that*

$$\|U(t, s)x\| \leq C\|x\|, \qquad x \in X, \ 0 \leq s \leq t \leq T$$

and

$$\|A(t)U(t, s)x\| \leq C(t - s)^{-1}\|x\|, \qquad x \in X, \ 0 \leq s < t \leq T.$$

The second estimate in (d) corresponds to estimate (1.6) in the case of analytic semigroups. It reflects smoothing properties of evolution systems of parabolic type. Such an estimate does in general not hold in the hyperbolic case. The next result collects norm estimates for parabolic evolution systems in interpolation spaces. It will be very useful in Section 4.2.1.

Proposition 1.60 ([Lun95, Corollary 6.1.8]). *Let $\{U(t,s)\}_{(t,s)\in\Lambda}$ be the evolution system from Proposition 1.59. Moreover, let $T > 0$, $1 \le p \le \infty$, and $0 < \theta < 1$. Then there exists a constant $C := C(T) > 0$ such that*

$$\|U(t,s)x\|_Y \le C\|x\|_Y, \qquad\qquad x \in X, \quad 0 \le s \le t \le T,$$
$$\|U(t,s)x\|_Y \le C(t-s)^{-(1-\theta)}\|x\|_{(X,Y)_{\theta,p}}, \qquad\qquad x \in X, 0 \le s < t \le T.$$

Moreover, denote by $\{T_s(t)\}_{t\ge 0}$ the semigroup generated by $A(s)$ for fixed $0 \le s \le T$. Assume that there exists a constant $M := M(T) > 0$ such that $\|T_s(t)x\|_{[X,Y]_\theta} \le M\|x\|_{[X,Y]_\theta}$ for all $x \in [X,Y]_\theta$ and all $0 \le s,t \le T$. Then there exists a constant $C := C(T) > 0$ such that

$$\|U(t,s)x\|_{[X,Y]_\theta} \le C\|x\|_{[X,Y]_\theta}, \qquad x \in [X,Y]_\theta, \quad 0 \le s \le t \le T.$$

To conclude this section we consider the inhomogeneous problem

$$\begin{cases} u'(t) - A(t)u(t) &= f(t), \quad t > s \ge 0, \\ u(s) &= x, \end{cases} \tag{1.10}$$

for $x \in X$ and $f \in L^1_{\text{loc}}(\mathbb{R}_+; X)$. If $f \in C([0,\infty); X)$, then as in the homogeneous case (1.9), by a *classical* solution to (1.10) we mean a function $u(\cdot; s, x) \in C([s,\infty), X) \cap C^1((s,\infty), X)$ such that $u(t; s, x) \in \mathcal{D}(A(t))$ for all $t > s$ and (1.10) holds. If there exists an evolution system $\{U(t,s)\}_{(t,s)\in\Lambda}$ associated to the family of operators $\{A(t), \mathcal{D}(A(t))\}_{t\ge 0}$ such that for every $x \in \mathcal{D}(A(s))$, $U(t,s)x \in \mathcal{D}(A(t))$ and $U(t,s)$ is differentiable with respect to t and s satisfying

$$\partial_t U(t,s)x = A(t)U(t,s)x,$$
$$\partial_s U(t,s)x = -U(t,s)A(s)x,$$

then every classical solution u to (1.10) is given via the *variation of constants formula*

$$u(t) = U(t,s)x + \int_s^t U(t,r)f(r)\,\mathrm{d}r, \tag{1.11}$$

see [Paz83, Section 5.1]. We say u is a *mild solution* to (1.10) if there is an evolution system $\{U(t,s)\}_{(t,s)\in\Lambda}$ associated to the family of operators $\{A(t)\}_{t\ge 0}$ and if u satisfies (1.11).

1.3 The Helmholtz decomposition

In the mathematical formulation of problems related to viscous, incompressible fluid flows, modeled by the Navier-Stokes equations, the notion of solenoidal (divergence-free) vector fields appears in a natural way. Of particular interest is the question if for a domain $\Omega \subset \mathbb{R}^d$ a given function $u \in L^p(\Omega)^d$, $1 < p < \infty$, can be decomposed in a *solenoidal* (divergence-free) and a *potential* part. Such a decomposition, called the *Helmholtz decomposition*, leads to a linear projection onto the space of solenoidal L^p-functions, called the *Helmholtz projection*. As it was briefly outlined in the introduction, the Helmholtz decomposition and the Helmholtz projection are essential for the semigroup approach to the Navier-Stokes equations and in particular for the definition of the Stokes operator. In this section we start with a brief discussion of the Helmholtz decomposition in the Hilbert space case $p = 2$, and then we shall state the main results for arbitrary $1 < p < \infty$. For more details and for proofs of the following results we refer to the monograph by Galdi [Gal94a, Chapter III].

For $\emptyset \neq \Omega \subset \mathbb{R}^d$, $d \in \mathbb{N}$, open and for $1 < p < \infty$ we define

$$C_{c,\sigma}^\infty(\Omega) := \{u \in C_c^\infty(\Omega)^d : \operatorname{div} u = 0\},$$

$$L_\sigma^p(\Omega) := \overline{C_{c,\sigma}^\infty(\Omega)}^{\|\cdot\|_{L^p(\Omega)}},$$

$$H_{0,\sigma}^1(\Omega) := \overline{C_{c,\sigma}^\infty(\Omega)}^{\|\cdot\|_{H^1(\Omega)}}.$$

The norms on $L_\sigma^p(\Omega)$ and $H_{0,\sigma}^1(\Omega)$ are simply denoted by $\|\cdot\|_{L^p(\Omega)}$ and $\|\cdot\|_{H^1(\Omega)}$, respectively.

The space $L_\sigma^p(\Omega)$ is the space of solenoidal vector fields in $L^p(\Omega)^d$. It is clear that $L_\sigma^p(\Omega)$ is a closed subspace of $L^p(\Omega)^d$ and thus in the case $p = 2$ it is even a Hilbert space with respect to the usual $L^2(\Omega)^d$-scalar product. Moreover, $H_{0,\sigma}^1(\Omega) \subset L_\sigma^2(\Omega)$ is also Hilbert space with respect to the $H^1(\Omega)^d$-scalar product. The dual space of $H_{0,\sigma}^1(\Omega)$ is denoted by $H_\sigma^{-1}(\Omega)$.

By defining the space $G_2(\Omega) := L_\sigma^2(\Omega)^\perp$ as the orthogonal complement we obtain the orthogonal decomposition

$$L^2(\Omega)^d = L_\sigma^2(\Omega) \overset{\perp}{\oplus} G_2(\Omega).$$

This representation of $L^2(\Omega)^d$ as the orthogonal sum of $L_\sigma^2(\Omega)$ and $G_2(\Omega)$ is called *Helmholtz decomposition*. There exists an orthogonal projection $\mathbb{P}_{\Omega,2} : L^2(\Omega)^d \to L_\sigma^2(\Omega)$ which is called the *Helmholtz projection*.

Before we proceed to the general situation $p \neq 2$, let us briefly derive an equivalent characterization of $G_2(\Omega)$. In the following let $\Omega \subset \mathbb{R}^d$ be a domain. Note that for $u \in G_2(\Omega)$ we have $u \in L_{\mathrm{loc}}^1(\Omega)^d$ and

$$\int_\Omega u \cdot \varphi \, \mathrm{d}x = 0 \qquad \text{for all } \varphi \in C_{c,\sigma}^\infty(\Omega).$$

By [Gal94a, Lemma III 1.1] it directly follows that there exists a function $\mathrm{p} \in W_{\mathrm{loc}}^{1,1}(\Omega)$ such that $u = \nabla \mathrm{p}$. Thus all functions in $G_2(\Omega)$ have a representation as gradient fields. This observation motivates to define

$$G_p(\Omega) := \{u = \nabla \mathrm{p} : \mathrm{p} \in \widehat{W}^{1,p}(\Omega)\}, \qquad 1 < p < \infty.$$

In the following we say that the Helmholtz decomposition of $L^p(\Omega)^d$, $1 < p < \infty$, holds if the decomposition

$$L^p(\Omega)^d = L_\sigma^p(\Omega) \oplus G_p(\Omega) \tag{1.12}$$

is valid. In this case there exists a linear projection $\mathbb{P}_{\Omega,p} : L^p(\Omega)^d \to L_\sigma^p(\Omega)$.

To show the validity of (1.12) for arbitrary $1 < p < \infty$ is much more involved as the in the Hilbert space case $p = 2$ since not every closed subspace in $L^p(\Omega)^d$ may be complemented in general. It turns out that the decomposition (1.12) holds if and only if the weak Neumann problem

$$\int_\Omega \nabla \mathrm{p} \cdot \nabla \varphi \, dx = \int_\Omega u \cdot \nabla \varphi \, dx, \qquad \varphi \in \widehat{W}^{1,p'}(\Omega) \quad \text{with} \quad \frac{1}{p} + \frac{1}{p'} = 1 \qquad (1.13)$$

has a unique solution $\mathrm{p} \in \widehat{W}^{1,p}(\Omega)$ for every $u \in L^p(\Omega)^d$, see [Gal94a, Lemma III 1.2]. In this case $u = \mathbb{P}_{\Omega,p} u + \nabla \mathrm{p}$, and there exists a $C > 0$ such that $\|\nabla \mathrm{p}\|_{L^p(\Omega)} \leq C\|u\|_{L^p(\Omega)}$.

Note that (1.13) cannot be uniquely solved in general for all $1 < p < \infty$ and all domains Ω. In [MB86] Maslennikova and Bogovskiĭ gave examples of unbounded domains Ω with smooth boundary for which the weak Neumann problem (1.13) loses either existence or uniqueness for certain values of p. Thus, by the equivalence of unique solvability of the weak Neumann problem (1.13) and the validity of (1.12), the same examples can be used to show that the Hemholtz decomposition does not hold for some domains Ω and certain values of p, see [Bog86]. However, for a large class of domains the Helmholtz decomposition holds true.

Proposition 1.61 ([Gal94a, Theorem III 1.2]). *Let $\Omega \subset \mathbb{R}^d$, $d \geq 2$, be either the whole space $\mathbb{R}^d$, the half-space $\mathbb{R}^d_+$ or a domain with a compact boundary that is of class C^2. Then (1.12) holds for all $1 < p < \infty$. The associated Helmholtz projections*

$$\mathbb{P}_{\Omega,p} : L^p(\Omega)^d \to L^p_\sigma(\Omega)$$

are linear and bounded and satisfy $\mathbb{P}^*_{\Omega,p} = \mathbb{P}_{\Omega,p'}$ with $\frac{1}{p} + \frac{1}{p'} = 1$. Moreover, $L^p_\sigma(\Omega)' = L^{p'}_\sigma(\Omega)$ and $G_p(\Omega)' = G_{p'}(\Omega)$

Remarks 1.62. 1. Let $\Omega \subset \mathbb{R}^d$ be as in Proposition 1.61. Then $\{\mathbb{P}_{\Omega,p}\}_{1<p<\infty}$ is a compatible family of bounded linear operators, i.e. for $u \in L^p(\Omega)^d \cap L^q(\Omega)^d$ with $1 < p, q < \infty$ we have $\mathbb{P}_{\Omega,p} u = \mathbb{P}_{\Omega,q} u$.

2. For convenience we shall frequently just write $\mathbb{P}$ instead of $\mathbb{P}_{\Omega,p}$ if no confusion arises and if it is clear from the context what the actual value of $1 < p < \infty$ is and on which domain we are working.

3. In the whole space $\mathbb{R}^d$ there is an explicit formula for the Helmholtz projection $\mathbb{P}_{\mathbb{R}^d}$. More precisely, for $u \in L^p(\mathbb{R}^d)^d$ we have

$$\mathbb{P}_{\mathbb{R}^d} u = \mathcal{F}^{-1}\left(\delta_{ij} - \frac{\xi_i \xi_j}{|\xi|^2}\right)^d_{i,j=1} \mathcal{F}u,$$

where δ_{ij} is the Kronecker delta[3]. By using this formula it is easy to see that $\mathbb{P}_{\mathbb{R}^d}\Delta u = \Delta \mathbb{P}_{\mathbb{R}^d} u$ holds for all $u \in W^{2,p}(\mathbb{R}^d)^d$.

Lemma 1.63 ([Gal94a, Lemma III.2.1]). *Let $\Omega \subset \mathbb{R}^d$, $d \geq 2$, be a domain and $u \in L^p(\Omega^d)$, $1 < p < \infty$. Then $u \in L^p_\sigma(\Omega)$ if and only if $\langle u, \nabla \mathrm{p}\rangle = 0$ for all $\mathrm{p} \in \widehat{W}^{1,p'}(\Omega)$ with $\frac{1}{p} + \frac{1}{p'} = 1$.*

---

[3]This means, that $\delta_{ij} = 1$ if $i = j$ and $\delta_{ij} = 0$ if $i \neq j$.

Remarks 1.64. Let $\Omega \subset \mathbb{R}^d$, $d \geq 2$, be either the whole space $\mathbb{R}^d$, the half-space $\mathbb{R}^d_+$ or a domain with compact boundary that is locally Lipschitz.

1. It holds that
$$L^p_\sigma(\Omega) = \left\{ u \in L^p(\Omega)^d : \operatorname{div} u = 0, \quad u \cdot \nu|_{\partial\Omega} = 0 \right\}. \tag{1.14}$$

   Here $\operatorname{div} u$ is understood in the sense of distributions and the trace $u \cdot \nu|_{\partial\Omega} = 0$ has to be understood in the sense of the generalized Gauss Theorem. We refer to [Gal94a, Section III.2] for details. Note that (1.14) does not hold for certain domains with non-compact boundaries.

2. As a consequence of (1.14) wee see that $W_0^{1,p}(\Omega)^d \cap L^p_\sigma(\Omega) = \{u \in W_0^{1,p}(\Omega)^d : \operatorname{div} u = 0\}$. Moreover, under the assumptions on Ω it also holds that
$$\overline{C_{c,\sigma}^\infty(\Omega)}^{\|\cdot\|_{W^{1,p}(\Omega)}} = \{u \in W_0^{1,p}(\Omega)^d : \operatorname{div} u = 0\} = W_0^{1,p}(\Omega)^d \cap L^p_\sigma(\Omega),$$

   see [Gal94a, Chapter III.4]. In particular, $H_{0,\sigma}^1(\Omega) = H_0^1(\Omega)^d \cap L^2_\sigma(\Omega)$.

In this thesis we sometimes work on balls B_R for a radius $R > 1$. In these situations it will be important to estimate the norm of the Helmholtz projection $\mathbb{P}_{B_R,p}$ independently of the actual value of $R > 1$. We know that $\mathbb{P}_{B_R,p} u = u - \nabla\mathrm{p}$, where p is the solution to (1.13). Thus, it suffices to estimate the L^p-norm of $\nabla\mathrm{p}$ independently of R. This can be done by using a scaling argument. For $x \in B_R$ we set
$$y := \frac{x}{R}, \qquad \mathrm{q}(y) := \mathrm{p}(x), \qquad \widetilde{u}(y) := R\,u(x).$$

For a given $\varphi \in \widehat{W}^{1,p'}(B_1)$ with $\frac{1}{p} + \frac{1}{p'} = 1$ set $\widetilde{\varphi}(x) := \varphi(y)$. It is clear that $\widetilde{\varphi} \in \widehat{W}^{1,p'}(B_R)$. A simple substitution shows that
$$\int_{B_1} \nabla\mathrm{q} \cdot \nabla\varphi \, dy = \frac{R^2}{R^d} \int_{B_R} \nabla\mathrm{p} \cdot \nabla\widetilde{\varphi} \, dx = \frac{R^2}{R^d} \int_{B_R} u \cdot \nabla\widetilde{\varphi} \, dx = \int_{B_1} \widetilde{u} \cdot \nabla\varphi \, dy.$$

Thus q solves the weak Neumann problem on the unit ball B_1 with right-hand-side $\widetilde{u}$ and there exists a constant $C := C(B_1) > 0$ such that
$$\|\nabla\mathrm{q}\|_{L^p(B_1)} \leq C\|\widetilde{u}\|_{L^p(B_1)}.$$

Again by a substitution, we see that
$$\|\nabla\mathrm{p}\|_{L^p(B_R)}^p = \frac{R^d}{R^p}\|\nabla\mathrm{q}\|_{L^p(B_1)}^p \leq C\frac{R^d}{R^p}\|\widetilde{u}\|_{L^p(B_1)}^p = C\|u\|_{L^p(B_R)}^p.$$

Hence we proved the following result:

Proposition 1.65. *Let $1 < p < \infty$ and $R > 1$. Then there exists a constant $C > 0$, independent of R, such that*
$$\|\mathbb{P}_{B_R,p} u\|_{L^p(B_R)} \leq C\|u\|_{L^p(B_R)}$$

for all $u \in L^p(B_R)^d$.

1.4 The Stokes operator and the Stokes semigroup

As it was outlined in the introduction, large parts of this thesis are based on the the the *Stokes operator* with Dirichlet boundary conditions on $L_\sigma^p(\Omega)$, $1 < p < \infty$. Therefore, having the Helmholtz decomposition at hand, we shall now introduce the Stokes operator and the associated Stokes semigroup on $L_\sigma^p(\Omega)$. As we have seen in Section 1.2.1, there are very effective abstract tools for evolution equations in the framework of Hilbert spaces. Therefore we shall start with the case of $L_\sigma^2(\Omega)$ before turning our attention to the general case $L_\sigma^p(\Omega)$.

1.4.1 L^2-theory

Let $\emptyset \neq \Omega \subset \mathbb{R}^d$, $d \in \mathbb{N}$, be open. In this section we apply the form methods introduced in Section 1.2.1. For employing these methods it is actually necessary to consider complex-valued function spaces. Therefore in this section all function spaces are assumed to be complex-valued. As usual, $L_\sigma^2(\Omega)$ is equipped with the scalar product

$$\langle u, v \rangle_{L^2(\Omega)} := \int_\Omega u \cdot \overline{v}\, dx, \qquad u, v \in L_\sigma^2(\Omega).$$

In $L_\sigma^2(\Omega)$ we define a closed symmetric form $a : H_{0,\sigma}^1(\Omega) \times H_{0,\sigma}^1(\Omega) \to \mathbb{C}$ by setting

$$a(u, v) := \int_\Omega \nabla u : \overline{\nabla v}\, dx := \sum_{j=1}^d \int_\Omega \nabla u_j \cdot \overline{\nabla v_j}\, dx, \qquad u, v \in H_{0,\sigma}^1(\Omega). \tag{1.15}$$

The form $a : H_{0,\sigma}^1(\Omega) \times H_{0,\sigma}^1(\Omega) \to \mathbb{C}$ has the following properties:

(i) The form a is continuous and elliptic.

(ii) The form a is symmetric.

(iii) $\{u \in H_{0,\sigma}^1(\Omega) : a(u, u) = 0\} = \{0\}$.

(iv) $a(u, u) > 0$ for all $u \in H_{0,\sigma}^1(\Omega) \setminus \{0\}$.

Based on this form $a : H_{0,\sigma}^1(\Omega) \times H_{0,\sigma}^1(\Omega) \to \mathbb{C}$ we define the Stokes operator on $L_\sigma^2(\Omega)$.

Definition 1.66. Let $\emptyset \neq \Omega \subset \mathbb{R}^d$, $d \in \mathbb{N}$, be open and let $a : H_{0,\sigma}^1(\Omega) \times H_{0,\sigma}^1(\Omega) \to \mathbb{C}$ be the form defined in (1.15). Then the *Stokes operator* with Dirichlet boundary conditions

$$A_{\Omega,2} : H_{0,\sigma}^1(\Omega) \supseteq \mathcal{D}(A_{\Omega,2}) \to L_\sigma^2(\Omega)$$

is defined as the negative of the operator induced by the form $a : H_{0,\sigma}^1(\Omega) \times H_{0,\sigma}^1(\Omega) \to \mathbb{C}$, i.e. $-A_{\Omega,2}$ is induced by a.

Remark 1.67. Note that there is a certain inconsistency in the literature. For example in the monograph by Sohr [Soh01] the operator induced by the form $a : H_{0,\sigma}^1(\Omega) \times H_{0,\sigma}^1(\Omega) \to \mathbb{C}$ is called Stokes operator. In this case the negative of the Stokes operator generates an analytic semigroup. However, in our case we want the Stokes operator itself to be the generator of an analytic semigroup. This is more suitable for our purposes.

The following basic properties of the Stokes operator $A_{\Omega,2}$ are a direct consequence of the definition via the form in (1.15) and of the results stated in Section 1.2.1, see Proposition 1.40, Proposition 1.42 and Remarks 1.43.

Proposition 1.68. *Let* $\emptyset \neq \Omega \subset \mathbb{R}^d$, $d \in \mathbb{N}$, *be open and let* $(A_{\Omega,2}, \mathcal{D}(A_{\Omega,2}))$ *be the Stokes operator. Then the following assertions hold:*

(a) $A_{\Omega,2}$ *is injective and self-adjoint with dense domain* $\mathcal{D}(A_{\Omega,2}) \subset L^2_\sigma(\Omega)$.

(b) *There exists an operator* $A_{\Omega,2}^{-1} : \mathcal{D}(A_{\Omega,2}^{-1}) \to L^2_\sigma(\Omega)$ *with domain* $\mathcal{D}(A_{\Omega,2}^{-1}) := \mathcal{R}(A_{\Omega,2})$ *such that* $A_{\Omega,2}^{-1}A_{\Omega,2}x = x$ *for all* $x \in \mathcal{D}(A_{\Omega,2})$. *In particular,* $A_{\Omega,2}^{-1}$ *is again self-adjoint.*

(c) *The numerical range* $W(-A_{\Omega,2})$ *is contained in* $[0, \infty)$.

(d) *The operator* $A_{\Omega,2}$ *generates a bounded analytic semigroup* $\{T_{\Omega,2}(z)\}_{z \in \Sigma_{\pi/2}}$ *of angle* $\frac{\pi}{2}$ *and there exists a constant* $C > 0$ *such that*

$$\|T_{\Omega,2}(z)\|_{\mathscr{L}(L^2_\sigma(\Omega))} \leq 1, \qquad\qquad z \in \Sigma_{\frac{\pi}{2}}.$$
$$\|A_{\Omega,2}T_{\Omega,2}(t)\|_{\mathscr{L}(L^2_\sigma(\Omega))} \leq Ct^{-1}, \qquad\qquad t > 0.$$

Since $A_{\Omega,2}$ is injective, the inverse $A_{\Omega,2}^{-1}$ is well-defined on $\mathcal{R}(A_{\Omega,2})$ and since $A_{\Omega,2}$ is self-adjoint, it immediately follows that $A_{\Omega,2}^{-1}$ is also self-adjoint, see e.g. [Haa06, Section C.4].

The semigroup $\{T_{\Omega,2}(t)\}_{t\geq 0}$ is called the *Stokes semigroup* on $L^2_\sigma(\Omega)$. For the definition of the Stokes operator and for the generation of the Stokes semigroup no assumptions on the open set Ω are needed. This is a particular feature of working on $L^2_\sigma(\Omega)$ since here the Helmholtz decomposition is valid for all open sets and we have form methods at hand which directly yield the generation of an analytic semigroup. However, in general no explicit characterization of the domain $\mathcal{D}(A_{\Omega,2})$ is available. If one imposes more assumptions on Ω, then one obtains an easier characterization of $(A_{\Omega,2}, \mathcal{D}(A_{\Omega,2}))$.

Proposition 1.69. *Let* $\Omega \subset \mathbb{R}^d$, $d \geq 2$, *be either the whole space* $\mathbb{R}^d$, *the half-space* $\mathbb{R}^d_+$ *or a domain with compact boundary that is of class* C^2. *Then*

$$\mathcal{D}(A_{\Omega,2}) = H^2(\Omega)^d \cap H^1_{0,\sigma}(\Omega) = H^2(\Omega)^d \cap H^1_0(\Omega)^d \cap L^2_\sigma(\Omega),$$
$$A_{\Omega,p}u = \mathbb{P}_{\Omega,2}\Delta u.$$

For a proof of this proposition we refer to [Soh01, Theorem III.2.1.1 (d)], however one shall keep in mind that Sohr defines the Stokes operator as the negative of our operator $A_{\Omega,2}$.

Later in Chapter 5 we shall mainly work on bounded domains. In this case one has additional properties of the Stokes operator and the Stokes semigroup which mainly follow from Poincaré's inequality and from Rellich's result on compact embeddings.

Proposition 1.70. *Let $\Omega \subset \mathbb{R}^d$ be a bounded domain.*

(a) *The Stokes semigroup $\{T_{\Omega,2}(t)\}_{t \geq 0}$ is exponentially stable, i.e. there exists an $\varepsilon > 0$ such that*
$$\|T_{\Omega,2}(t)\|_{\mathscr{L}(L^2_\sigma(\Omega))} \leq e^{-\varepsilon t}, \qquad t > 0.$$

(b) *It holds that $0 \in \rho(A_{\Omega,2})$. In particular, it holds that $\mathcal{D}(A_{\Omega,2}^{-1}) = \mathcal{R}(A_{\Omega,2}) = L^2_\sigma(\Omega)$ and $A_{\Omega,2}^{-1} \in \mathscr{L}(L^2_\sigma(\Omega))$.*

(c) *The operator $A_{\Omega,2}^{-1} : L^2_\sigma(\Omega) \to L^2_\sigma(\Omega)$ is compact.*

(d) *The space $L^2_\sigma(\Omega)$ has an orthonormal basis $\{\psi_k\}_{k \in \mathbb{N}} \subset \mathcal{D}(A_{\Omega,2})$ consisting of eigenfunctions of the Stokes operator corresponding to a sequence of eigenvalues $\{-\lambda_k\}_{k \in \mathbb{N}}$ such that*
$$0 < \lambda_1 \leq \lambda_2 \leq \ldots \to \infty.$$

Proof. Let $\varepsilon > 0$ and consider the shifted operator $A_{\Omega,2} + \varepsilon$. Then, by applying Poincaré's inequality, we see that

$$\mathrm{Re}\,\langle -(A_{\Omega,2}+\varepsilon)u, u\rangle_{L^2(\Omega)} = -\mathrm{Re}\,\langle A_{\Omega,2}u, u\rangle_{L^2(\Omega)} - \varepsilon\|u\|^2_{L^2(\Omega)} \geq \|\nabla u\|^2_{L^2(\Omega)} - C^2\varepsilon\|\nabla u\|^2_{L^2(\Omega)} > 0$$

if $\varepsilon > 0$ is sufficiently small. Here $C := C(\Omega) > 0$ denotes the Poincaré constant of the bounded domain Ω. Thus $-(A_{\Omega,2} + \varepsilon)$ is m-accretive and as a consequence of the Lumer-Phillips Theorem $A_{\Omega,2} + \varepsilon$ generates a contraction semigroup on $L^2_\sigma(\Omega)$. By standard scaling arguments for semigroup generators, see [EN00, Chapter II.2], assertion (a) follows. By Remark 1.33 (ii), $0 \in \rho(A_{\Omega,2})$ directly follows from (a). Note that $A_{\Omega,2}^{-1} : L^2_\sigma(\Omega) \to \mathcal{D}(A_{\Omega,2})$. The compact embedding $H^1_{0,\sigma}(\Omega) \hookrightarrow L^2_\sigma(\Omega)$ yields assertion (c). The spectral theorem for compact self-adjoint operators, see e.g. [Rud73, Theorem 12.29 and 12.30], states that the operator $A_{\Omega,2}^{-1} : L^2_\sigma(\Omega) \to L^2_\sigma(\Omega)$ has a null-sequence of eigenvalues $\{-\mu_k\}_{k \in \mathbb{N}} \subset \mathbb{R} \setminus \{0\}$ and the corresponding eigenfunctions $\{\psi_k\}_{k \in \mathbb{N}}$ form an orthonormal basis of $L^2_\sigma(\Omega)$. If we set $\lambda_k = 1/\mu_k$, then $\{\psi_k\}_{k \in \mathbb{N}}$ is a sequence of eigenfunctions of the Stokes operator $A_{\Omega,2}$ corresponding to the eigenvalues $\{-\lambda_k\}_{k \in \mathbb{N}}$. Since $\mu_k \to 0$ as $k \to \infty$ and since

$$0 < \langle \nabla\psi_k, \nabla\psi_k\rangle_{L^2(\Omega)} = -\langle \psi_k, A_{\Omega,2}\psi_k\rangle_{L^2(\Omega)} = \lambda_k\|\psi_k\|^2_{L^2(\Omega)}$$

we can assume without loss of generality that $0 < \lambda_1 \leq \lambda_2 \leq \ldots \to \infty$. $\qquad\square$

Remark 1.71. Let $\{\psi_k\}_{k \in \mathbb{N}} \subset \mathcal{D}(A_{\Omega,2})$ be the orthonormal basis of $L^2_\sigma(\Omega)$ consisting of eigenfunctions of the Stokes operator, corresponding to the sequence of eigenvalues $\{-\lambda_k\}_{k \in \mathbb{N}}$. Then $\{\lambda_k^{-1/2}\psi_k\}_{k \in \mathbb{N}}$ is an orthonormal basis of $H^1_{0,\sigma}(\Omega)$ with respect to the scalar product $\langle \nabla\cdot, \nabla\cdot\rangle_{L^2(\Omega)}$ and $\{\lambda_k^{-1}\psi_k\}_{k \in \mathbb{N}}$ is an orthonormal basis of $\mathcal{D}(A_{\Omega,2})$ with respect to the scalar product $\langle A_{\Omega,2}\cdot, A_{\Omega,2}\cdot\rangle_{L^2(\Omega)}$.

Since the resolvent $R(\lambda, A_{\Omega,2})$, for large $\lambda > 0$, leaves real-valued functions invariant, the same also holds true for the semigroup $\{T_{\Omega,2}(t)\}_{t \geq 0}$. Similarly, as the eigenvalues $\{-\lambda_k\}_{k \in \mathbb{N}}$ are real, one can also choose the basis $\{\psi_k\}_{k \in \mathbb{N}} \subset \mathcal{D}(A_{\Omega,2})$ to be real-valued. Therefore, in the following we restrict our attention to the real-valued case.

1.4.2 L^p-theory

We consider the situation of $L^p_\sigma(\Omega)$ for $1 < p < \infty$. In this case, unlike for $L^2_\sigma(\Omega)$, we do not longer have form methods at hand to define the Stokes operator for general open sets $\Omega \subset \mathbb{R}^d$. Moreover, as it was discussed in Section 1.3, the Helmholtz decomposition does in general not hold for all domains $\Omega \subset \mathbb{R}^d$ and all values of $1 < p < \infty$. Therefore we shall only define the Stokes operator on $L^p_\sigma(\Omega)$ for certain domains in which the Helmholtz decomposition is known to exist.

Definition 1.72. Let $\Omega \subset \mathbb{R}^d$, $d \geq 2$, be either the whole space $\mathbb{R}^d$, the half-space $\mathbb{R}^d_+$ or a domain with compact boundary that is of class C^2. Then for $1 < p < \infty$ the *Stokes operator with Dirichlet boundary conditions* on $L^p_\sigma(\Omega)$ is defined by

$$
\begin{aligned}
\mathcal{D}(A_{\Omega,p}) &:= W^{2,p}(\Omega) \cap W_0^{1,p}(\Omega) \cap L^p_\sigma(\Omega), \\
A_{\Omega,p} u &:= \mathbb{P}_{\Omega,p} \Delta u.
\end{aligned}
$$

Remarks 1.73. 1. Note that this definition is consistent with definition in the case $p = 2$ by Proposition 1.69.

2. For convenience we shall frequently just write A instead of $A_{\Omega,p}$ if it is clear from the context what the actual value of $1 < p < \infty$ is and on which domain we are working.

As we have seen, in $L^2_\sigma(\Omega)$ it was rather easy to show that the Stokes operator $A_{\Omega,2}$ generates a bounded analytic semigroup of angle $\frac{\pi}{2}$. To show that the Stokes operator $A_{\Omega,p}$ generates an analytic semigroup on $L^p_\sigma(\Omega)$ is more involved and requires more technical work. In this thesis we use of the following result which basically follows from the classical results by Solonnikov, see [Sol77]. See also [Gig81], [GHH+10] for bounded and exterior domains, [BS87], [BV93], [MS97] for exterior domains and [McC81] [Uka87], [DHP01] for the half-space $\mathbb{R}^d_+$.

Proposition 1.74. *Let $\Omega \subset \mathbb{R}^d$, $d \geq 2$, be either the whole space $\mathbb{R}^d$, the half-space $\mathbb{R}^d_+$, or a domain with compact boundary that is of class C^2.*

(a) *For $1 < p < \infty$ the Stokes operator $(A_{\Omega,p}, \mathcal{D}(A_{\Omega,p}))$ generates an analytic semigroup $\{T_{\Omega,p}(t)\}_{t \geq 0}$ on $L^p_\sigma(\Omega)$.*

(b) *For $1 < p, q < \infty$ and $T_0 > 0$ the Stokes operator $(A_{\Omega,p}, \mathcal{D}(A_{\Omega,p}))$ admits maximal L^q-regularity on $J := (0, T_0)$ in $L^p_\sigma(\Omega)$.*

The semigroup $\{T_{\Omega,p}(t)\}_{t \geq 0}$ from Proposition 1.74 (a) is called *Stokes semigroup* on $L^p_\sigma(\Omega)$. If the value of $1 < p < \infty$ is clear from the context, or if no confusion arises, then we simply write $\{T_\Omega(t)\}_{t \geq 0}$.

Remarks 1.75.   1. Let $1 < p, q < \infty$. The family of semigroups $\{T_{\Omega,p}(t)\}_{t \geq 0}$ is consistent in the sense that $T_{\Omega,p}(t)f = T_{\Omega,q}(t)f$ for all $t \geq 0$ and all $f \in L^p_\sigma(\Omega) \cap L^q_\sigma(\Omega)$.

2. Let $1 < p, p' < \infty$ with $\frac{1}{p} + \frac{1}{p'} = 1$. The adjoint of the Stokes operator $(A_{\Omega,p}, \mathcal{D}(A_{\Omega,p}))$ is given by the Stokes operator $(A_{\Omega,p'}, \mathcal{D}(A_{\Omega,p'}))$. Moreover, $T^*_{\Omega,p}(t) = T_{\Omega,p'}(t)$ holds for all $t \geq 0$.

3. If Ω is bounded, then in Proposition 1.74 (b) we may choose $J = (0, \infty)$, see e.g. [GHH$^+$10, Theorem 1.1].

For the formulation of Definition 1.72 and Proposition 1.74 we restricted our attention to classes of domains for which the Helmholtz decomposition is known to exist. In fact, Definition 1.72 and Proposition 1.74 can be generalized to domains $\Omega \subset \mathbb{R}^d$ with sufficiently smooth boundary under the assumption that the Helmholtz decomposition in $L^p(\Omega)^d$ holds. This is done by Geissert, Heck, Hieber and Sawada in [GHHS10].

Proposition 1.76. *Let $\Omega \subset \mathbb{R}^d$, $d \geq 2$, be either the whole space $\mathbb{R}^d$, the half-space $\mathbb{R}^d_+$ or a domain with compact boundary that is of class C^2. Moreover, let $J := (0, T_0)$ for some $T_0 > 0$ and $1 < p, q < \infty$. For every $f \in L^q(J; L^p_\sigma(\Omega))$ and every $u_0 \in (L^p_\sigma(\Omega), \mathcal{D}(A_{\Omega,p}))_{1-1/q,q}$ there exists a unique solution $u \in W^{1,q}(J; L^p_\sigma(\Omega)) \cap L^q(J; \mathcal{D}(A_{\Omega,p}))$ to the inhomogeneous Stokes problem*

$$\begin{cases} u'(t) - A_{\Omega,p}u(t) &= f(t), \quad t \in J, \\ u(0) &= u_0. \end{cases}$$

Now let us assume that $u_0 = 0$. There exists a constant $C > 0$, independent of T_0, such that

$$\|u'\|_{L^q(J;L^p_\sigma(\Omega))} + \|A_{\Omega,p}u\|_{L^q(J;L^p_\sigma(\Omega))} \leq C\|f\|_{L^q(J;L^p_\sigma(\Omega))}.$$

The fact that the constant $C > 0$ in Proposition 1.76 is independent of T_0 does not automatically follow from the abstract result in Proposition 1.47, but this has to be shown directly in each case. In the whole space case $\mathbb{R}^d$ one can closely check the proof of [Lad69, Chapter 4, Theorem 10] to see that $C > 0$ is independent of T_0. The half-space case is sketched e.g. in [MS97, Section 2] and there it is stated that the constant does not depend on T_0. For bounded domains we refer e.g. to [GHH$^+$10, Theorem 1.1] and for exterior domains to [MS97, Theorem 1.4].

Remarks 1.77.   1. Let Ω be as in Proposition 1.76 and $f \in L^p_\sigma(\Omega)$. Then there exists a unique solution

$$(u, \mathrm{p}) \in W^{1,q}(J; L^p_\sigma(\Omega)) \cap L^q(J; \mathcal{D}(A_{\Omega,p})) \times L^q(J; \widehat{W}^{1,p}(\Omega))$$

to

$$\begin{cases} u_t - \Delta u + \nabla \mathrm{p} &= f \quad \text{in } J \times \Omega, \\ \operatorname{div} u &= 0 \quad \text{in } J \times \Omega, \\ u &= 0 \quad \text{on } J \times \partial\Omega, \\ u|_{t=0} &= 0 \quad \text{in } \Omega. \end{cases}$$

Moreover, there exists a constant $C > 0$ independent of T_0 such that

$$\|\nabla\mathrm{p}\|_{L^q(J;L^p(\Omega))} \le C\|f\|_{L^q(J;L^p_\sigma(\Omega))}.$$

Here $u \in W^{1,q}(J;L^p_\sigma(\Omega)) \cap L^q(J;\mathcal{D}(A_{\Omega,p}))$ is the function from Proposition 1.76 (a) and $\mathrm{p} \in L^q(J;\widehat{W}^{1,p}(\Omega))$ is chosen such that $\nabla\mathrm{p} = (\mathrm{Id} - \mathbb{P}_{\Omega,p})\Delta u$.

2. Let Ω be as in Proposition 1.76 and let $f \in L^p(\Omega)^d$ and $u_0 \in (L^p_\sigma(\Omega), \mathcal{D}(A_{\Omega,p}))_{1-1/q,q}$. Then there exists a unique solution

$$(u,\mathrm{p}) \in W^{1,q}(J;L^p_\sigma(\Omega)) \cap L^q(J;\mathcal{D}(A_{\Omega,p})) \times L^q(J;\widehat{W}^{1,p}(\Omega))$$

to

$$\begin{cases} u_t - \Delta u + \nabla\mathrm{p} &=\ f \quad \text{in } J \times \Omega, \\ \operatorname{div} u &=\ 0 \quad \text{in } J \times \Omega, \\ u &=\ 0 \quad \text{on } J \times \partial\Omega, \\ u|_{t=0} &=\ u_0 \quad \text{in } \Omega. \end{cases}$$

Here $u \in W^{1,q}(J;L^p_\sigma(\Omega)) \cap L^q(J;\mathcal{D}(A_{\Omega,p}))$ is the function from Proposition 1.76 (b) and $\mathrm{p} \in L^q(J;\widehat{W}^{1,p}(\Omega))$ is chosen such that $\nabla\mathrm{p} = (\mathrm{Id} - \mathbb{P}_{\Omega,p})\Delta u + (\mathrm{Id} - \mathbb{P}_{\Omega,p})f$.

From general properties of analytic semigroups and the Gagliardo-Nirenberg inequality we can directly derive the following *local*[4] estimates for the Stokes semigroup $\{T_{\Omega,p}(t)\}_{t\ge 0}$.

Proposition 1.78. *Let $\Omega \subset \mathbb{R}^d$, $d \ge 2$, be either the whole space $\mathbb{R}^d$, the half-space $\mathbb{R}^d_+$ or a domain with compact boundary that is of class C^2. Moreover, let $1 < p \le q < \infty$, $m = 1,2$ and $T_0 > 0$. Then there exists a constant $C := C(T_0) > 0$ such that for $f \in L^p_\sigma(\Omega)$*

(i) $\|T_{\Omega,p}(t)f\|_{L^q(\Omega)} \le Ct^{-\frac{d}{2}\left(\frac{1}{p}-\frac{1}{q}\right)}\|f\|_{L^p(\Omega)}, \quad t \in (0, T_0],$

(ii) $\|T_{\Omega,p}(t)f\|_{L^\infty(\Omega)} \le Ct^{-\frac{d}{2p}}\|f\|_{L^p(\Omega)}, \quad t \in (0, T_0],$

(iii) $\|\nabla^m T_{\Omega,p}(t)f\|_{L^q(\Omega)} \le Ct^{-\frac{d}{2}\left(\frac{1}{p}-\frac{1}{q}\right)-\frac{m}{2}}\|f\|_{L^p(\Omega)}, \quad t \in (0, T_0],$

(iv) $\|A_{\Omega,p}T_{\Omega,p}(t)f\|_{L^q(\Omega)} \le Ct^{-\frac{d}{2}\left(\frac{1}{p}-\frac{1}{q}\right)-1}\|f\|_{L^p(\Omega)}, \quad t \in (0, T_0].$

Proof. From (1.5) it follows that (i) holds for $p = q$. Moreover, Remarks 1.36 directly yields estimate (iv) for $p = q$. Next we take $\lambda > 0$ such that $\lambda \in \rho(A_{\Omega,p})$. Then we see that[5]

$$\|\nabla^2 T_{\Omega,p}(t)f\|_{L^p(\Omega)} = \|\nabla^2 R(\lambda, A_{\Omega,p})(\lambda - A_{\Omega,p})T_{\Omega,p}(t)f\|_{L^p(\Omega)}$$
$$\le C\|(\lambda - A_{\Omega,p})T_{\Omega,p}(t)f\|_{L^p(\Omega)} \le Ct^{-1}\|f\|_{L^p(\Omega)}.$$

---

[4]The estimates in Proposition 1.78 are called local as they hold only on finite time-intervals and they express a certain behaviour of the Stokes semigroup at $t = 0$.

[5]Here we use that $\nabla^2 R(\lambda, A_{\Omega,p})$ is a closed operator defined on the whole $L^p_\sigma(\Omega)$, and thus it is bounded as a consequence of the closed graph theorem.

This proves (iii) for $m = 2$ and $p = q$. The other assertions follow now by applying the Gagliardo-Nirenberg inequality, see Proposition 1.17, and the semigroup property. See also the proof of Proposition 4.15 where this argumentation is presented in full detail. $\qquad\square$

To conclude this section we collect some decay estimates for the Stokes semigroup $\{T_{\Omega,p}(t)\}_{t\geq 0}$ that show the behaviour of the semigroup as $t \to \infty$. Such estimates are also referred to as *global* estimates. Here we only consider the cases $\Omega = \mathbb{R}^d$ and $\Omega \subset \mathbb{R}^d$ a bounded domain. The case of exterior domains is discussed in more detail in Chapter 3.

In the whole space $\mathbb{R}^d$ the Stokes semigroup $\{T_{\mathbb{R}^d}(t)\}_{t\geq 0}$ can be explicitly given via the classical heat (also called Gaussian) kernel:

$$T_{\mathbb{R}^d}(t)f := \frac{1}{(4\pi t)^{\frac{d}{2}}} \int_{\mathbb{R}^n} f(x - y)e^{-\frac{|y|^2}{4t}} \, \mathrm{d}y, \qquad t > 0, \ f \in L^p_\sigma(\mathbb{R}^d). \tag{1.16}$$

We also define a semigroup $\{\tilde{T}_{\mathbb{R}^d}(t)\}_{t\geq 0}$ on $L^p(\mathbb{R}^d)^d$, $1 \leq p < \infty$, by (1.16) for all $f \in L^p(\mathbb{R}^d)^d$. This is the classical heat semigroup. Then an easy calculation shows that $\{\tilde{T}_{\mathbb{R}^d}(t)\}_{t\geq 0}$ leaves $L^p_\sigma(\mathbb{R}^d)$ invariant and this shows that $T_{\mathbb{R}^d}(t) = \tilde{T}_{\mathbb{R}^d}(t)|_{L^p_\sigma(\mathbb{R}^d)}$. Moreover, by using the explicit representation of the Helmholtz projection $\mathbb{P}_{\mathbb{R}^d}$ we can conclude that for $f \in L^p(\mathbb{R}^d)^d$ we have

$$T_{\mathbb{R}^d}(t)\mathbb{P}_{\mathbb{R}^d}f = \mathbb{P}_{\mathbb{R}^d}\tilde{T}_{\mathbb{R}^d}(t)f.$$

The explicit formula and Young's inequality for convolutions directly yield the following results. The proofs are very similar to the one of Theorem 4.9 and therefore are omitted.

Proposition 1.79. *Let $1 < p < q \leq \infty$ or $1 < p \leq q < \infty$, and let $m = 1, 2$. Then there exists a constant $C := C(p, q, d) > 0$ such that for $f \in L^p(\mathbb{R}^d)^d$*

(i) $\|T_{\mathbb{R}^d}(t)\mathbb{P}_{\mathbb{R}^d}f\|_{L^q(\mathbb{R}^d)} \leq Ct^{-\frac{d}{2}\left(\frac{1}{p}-\frac{1}{q}\right)}\|f\|_{L^p(\mathbb{R}^d)}, \quad t > 0,$

(ii) $\|\nabla^m T_{\mathbb{R}^d}(t)\mathbb{P}_{\mathbb{R}^d}f\|_{L^q(\mathbb{R}^d)} \leq Ct^{-\frac{d}{2}\left(\frac{1}{p}-\frac{1}{q}\right)-\frac{m}{2}}\|f\|_{L^p(\mathbb{R}^d)}, \quad t > 0,$

(iii) $\|A_{\mathbb{R}^d} T_{\mathbb{R}^d}(t)\mathbb{P}_{\mathbb{R}^d}f\|_{L^q(\mathbb{R}^d)} \leq Ct^{-\frac{d}{2}\left(\frac{1}{p}-\frac{1}{q}\right)-1}\|f\|_{L^p(\mathbb{R}^d)}, \quad t > 0.$

Proposition 1.80. *If $f \in L^p_\sigma(\mathbb{R}^d) \cap L^\infty(\mathbb{R}^d)^d$ for some $1 < p < \infty$, then there exists a constant $C := C(d) > 0$ such that*

$$\|T_{\mathbb{R}^d}(t)f\|_{L^\infty(\mathbb{R}^d)} \leq C\|f\|_{L^\infty(\mathbb{R}^d)}, \qquad t > 0.$$

Moreover, let $1 < q \leq \infty$. If $f \in L^p(\mathbb{R}^d)^d \cap L^1(\mathbb{R}^d)^d$ for some $1 < p < \infty$, then there exists a constant $C := C(q, p, d) > 0$ such that

$$\|T_{\mathbb{R}^d}(t)\mathbb{P}_{\mathbb{R}^d}f\|_{L^q(\mathbb{R}^d)} \leq C\|f\|_{L^1(\mathbb{R}^d)}, \qquad t > 0.$$

Proposition 1.81. *Let $1 < p < \infty$ and $m = 1, 2$. Then there exists a constant $C := C(p) > 0$ such that for all $f \in W^{m,p}(\mathbb{R}^d)^d$*

$$\|T_{\mathbb{R}^d,p}(t)\mathbb{P}_{\mathbb{R}^d,p}f\|_{\widehat{W}^{m,p}(\mathbb{R}^d)} \leq C\|f\|_{\widehat{W}^{m,p}(\mathbb{R}^d)}, \qquad t > 0.$$

For bounded domains we have the following estimates.

Proposition 1.82. *Let $1 < p \leq q < \infty$, $m = 1, 2$, and let $\Omega \subset \mathbb{R}^d$ be a bounded domain of class C^2. Then there exist constants $\delta := \delta(p, \Omega) > 0$, $C := C(p, q, \Omega) > 0$ such that for $f \in L^p_\sigma(\Omega)$ we have*

(i) $\|T_{\Omega,p}(t)f\|_{L^p(\Omega)} \leq Ce^{-\delta t}\|f\|_{L^p(\Omega)}, \quad t > 0,$

(ii) $\|T_{\Omega,p}(t)f\|_{L^q(\Omega)} \leq Ce^{-\delta t}t^{-\frac{d}{2}\left(\frac{1}{p}-\frac{1}{q}\right)}\|f\|_{L^p(\Omega)}, \quad t > 0,$

(iii) $\|\nabla^m T_{\Omega,p}(t)f\|_{L^q(\Omega)} \leq Ce^{-\delta t}t^{-\frac{d}{2}\left(\frac{1}{p}-\frac{1}{q}\right)-\frac{m}{2}}\|f\|_{L^p(\Omega)}, \quad t > 0,$

(iv) $\|A_{\Omega,p}T_{\Omega,p}(t)f\|_{L^q(\Omega)} \leq Ce^{-\delta t}t^{-\frac{d}{2}\left(\frac{1}{p}-\frac{1}{q}\right)-1}\|f\|_{L^p(\Omega)}, \quad t > 0.$

Proof. By Proposition 1.70 we know that $0 \in \rho(A_{\Omega,2})$. Since $W^{1,p}(\Omega)^d$ is compactly embedded into $L^p(\Omega)^d$, the operator $A_{\Omega,p}$ has compact resolvents for all $1 < p < \infty$. By [Are94, Proposition 2.6] it follows that $\sigma(A_{\Omega,2}) = \sigma(A_{\Omega,p})$ for $1 < p < \infty$. This shows $0 \in \rho(A_{\Omega,p})$. Remark 1.33 yields that there exist constants $\delta := \delta(p, \Omega) > 0$, $C := C(p, \Omega) > 0$ such that

$$\|T_{\Omega,p}(t)f\|_{L^p(\Omega)} \leq Ce^{-\delta t}\|f\|_{L^p(\Omega)}, \qquad t > 0. \tag{1.17}$$

Moreover, since $\{T_{\Omega,p}(t)\}_{t \geq 0}$ is an analytic semigroup, it follows from Remarks 1.36 that

$$\|A_{\Omega,p}T_{\Omega,p}(t)f\|_{L^p(\Omega)} \leq Ct^{-1}\|f\|_{L^p(\Omega)}, \qquad t \in (0, 1]$$

for $C := C(p, \Omega) > 0$. By applying the semigroup property and by using (1.17) we see that

$$\|A_{\Omega,p}T_{\Omega,p}(t)f\|_{L^p(\Omega)} \leq C\|T_{\Omega,p}(t-1)f\|_{L^p(\Omega)} \leq Ce^{-\delta(t-1)}\|f\|_{L^p(\Omega)}, \qquad t > 1$$

for $C := C(\Omega, p) > 0$. This proves (iv) in the case $p = q$. Since $A_{\Omega,p}$ is invertible, we obtain

$$\|\nabla^2 T_{\Omega,p}(t)f\|_{L^p(\Omega)} \leq C\|A_{\Omega,p}T_{\Omega,p}(t)f\|_{L^p(\Omega)}, \qquad t > 0$$

for a constant $C := C(p, \Omega) > 0$. This shows *(iii)* for $m = 2$ and $p = q$. The other assertions follow now by applying the Gagliardo-Nirenberg inequality and the semigroup property. $\square$

1.5 The Stokes operator with rotating effect

In the previous section we introduced the classical Stokes operator and the associated semigroup. As it was explained in the introduction, in order to study Navier-Stokes flows around a rotating obstacle the classical Stokes operator has to be replaced by a modified operator which has unbounded coefficients. It is the purpose of this section to introduce this operator which we shall call *Stokes operator with rotating effect*.

Let $\Omega \subset \mathbb{R}^d$, $d \geq 2$, denote an exterior domain of class C^2 or let $\Omega := \mathbb{R}^d$. We consider a linear operator formally defined on smooth vector fields $u = (u_1, ..., u_d)$ by

$$\mathcal{L}u(x) = \Delta u(x) + Mx \cdot \nabla u(x) + c \cdot \nabla u(x) - Mu(x), \quad x \in \Omega. \tag{1.18}$$

Here $M \in \mathbb{R}^{d \times d}$ is a given matrix and $c \in \mathbb{R}^d$ is a given vector. The particularity of this linear operator is that the coefficients of the drift term grow linearly in the space variable x and are therefore unbounded over Ω. Therefore, it is not possible to consider the first order term as a *small perturbation* of the Laplace operator Δ, and thus classical perturbation theory, see Proposition 1.37, cannot be applied. Operators with linearly growing drift coefficients, also referred to as *Ornstein-Uhlenbeck type operators*, have been studied intensively in the last two decades in various function spaces over $\mathbb{R}^d$ and exterior domains. We refer e.g. to [Met01, MPRS02, GHHW05, HDW05] and the references stated therein. See also the monograph [LB07] for an overview. Note that in all these references only scalar-valued operators were studied and not systems of equations as in (1.18). Parabolic systems with unbounded coefficients were studied in the Ph.D. thesis by Wiedl, see [Wie07, Chapter 5].

Let $1 < p < \infty$. We define the L^p-realizations of the formally defined operator $\mathcal{L}$ by setting

$$\mathcal{D}(\tilde{L}_{\Omega,p}) \; := \; \{u \in W^{2,p}(\Omega)^d \cap W_0^{1,p}(\Omega)^d : Mx \cdot \nabla u \in L^p(\Omega)^d\},$$
$$\tilde{L}_{\Omega,p} u \; := \; \mathcal{L}u.$$

Note that the condition $Mx \cdot \nabla u \in L^p(\Omega)^d$ has to be part of the domain, as in general this does not hold for an arbitrary L^p-function. The space $C_c^\infty(\Omega)^d$ is contained in $\mathcal{D}(\tilde{L}_{\Omega,p})$.

Definition 1.83. Let $\Omega \subset \mathbb{R}^d$, $d \geq 2$, be either the whole space $\mathbb{R}^d$ or an exterior domain of class C^2. Moreover, let $M \in \mathbb{R}^{d \times d}$ and $c \in \mathbb{R}^d$. Then the *Stokes operator with rotating effect* on $L_\sigma^p(\Omega)$, $1 < p < \infty$, is defined by

$$\mathcal{D}(L_{\Omega,p}) \; := \; \mathcal{D}(\tilde{L}_{\Omega,p}) \cap L_\sigma^p(\Omega),$$
$$L_{\Omega,p} u \; := \; \mathbb{P}_{\Omega,p} \tilde{L}_{\Omega,p} u.$$

Remark 1.84. For simplicity we often write L_Ω and $\tilde{L}_\Omega$ instead of $L_{\Omega,p}$ and $\tilde{L}_{\Omega,p}$, respectively, if it is clear from the context what the actual value of $1 < p < \infty$ is.

Proposition 1.85. *Let $1 < p < \infty$ and let $\Omega \subset \mathbb{R}^d$, $d \geq 2$, be either the whole space $\mathbb{R}^d$ or an exterior domain of class C^2. Then the Stokes operator with rotating effect $(L_{\Omega,p}, \mathcal{D}(L_{\Omega,p}))$ generates a C_0-semigroup on $L_\sigma^p(\Omega)$ which is not analytic.*

This generation result was first proved by Hishida [His99a] for $d = 3$, $Mx = (0, 0, 1)^\top \times x$, $c = 0$ and $p = 2$. Hishida also showed that the semigroup is not analytic. Also in the special situation $d = 3$, $Mx = (0, 0, 1)^\top \times x$, $c = 0$ and $p = 2$, it was shown by Farwig and Neustupa [FN07] that the essential spectrum of $L_{\Omega,2}$ consists of equally spaced half lines in the left half-plane parallel to the negative real-axis. This immediately yields that $L_{\Omega,2}$ cannot be the generator of an analytic semigroup. The more general generation result for arbitrary $M \in \mathbb{R}^{d \times d}$, for $c = 0$ and for $1 < p < \infty$ was proved for the whole space $\mathbb{R}^d$ by Hieber and Sawada [HS05] and for exterior domains by Geissert, Heck and Hieber [GHH06a]. The result for $c \neq 0$ then follows directly from these results by using classical perturbation theory for C_0-semigroups, e.g. [EN00, Theorem 2.7]. In the case $d = 3$, $Mx = (0, 0, k)^\top \times x$ for some $k \in \mathbb{R}$ and $c = 0$, it was shown by Hishida and Shibata [HS09] that the semigroup is even bounded and they derived global L^p-L^q-estimates and gradient estimates for the semigroup. See also [Shi08, Shi10] for the same result in the case $c = (0, 0, k)^\top$.

In the whole space situation $\mathbb{R}^d$ the following lemma is quite helpful.

Lemma 1.86. *Let $1 < p < \infty$. Then $L_{\mathbb{R}^d,p}u = \tilde{L}_{\mathbb{R}^d,p}u$ holds for all $u \in \mathcal{D}(L_{\mathbb{R}^d,p})$.*

Proof. Let $u \in C_{c,\sigma}^\infty(\mathbb{R}^d)$. As a consequence of Remarks 1.62 we know that $\mathbb{P}_{\mathbb{R}^d,p}\Delta u = \Delta u$ holds. Moreover, an easy computation shows that

$$\operatorname{div}(Mx \cdot \nabla u + c \cdot \nabla u - Mu) = Mx \cdot \nabla(\operatorname{div} u) + \operatorname{div} Mu - \operatorname{div} Mu + c \cdot \nabla(\operatorname{div} u) = 0.$$

Thus, $\tilde{L}_{\mathbb{R}^d,p}u \in L_\sigma^p(\mathbb{R}^d)$ for $u \in C_{c,\sigma}^\infty(\mathbb{R}^d)$. Density of $C_{c,\sigma}^\infty(\mathbb{R}^d)$ in $\mathcal{D}(L_{\Omega,p})$ yields the claim. $\square$

1.6 The Bogovskiĭ operator and applications

In this section we introduce the Bogovskiĭ operator, denoted by $\mathbb{B}_\Omega$, which is frequently used in this thesis. This operator is roughly speaking the right-inverse to the divergence operator, i.e. under suitable assumptions on the scalar-valued function g, the function $u = \mathbb{B}_\Omega g$ solves

$$\begin{cases} \operatorname{div} u = g & \text{in } \Omega, \\ u = 0 & \text{on } \partial\Omega, \end{cases} \tag{1.19}$$

where $\Omega \subset \mathbb{R}^d$, $d \geq 2$, is a bounded and locally Lipschitz domain. The following result collects properties and estimates for the Bogovskiĭ operator, see Bogovskiĭ [Bog79], Geissert, Heck and Hieber [GHH06b] and the monograph by Galdi [Gal94a, Section III.3].

Proposition 1.87. *Let $\Omega \subset \mathbb{R}^d$, $d \geq 2$, be a bounded and locally Lipschitz domain, let $1 < p < \infty$ and let $m \in \mathbb{Z}$. Then there exists a mapping $\mathbb{B}_\Omega : C_c^\infty(\Omega) \to C_c^\infty(\Omega)^d$ such that for all $g \in C_c^\infty(\Omega)$ with*

$$\int_\Omega g(x)\,\mathrm{d}x = 0, \tag{1.20}$$

it holds that $\operatorname{div}\mathbb{B}_\Omega g = g$. Moreover, $\mathbb{B}_\Omega$ can be extended continuously to a bounded operator

$$\mathbb{B}_\Omega : W_0^{m,p}(\Omega) \to W_0^{m+1,p}(\Omega)^d,$$

such that

$$\|\mathbb{B}_\Omega g\|_{W^{m+1,p}(\Omega)} \leq C\|g\|_{W^{m,p}(\Omega)}, \qquad g \in W_0^{m,p}(\Omega), \tag{1.21}$$

provided that $m > -2 + \frac{1}{p}$ holds.

Remark 1.88. Since $C_c^\infty(\Omega)$ is dense in $L^p(\Omega)$ it is clear $\operatorname{div}\mathbb{B}_\Omega g = g$ holds for all $g \in L^p(\Omega)$ satisfying the compatibility condition (1.20). The compatibility condition (1.20) on g is not needed for the estimate (1.21). However, if we do not require (1.20), then $\mathbb{B}_\Omega g$ is not a solution to problem (1.19).

In this thesis the Bogovskiĭ operator is mainly used for two purposes: First of all, this operator is used to keep the solenoidal (divergence-free) condition in cut-off and localization procedures, as the usual localization procedures for elliptic and parabolic problems, see e.g. [ADN59], do not automatically keep the divergence-free condition. Second, the Bogovskiĭ operator is used to deal with inhomogeneous Dirichlet boundary conditions for the Navier-Stokes equations.

1.6.1 Cut-off procedures

Let $\Omega \subset \mathbb{R}^d$ be an exterior domain of class C^2. In the following we consider the homogeneous Stokes equations

$$\begin{cases} u_t - \Delta u + \nabla p & = & 0 & \text{in } \mathbb{R}_+ \times \Omega, \\ \operatorname{div} u & = & 0 & \text{in } \mathbb{R}_+ \times \Omega, \\ u & = & 0 & \text{on } \mathbb{R}_+ \times \partial\Omega, \\ u|_{t=0} & = & u_0 & \text{in } \Omega. \end{cases} \tag{1.22}$$

Let (u, p) be a solution to (1.22). The solution u is given via the Stokes semigroup $\{T_{\Omega,p}(t)\}_{t \geq 0}$, i.e. $u(t) = T_{\Omega,p}(t)u_0$. If one wants to derive norm estimates for $u(t)$, then it is convenient to reduce the exterior domain problem to a problem in the whole space $\mathbb{R}^d$ and to a problem in a bounded domain close to the boundary $\partial\Omega$. Set $\mathcal{O} := \mathbb{R}^d \setminus \Omega$ and choose $R > 0$ sufficiently large such that $\mathcal{O} \subset B_R$. Moreover, set

$$D := \Omega \cap B_{R+4},$$
$$D_1 := \{x \in \Omega : R < |x| < R + 3\},$$

and choose a cut-off function $\varphi_1 \in C_c^\infty(\mathbb{R}^d)$ such that $0 \leq \varphi_1 \leq 1$ and

$$\varphi_1(x) := \begin{cases} 1, & |x| \leq R + 1, \\ 0, & |x| \geq R + 2. \end{cases}$$

Set $\varphi_2 := (1 - \varphi_1)$. Note that $\varphi_1 + \varphi_2 = 1$ and $\nabla\varphi_1 = -\nabla\varphi_2$. Thus, it is clear that $u(t) - \varphi_1 u(t) + \varphi_2 u(t)$.

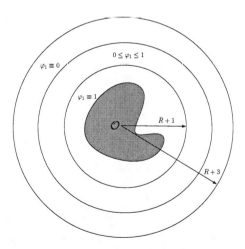

Figure 1.1: Definition of the cut-off function φ_1.

The functions $\varphi_1 u(t)$ and $\varphi_2 u(t)$ are not divergence-free anymore since a simple calculation yields $\operatorname{div}(\varphi_i u) = \varphi_i \operatorname{div} u + (\nabla\varphi_i) \cdot u$, $i = 1, 2$. In the following we shall modify the functions

$\varphi_1 u(t)$ and $\varphi_2 u(t)$ in such a way that they will become divergence-free. For $i = 1, 2$, set

$$
\begin{aligned}
u^{(i)}(t) &:= \varphi_i u(t) - \mathbb{B}_{D_1}((\nabla \varphi_i) \cdot u(t)), \\
\mathrm{p}^{(i)}(t) &:= \varphi_i \mathrm{p}(t).
\end{aligned}
\tag{1.23}
$$

It is clear that $u = u^{(1)} + u^{(2)}$ and $\mathrm{p} = \mathrm{p}^{(1)} + \mathrm{p}^{(2)}$. Here $\mathbb{B}_{D_1}$ is the Bogovskiĭ operator on the bounded domain D_1, and we interpret the term $\mathbb{B}_{D_1}((\nabla \varphi_i) \cdot u)$ as a function on D for $i = 1$ and on $\mathbb{R}^d$ for $i = 2$ by extending it by zero[6]. Integration by parts and the divergence theorem yield

$$
\int_{D_1} \nabla \varphi_i(x) \cdot u(t, x)\, \mathrm{d}x = \int_{\partial D_1} u(t, x) \cdot \nu\, \mathrm{d}\sigma - \int_{D_1} \varphi(x)\, \mathrm{div}\, u(t, x)\, \mathrm{d}x = 0
$$

for $i = 1, 2$, since $u(t) \in L^p_\sigma(\Omega)$. Hence, the use of the Bogovskiĭ operator ensures that $u^{(i)}$, $i = 1, 2$, are divergence-free. More precisely, by using Proposition 1.87 we see that

$$
u^{(1)} \in W^{2,p}(D)^d \cap W_0^{1,p}(D)^d \cap L^p_\sigma(D) \qquad \text{and} \qquad u^{(2)} \in W^{2,p}(\mathbb{R}^d)^d \cap L^p_\sigma(\mathbb{R}^d).
$$

A short calculation shows that $u^{(1)}$ solves the inhomogeneous Stokes problem

$$
\begin{cases}
u_t^{(1)} - \Delta u^{(1)} + \nabla \mathrm{p}^{(1)} &= F^{(1)} \quad \text{in } \mathbb{R}_+ \times D, \\
\mathrm{div}\, u^{(1)} &= 0 \quad \text{in } \mathbb{R}_+ \times D, \\
u^{(1)} &= 0 \quad \text{on } \mathbb{R}_+ \times \partial D, \\
u^{(1)}|_{t=0} &= u_0^{(1)} \quad \text{in } D,
\end{cases}
$$

in the bounded domain D, and $u^{(2)}$ solves the inhomogeneous Stokes problem

$$
\begin{cases}
u_t^{(2)} - \Delta u^{(2)} + \nabla \mathrm{p}^{(2)} &= F^{(2)} \quad \text{in } \mathbb{R}_+ \times \mathbb{R}^d, \\
\mathrm{div}\, u^{(2)} &= 0 \quad \text{in } \mathbb{R}_+ \times \mathbb{R}^d, \\
u^{(2)}|_{t=0} &= u_0^{(2)} \quad \text{in } \mathbb{R}^d,
\end{cases}
$$

in the whole space $\mathbb{R}^d$. Here the "error terms" $F^{(i)}$, $i = 1, 2$, are given by

$$
F^{(i)}(t) = -2(\nabla \varphi_i) \cdot \nabla u(t) - (\Delta \varphi_i) u(t) - (\partial_t - \Delta) \mathbb{B}_{D_1}((\nabla \varphi_i) \cdot u(t)) + (\nabla \varphi_i)\mathrm{p},
$$

and the initial values $u_0^{(i)}$, $i = 1, 2$, are given by $u_0^{(i)} = \varphi_i u_0 - \mathbb{B}_{D_1}((\nabla \varphi_i) \cdot u_0)$. Note that supports of $F^{(i)}(t)$, $i = 1, 2$, are contained in $D_1 \subset D$ for all $t \in \mathbb{R}_+$. With this technique we decomposed the function $u(t)$ into a function $u^{(1)}(t)$ "living" in the bounded domain D and into a function $u^{(2)}(t)$ "living" in the whole space $\mathbb{R}^d$. This now allows to use techniques and results for bounded domains and for the whole space $\mathbb{R}^d$. Since the "error terms" $F^{(i)}(t)$, $i = 1, 2$, are supported only close to the boundary $\partial \Omega$, they can be estimated by using e.g. Poincaré's inequality. Such a cut-off technique and variants of this technique are used frequently in this thesis. For example, such a construction is used in Chapter 3 and variant of it is applied to stationary Stokes problems in Chapter 2.

---

[6]This is possible as $\mathbb{B}_{D_1} : W_0^{m,p}(D_1) \to W_0^{m+1,p}(D_1)^d$.

1.6.2 Inhomogeneous Dirichlet boundary conditions

Let $\Omega \subset \mathbb{R}^d$ be an exterior domain of class C^2. Moreover, let h be a function defined on the boundary $\partial\Omega$. We consider the Navier-Stokes equations

$$\begin{cases} w_t - \Delta w + w \cdot \nabla w + \nabla p &= 0 & \text{in } \mathbb{R}_+ \times \Omega, \\ \operatorname{div} w &= 0 & \text{in } \mathbb{R}_+ \times \Omega, \\ w &= h & \text{on } \mathbb{R}_+ \times \partial\Omega, \\ w|_{t=0} &= w_0 & \text{in } \Omega. \end{cases} \tag{1.24}$$

Due to technical reasons it is often more convenient to work with zero boundary conditions. Therefore we shall construct a suitable solenoidal extension in Ω of the boundary velocity h. For that purpose, assume there exists an extension H of h onto $\mathbb{R}^d$ such that

$$H|_{\partial\Omega} = h, \qquad H \in C^1([0,\infty); C^2(\mathbb{R}^d)), \qquad \operatorname{div} H(t) = 0 \text{ for all } t \geq 0.$$

A typical situation is that $h(t,x) := M(t)x$, where $M \in C^1([0,\infty); \mathbb{R}^{d\times d})$ and $M(t)$ is a skew symmetric matrix for every $t \geq 0$, i.e. $M(t)^\top = -M(t)$. Then the extension is given by $H(t,x) = M(t)x$. Since $\operatorname{tr} M(t) = 0$ for all $t \geq 0$ it is clear that $\operatorname{div} H(t) = 0$ for all $t \geq 0$.

Next, we shall cut-off the function H away from $\partial\Omega$. Set $\mathcal{O} := \mathbb{R}^d \setminus \Omega$ and choose $R > 0$ such that $\mathcal{O} \subset B_R$. Moreover, choose a cut-off function $\xi \in C_c^\infty(\mathbb{R}^d)$ such that $0 \leq \xi \leq 1$ and

$$\xi(x) := \begin{cases} 1, & |x| \leq R+1, \\ 0, & |x| \geq R+2. \end{cases}$$

Set $\Omega_b := \{x \in \Omega : R < |x| < R+3\}$. Then we define $b : [0,\infty) \times \mathbb{R}^d \to \mathbb{R}^d$ by

$$b(t,x) := \xi(x)H(t,x) - \mathbb{B}_{\Omega_b}((\nabla\xi) \cdot H(t,x)),$$

where $\mathbb{B}_{\Omega_b}$ is the Bogovskiĭ operator on the bounded domain Ω_b. Note that we can interpret $\mathbb{B}_{\Omega_b}((\nabla\xi) \cdot H(t,x))$ as a function on $\mathbb{R}^d$ by extending it by zero. Moreover, integration by parts and the divergence theorem yield

$$\int_{\Omega_b} \nabla\xi(x) \cdot H(t,x) \, dx = \int_{\partial\Omega_b} H(t,x) \cdot \nu \, d\sigma - \int_{\Omega_b} \xi(x) \operatorname{div} H(t,x) \, dx = 0$$

since $\operatorname{div} H = 0$. Hence, by applying Proposition 1.87, we see that $b \in C^1([0,\infty); C^2(\mathbb{R}^d))$ and $\operatorname{div} b(t) = 0$ for all $t \geq 0$. Furthermore, $b|_{\partial\Omega} = h$. If we set $w := u + b$, then it is easily seen that problem (1.24) is equivalent to

$$\begin{cases} u_t - \Delta u + u \cdot \nabla u + b \cdot \nabla u + u \cdot \nabla b + \nabla p &= \Delta b - b_t - b \cdot \nabla b & \text{in } \mathbb{R}_+ \times \Omega, \\ \operatorname{div} u &= 0 & \text{in } \mathbb{R}_+ \times \Omega, \\ u &= u & \text{on } \mathbb{R}_+ \times \partial\Omega, \\ u|_{t=0} &= w_0 - b(0) & \text{in } \Omega. \end{cases}$$

This construction is used several times in this thesis in different situations, but the idea is always as described above.

1.7 Classical compactness results

In this section we collect classical compactness and embedding results for time-dependent functions with values in Banach spaces. The results stated here are mainly used in Chapter 5. In the entire section $T > 0$ denotes a fixed positive time.

Proposition 1.89 ([Sim87, page 85, Corollary 4]). *Let X, E and Y be Banach spaces such that $X \subset E \subset Y$ with a compact embedding $X \hookrightarrow E$. The following embeddings are compact:*

(i) $L^p((0,T); X) \cap \left\{ \phi : \frac{\partial}{\partial t} \phi \in L^1((0,T); Y) \right\} \hookrightarrow L^p((0,T); E)$ *for* $1 \leq p < \infty$,

(ii) $L^\infty((0,T); X) \cap \left\{ \phi : \frac{\partial}{\partial t} \phi \in L^r((0,T); Y) \right\} \hookrightarrow C([0,T]; E)$ *for* $1 < r \leq \infty$.

Definition 1.90. Let X be a Banach space. We say a function $\phi : [0,T] \to X$ is weakly continuous in $[0,T]$ if for all $x' \in X'$ the scalar function $t \mapsto \langle \phi(t), x' \rangle_{X,X'}$ is continuous in $[0,T]$. The space of all such functions is denoted by $C([0,T]; X - weak)$.

The next lemma gives a criterion for a function $\phi : [0,T] \to X$ to belong to the space $C([0,T]; X - weak)$.

Lemma 1.91 ([Tem01, Chapter III Lemma 1.4]). *Let X and Y be two Banach spaces with $X \hookrightarrow Y$. If $\phi \in L^\infty((0,T); X) \cap C([0,T]; Y - weak)$, then $\phi \in C([0,T]; X - weak)$.*

In particular, Lemma 1.91 states that if $\phi \in L^\infty((0,T); X) \cap C([0,T]; Y - weak)$, then $\phi(t) \in X$ for all $t \in [0,T]$.

In the following let X be a separable reflexive Banach space such that $X \hookrightarrow Y$ for a Banach space Y. Assume that Y' is separable and dense in X'. Let $\{\phi_n\}_{n \in \mathbb{N}} \subset L^\infty((0,T); X)$ be a bounded sequence such that $\{\phi_n\}_{n \in \mathbb{N}} \subset C([0,T]; Y)$ and for all $y' \in Y'$ the map $t \mapsto \langle \phi_n(t), y' \rangle_{Y,Y'}$ is uniformly equicontinuous for $n \in \mathbb{N}$. As a consequence of Lemma 1.91 we know that $\phi_n(t) \in X$ for all $t \in [0,T]$ and all $n \in \mathbb{N}$. Moreover, we can find a ball B of X that contains $\phi_n(t)$ for all $t \in [0,T]$ and all $n \in \mathbb{N}$. Since X' is separable, the weak topology makes B a compact metric space. Thus, in the following we can consider B as a complete metric space and so is the space $C([0,T]; B)$. Now we are in position to state the following lemma.

Lemma 1.92 ([Lio96, Appendix C]). *Let the sequence $\{\phi_n\}_{n \in \mathbb{N}}$ be as described above. Then $\{\phi_n\}_{n \in \mathbb{N}}$ is relatively compact in $C([0,T]; B) \subset C([0,T]; X - weak)$.*

2 Stationary Stokes equations and elliptic L^p-estimates

As outlined in the introduction, the main objective of this thesis is to study non-stationary Navier-Stokes flows, mainly with rotating effect. For this purpose we frequently need to control the L^p-norm of higher order derivatives of the velocity field. This can usually be achieved by using elliptic theory for the stationary Stokes system. Therefore, in this chapter we consider the stationary Stokes equations and collect and prove elliptic L^p-estimates which are needed in the course of this thesis. Most of the results stated here are well-known or even classical, however we place a particular emphasis on the dependence or independence of the constants on the actual form and size of the domain. This is crucial for our purposes. Moreover, the result for the modified Stokes system with rotating effect, stated in Theorem 2.12, seems to be new. This result is a non-trivial variant of a the result of Farwig, Hishida and Müller [FHM04] on a similar problem; see also Galdi and Kyed [GK11]. Theorem 2.12 is interesting in its own right and it also serves as an important tool in Chapter 5.

2.1 The classical Stokes system

In this section we consider the classical stationary Stokes system

$$
\begin{cases}
-\Delta u + \nabla \mathrm{p} &= g \quad \text{in } \Omega, \\
\operatorname{div} u &= 0 \quad \text{in } \Omega, \\
u &= 0 \quad \text{on } \partial\Omega,
\end{cases}
\tag{2.1}
$$

where $g \in L^p(\Omega)^d$, $1 < p < \infty$, is a given vector field and $\Omega \subset \mathbb{R}^d$, $d \geq 2$, is a domain specified later. In the following, we are mainly interested in estimates of the form

$$
\|\nabla^2 u\|_{L^p(\Omega)} \leq C \|g\|_{L^p(\Omega)},
\tag{2.2}
$$

which are referred to as *elliptic estimates*. Estimate (2.2) states that second order derivatives of u can be controlled by the right-hand side g. Such elliptic estimates will be frequently applied when dealing with non-stationary Stokes equations:

$$
\begin{cases}
u_t - \Delta u + \nabla \mathrm{p} &= 0 \quad \text{in } \mathbb{R}_+ \times \Omega, \\
\operatorname{div} u &= 0 \quad \text{in } \mathbb{R}_+ \times \Omega, \\
u &= 0 \quad \text{on } \partial\Omega, \\
u|_{t=0} &= u_0 \quad \text{in } \Omega.
\end{cases}
\tag{2.3}
$$

Then for fixed $t \in \mathbb{R}_+$ we may consider the solution u of the non-stationary Stokes problem (2.3) as a solution to the stationary Stokes problem (2.1) with right-hand side $g := -u_t$. Thus, elliptic theory allows us to compare second order derivatives of a non-stationary solution u with the time derivative u_t, or equivalently with $A_{\Omega,p} u$, where $A_{\Omega,p}$ is the Stokes operator.

2.1.1 The whole space $\mathbb{R}^d$ and the half-space $\mathbb{R}^d_+$

We consider the whole space situation $\Omega := \mathbb{R}^d$. Clearly, in this case we do not have any boundary condition in (2.1). The particular feature of the whole space $\mathbb{R}^d$ is that we can derive explicit formulas for the solution (u, p). More precisely, formally applying the divergence operator to (2.1) yields

$$\Delta \mathrm{p} = \operatorname{div} g.$$

Now we formally apply the Fourier transform to this equation and obtain

$$\widehat{\mathrm{p}}(\xi) = -i\frac{1}{|\xi|^2}\xi \cdot \widehat{g}, \qquad \xi \in \mathbb{R}^d. \tag{2.4}$$

Moreover, by formally applying the Fourier transform to (2.1) we see that

$$\widehat{u}(\xi) = \frac{1}{|\xi|^2}\left(\delta_{ij} - \frac{\xi_i \xi_j}{|\xi|^2}\right)^d_{i,j=1}\widehat{g}, \qquad \xi \in \mathbb{R}^d. \tag{2.5}$$

By using formulas (2.4) and (2.5) and by applying multiplier theory one can show the following result. See also [Gal94a, Theorem IV.2.1], where potential theory is used for the proof.

Proposition 2.1. *Let $1 < p < \infty$, $m \in \mathbb{N}_0$, and $g \in W^{m,p}(\mathbb{R}^d)^d$. Then there exists a solution $(u, \mathrm{p}) \in D^{m+2,p}(\mathbb{R}^d)^d \times D^{m+1,p}(\mathbb{R}^d)$ to (2.1) for $\Omega := \mathbb{R}^d$ which has the following properties:*

(a) For all $k \in [0, m]$ there exists a constant $C := C(d, p, k) > 0$ such that

$$\|\nabla^2 u\|_{W^{k,p}(\mathbb{R}^d)} + \|\nabla \mathrm{p}\|_{W^{k,p}(\mathbb{R}^d)} \leq C\|g\|_{W^{k,p}(\mathbb{R}^d)}.$$

(b) Let $1 < p < d$. For all $k \in [0, m]$ there exists a constant $C := C(d, p, k) > 0$ such that

$$\|\nabla u\|_{W^{k,r}(\mathbb{R}^d)} \leq C\|g\|_{W^{k,p}(\mathbb{R}^d)},$$

where $r = dp/(d - p)$.

(c) Let $1 < p < \frac{d}{2}$. For all $k \in [0, m]$ there exists a constant $C := C(d, p, k) > 0$ such that

$$\|u\|_{W^{k,s}(\mathbb{R}^d)} \leq C\|g\|_{W^{k,p}(\mathbb{R}^d)},$$

where $s = dp/(d - 2p)$.

If (v, q) is another solution to (2.1) for $\Omega := \mathbb{R}^d$ with $\|v\|_{\widehat{W}^{k+2,p}(\mathbb{R}^d)}$ finite for some $k \in [0, m]$, then $\|u - v\|_{\widehat{W}^{k+2,p}(\mathbb{R}^d)} = 0$ and $\|\mathrm{p} - \mathrm{q}\|_{\widehat{W}^{k+1,p}(\mathbb{R}^d)} = 0$.

The half-space situation $\mathbb{R}_+^d$ is already more complicated caused by the fact that the domain has a boundary. Nevertheless, it is still possible to derive explicit formulas by applying the Fourier transform in the tangential components or by using potential theory. Here we shall cite the following result, see [Gal94a, Section IV.3].

Proposition 2.2. *Let* $1 < p < \infty$, $m \in \mathbb{N}_0$, *and* $g \in W^{m,p}(\mathbb{R}^d)^d$. *Then there exists a solution* $(u, p) \in D^{m+2,p}(\mathbb{R}_+^d)^d \times D^{m+1,p}(\mathbb{R}_+^d)$ *to* (2.1) *for* $\Omega := \mathbb{R}_+^d$. *Moreover, for all* $k \in [0, m]$ *there exists a constant* $C := C(d, p, k) > 0$ *such that*

$$\|\nabla^2 u\|_{W^{k,p}(\mathbb{R}_+^d)} + \|\nabla p\|_{W^{k,p}(\mathbb{R}_+^d)} \leq C\|g\|_{W^{k,p}(\mathbb{R}_+^d)}.$$

If (v, q) *is another solution to* (2.1) *for* $\Omega := \mathbb{R}_+^d$ *with* $\|v\|_{\widehat{W}^{k+2,p}(\mathbb{R}_+^d)}$ *finite for some* $k \in [0, m]$, *then* $\|u - v\|_{\widehat{W}^{k+2,p}(\mathbb{R}_+^d)} = 0$ *and* $\|p - q\|_{\widehat{W}^{k+1,p}(\mathbb{R}_+^d)} = 0$.

2.1.2 The case of bounded domains

Next we consider problem (2.1) in bounded domains $\Omega \subset \mathbb{R}^d$ having a smooth boundary. The underlying idea for bounded domains is to derive suitable estimates in the interior of Ω, for that Proposition 2.1 is used, and then estimates near the boundary. For deriving the boundary estimates one usually follows the methods of Cattabriga [Cat61] which are based on the work of Agmon, Douglas and Nirenberg [ADN59]. The strategy is to use a suitable change of variables such that locally close to the boundary the problem becomes a half-space problem, and then Proposition 2.2 can be used. This method is also outlined in [Gal94a, Section IV.5]. Here we shall cite the following well-known result, see [Gal94a, Theorem IV.6.1].

Proposition 2.3. *Let* $1 < p < \infty$, $m \in \mathbb{N}_0$, $g \in W^{m,p}(\Omega)^d$ *and* $\Omega \subset \mathbb{R}^d$ *be a bounded domain of class* C^{m+2}. *Then there exists a solution* $(u, p) \in W^{m+2,p}(\Omega)^d \times W^{m+1,p}(\Omega)$ *to* (2.1). *Here* u *is unique and* p *is unique up to an additive constant. Moreover, for all* $k \in [0, m]$ *there exists a constant* $C := C(p, k, \Omega) > 0$ *such that*

$$\|u\|_{W^{k+2,p}(\Omega)} + \|\nabla p\|_{W^{k}(\Omega)} \leq C\|g\|_{W^{k,p}(\Omega)}.$$

The constant in Proposition 2.3 depends on the domain Ω. Later in Chapter 5 it is important to have estimates with constants independent of the actual size of the domain. Therefore we shall make use of the following estimate; see [Hey80, Lemma 1].

Proposition 2.4. *Let* $\Omega \subset \mathbb{R}^3$ *be a bounded domain of class* C^3 *and let* $(u, p) \in H^2(\Omega)^d \times H^1(\Omega)$ *be a solution to* (2.1) *with* $g \in L^2(\Omega)^d$. *Then there exists a constant* $C > 0$ *such that*

$$\|\nabla^2 u\|_{L^2(\Omega)} \leq C\left(\|\mathbb{P}_{\Omega,2}g\|_{L^2(\Omega)} + \|\nabla u\|_{L^2(\Omega)}\right),$$

and

$$\|\nabla u\|_{L^3(\Omega)} \leq C\left(\|\mathbb{P}_{\Omega,2}g\|_{L^2(\Omega)}^{\frac{1}{2}}\|\nabla u\|_{L^2(\Omega)}^{\frac{1}{2}} + \|\nabla u\|_{L^2(\Omega)}\right).$$

Here the constant C *depends only on the* C^3-*regularity of the boundary* $\partial\Omega$ *and not on the size of* $\partial\Omega$ *or* Ω.

Remark 2.5. By closely checking the proof of [Hey80, Lemma 1] and the explanations and results in [Gal94a, Section IV.5] one sees that a C^2-regularity for the boundary is sufficient for proving Proposition 2.4. However, for our purposes a C^3-regularity is sufficient and therefore we stated Proposition 2.4 in the original form of Heywood.

In some parts of this thesis, cf. Chapter 5, we work on special bounded domains, namely on balls $B_R \subset \mathbb{R}^d$ for $R > 1$. It is necessary for our purposes to obtain estimates independent of the actual value of $R > 1$. The advantage of working with balls B_R is that we can use a scaling argument to reduce the situation to the unit ball B_1, and then apply Proposition 2.3 for B_1. By a simple substitution we can get back to B_R without changing the constant.

Proposition 2.6. *Let $R > 1$, $1 < p < \infty$, and $m \in \mathbb{N}_0$. Let $(u, \mathrm{p}) \in W^{2,p}(B_R)^d \times W^{1,p}(B_R)$ be a solution to (2.1) for $\Omega := B_R$ and $g \in W^{m,p}(B_R)^d$. Then $u \in W^{m+2,p}(B_R)^d$, $\mathrm{p} \in W^{m+1}(B_R)$ and there exists a constant $C := C(m,p) > 0$, independent of R, such that*

$$\|\nabla^2 u\|_{W^{m,p}(B_R)} + \|\nabla \mathrm{p}\|_{W^{m,p}(B_R)} \leq C \|g\|_{W^{m,p}(B_R)}.$$

Proof. By Proposition 2.3 it is clear that $u \in W^{m+2,p}(B_R)^d$ and $\mathrm{p} \in W^{m+1}(B_R)$ whenever $g \in W^{m,p}(B_R)^d$. So it remains to prove the estimate with a constant independent of $R > 1$. The idea of the proof is to use a standard scaling argument. Let $x \in B_R$ and set

$$y := \frac{x}{R}, \qquad v(y) := u(x), \qquad \mathrm{q}(y) := R\,\mathrm{p}(x), \qquad \widetilde{g}(y) := R^2\,g(x).$$

Then it is easy to see that (v, q) solves the system

$$\begin{cases} -\Delta v + \nabla \mathrm{q} &= \widetilde{g} \quad \text{in } B_1, \\ \operatorname{div} v &= 0 \quad \text{in } B_1, \\ v &= 0 \quad \text{on } \partial B_1, \end{cases}$$

in the unit ball B_1. By Proposition 2.3 there exists a constant $C := C(m,p) > 0$ such that

$$\|\nabla^2 v\|_{W^{m,p}(B_1)} + \|\nabla \mathrm{q}\|_{W^{m,p}(B_1)} \leq C \|\widetilde{g}\|_{W^{m,p}(B_1)}. \tag{2.6}$$

Let $1 \leq i, j \leq d$. By using the above estimate and a simple substitution, we see that

$$\|\partial_{ij} u\|_{L^p(B_R)}^p = \frac{R^d}{R^{2p}} \|\partial_{ij} v\|_{L^p(B_1)}^p \leq C \frac{R^d}{R^{2p}} \|\widetilde{g}\|_{L^p(B_1)}^p = C \|g\|_{L^p(B_R)}^p.$$

This proves the claim for $m = 0$. Now we proceed inductively. Let $m > 0$ and assume that the assertion holds for $m - 1$. Let $\alpha \in \mathbb{N}_0^d$ be a multi-index with $|\alpha| = m$. Then, similarly as above, we obtain

$$\|\partial^\alpha \partial_{ij} u\|_{L^p(B_R)}^p = \frac{R^d}{R^{(|\alpha|+2)p}} \|\partial^\alpha \partial_{ij} v\|_{L^p(B_1)}^p \leq C \frac{R^d}{R^{(|\alpha|+2)p}} \|\widetilde{g}\|_{W^{m,p}(B_1)}^p$$

$$= C \frac{R^d}{R^{(|\alpha|+2)p}} \sum_{|\beta| \leq m} \|\partial^\beta \widetilde{g}\|_{L^p(B_1)}^p = C \sum_{|\beta| \leq m} \frac{R^{|\beta|p}}{R^{|\alpha|p}} \|\partial^\beta g\|_{L^p(B_R)}^p$$

$$\leq C \|g\|_{W^{m,p}(B_R)}^p.$$

This concludes the proof. $\qquad \square$

2.1.3 The case of exterior domains

In this section we consider the case of exterior domains $\Omega \subset \mathbb{R}^d$. This situation is presented in detail in [Gal94a, Chapter V]. See also [MS90].

Proposition 2.7. *Let $1 < p < \infty$, $m \in \mathbb{N}_0$, and $\Omega \subset \mathbb{R}^d$ be an exterior domain of class C^{m+2}. Moreover, let $g \in W^{m,p}(\Omega)^d$ and $(u, \mathrm{p}) \in D^{2,p}(\Omega)^d \times D^{1,p}(\Omega)$ be a solution to (2.1). Then for all $k \in [0, m]$ there exists a constant $C := C(p, k, \Omega) > 0$ such that*

$$\|\nabla^2 u\|_{W^{k,p}(\Omega)} + \|\nabla \mathrm{p}\|_{W^{k,p}(\Omega)} \leq C \left(\|g\|_{W^{k,p}(\Omega)} + \|u\|_{L^p(D)} \right), \tag{2.7}$$

where $D := \Omega \cap B_R$ for some sufficiently large $R > 0$. In particular, there exists a constant $C := C(p, k, \Omega) > 0$ such that

$$\|\nabla^2 u\|_{W^{k,p}(\Omega)} + \|\nabla \mathrm{p}\|_{W^{k,p}(\Omega)} \leq C \left(\|g\|_{W^{k,p}(\Omega)} + \|\nabla u\|_{L^p(\Omega)} \right). \tag{2.8}$$

Our basic strategy to prove Proposition 2.7 is to use a suitable localization or more precisely a cut-off technique to reduce the exterior domain problem to a problem in the whole space $\mathbb{R}^d$ and to a problem in a bounded domain close to the boundary $\partial \Omega$. The general approach of such a cut-off technique is outlined in Section 1.6.1.

Before we state the actual proof of Proposition 2.7 we need to show a certain local estimate for the pressure term. For this we need some preparations.

Definition 2.8. For a bounded domain $D \subset \mathbb{R}^d$ we define the space $L_0^p(D)$ of all L^p-functions having mean zero by setting

$$L_0^p(D) := \left\{ \mathrm{q} \in L^p(D) : \int_D \mathrm{q} \, dx = 0 \right\}.$$

If $D \subset \Omega$ is bounded, then we may choose, or modify, the pressure p of the solution (u, p) to (2.1) in such a way that we have $\mathrm{p} \in L_0^p(D)$. This is possible since we may set $\tilde{\mathrm{p}} := \mathrm{p} - |D|^{-1} \int_D \mathrm{p} \, dx$, and then $(u, \tilde{\mathrm{p}})$ is still a solution to (2.1) and $\nabla \tilde{\mathrm{p}} = \nabla \mathrm{p}$.

Lemma 2.9. *Let $1 < p < \infty$ and let $\Omega \subset \mathbb{R}^d$ be an exterior domain of class C^2. Fix some $R > 0$ sufficiently large such that $\mathcal{O} := \mathbb{R}^d \backslash \Omega \subset B_R$. Moreover, let $(u, \mathrm{p}) \in D^{2,p}(\Omega)^d \times D^{1,p}(\Omega)$ be a solution to (2.1) such that $\mathrm{p} \in L_0^p(D)$, where $D := \Omega \cap B_R$. Then for $\gamma \in (\frac{1}{p}, 1]$ there exists a constant $C := C(p, \gamma, D) > 0$ such that*

$$\|\mathrm{p}\|_{L^p(D)} \leq C \left(\|\nabla u\|_{W_p^\gamma(D)} + \|g\|_{L^p(D)} \right).$$

Recall that $W_p^\gamma(D)$ is the Sobolev-Slobodeckiĭ space defined in Definition 1.9. By using Poincaré's inequality[1] on D and the fact that $\nabla \mathrm{p} = g + \Delta u$ we can easily prove the lemma for $\gamma = 1$. However, for later purposes we are actually interested in the case $\gamma < 1$. For proving this we follow the ideas of Shibata and Shimada presented in [SS07a]. See also the explanations of Shibata in [Shi08, Section 4].

---

[1] Poincaré's inequality can be applied here since $\mathrm{p} \in L_0^p(D)$.

Proof. Let $\gamma \in (\frac{1}{p}, 1]$. We set $D := \Omega \cap B_R$ for some $R > 0$ sufficiently large. Moreover, let $1 < p' < \infty$ with $\frac{1}{p} + \frac{1}{p'} = 1$, $\varphi \in C_c^\infty(D)$ and set $\tilde{\varphi} = \varphi - |D|^{-1} \int_D \varphi \, \mathrm{d}x$. We consider the classical Neumann problem (see [ADN59])

$$\begin{cases} \Delta \psi &= \tilde{\varphi} \quad \text{in } D, \\ \nu \cdot \nabla \psi &= 0 \quad \text{on } \partial D. \end{cases} \tag{2.9}$$

It is well-known that there exits a unique solution $\psi \in W^{2,p'}(D)$ to (2.9) which satisfies the estimate

$$\|\psi\|_{W^{2,p'}(D)} \leq C \|\tilde{\varphi}\|_{L^{p'}(D)} \leq 2C \|\varphi\|_{L^{p'}(D)}$$

for some constant $C := C(p', D) > 0$. Now we obtain

$$\langle \mathrm{p}, \varphi \rangle_D = \langle \mathrm{p}, \tilde{\varphi} \rangle_D = \langle \mathrm{p}, \Delta \psi \rangle_D = -\langle \nabla \mathrm{p}, \nabla \psi \rangle_D = -\langle \Delta u + g, \nabla \psi \rangle_D$$
$$= \langle \nabla u, \nabla^2 \psi \rangle_D - \langle \nu \cdot \nabla u, \nabla \psi \rangle_{\partial D} - \langle g, \nabla \psi \rangle_D.$$

By using the embedding $W_p^{\frac{1}{p}+\delta}(D) \hookrightarrow L^p(\partial D)$ for $0 < \delta \leq 1 - \frac{1}{p}$, see Proposition 1.10, we obtain

$$|\langle \mathrm{p}, \varphi \rangle_D| \leq C \left(\|\nu \cdot \nabla u\|_{L^p(\partial D)} + \|\nabla u\|_{L^p(D)} + \|g\|_{L^p(\Omega_R)} \right) \|\psi\|_{W^{2,p'}(D)}$$
$$\leq C \left(\|\nabla u\|_{W_p^{1/p+\delta}(D)} + \|g\|_{L^p(D)} \right) \|\varphi\|_{L^{p'}(D)}.$$

If we choose δ such that $\gamma = \frac{1}{p} + \delta$, the assertion follows. $\qquad \square$

We are now in position to state the proof of Proposition 2.7.

Proof of Proposition 2.7. We shall apply a variant of the cut-off procedure explained in Section 1.6.1. Let $\mathcal{O} := \mathbb{R}^d \setminus \Omega$ and choose $R > 0$ sufficiently large such that $\mathcal{O} \subset B_R$. Set

$$D := \Omega \cap B_{R+4},$$
$$D_1 := \{x \in \Omega : R < |x| < R+3\},$$

and choose a cut-off function $\varphi_1 \in C_c^\infty(\mathbb{R}^d)$ such that $0 \leq \varphi_1 \leq 1$ and

$$\varphi_1(x) := \begin{cases} 1, & |x| \leq R+1, \\ 0, & |x| \geq R+2. \end{cases}$$

Set $\varphi_2 := (1 - \varphi_1)$. Let $(u, \mathrm{p}) \in D^{2,p}(\Omega) \times D^{1,p}(\Omega)$ be a solution to (2.1). Without loss of generality we may assume that $\mathrm{p} \in L_0^p(D)$. Next, set

$$u^{(i)} := \varphi_i u - \mathbb{B}_{D_1}((\nabla \varphi_i) \cdot u),$$
$$\mathrm{p}^{(i)} := \varphi_i \mathrm{p},$$

where $i = 1, 2$. Note that $u = u^{(1)} + u^{(2)}$ and $\mathrm{p} = \mathrm{p}^{(1)} + \mathrm{p}^{(2)}$. A short calculation shows that $u^{(1)}$ solves the stationary problem (2.1) in the bounded domain D with right hand side $g^{(1)} := \varphi_1 g + G^{(1)}$, and $u^{(2)}$ solves the stationary problem (2.1) in $\mathbb{R}^d$ with right hand side

$g^{(2)} := \varphi_2 g + G^{(2)}$. Here the "error terms" $G^{(i)}$, $i = 1, 2$, are given by

$$G^{(i)} = -2(\nabla\varphi_i) \cdot \nabla u - (\Delta\varphi_i)u + \Delta\mathbb{B}_{D_1}((\nabla\varphi_i) \cdot u) + (\nabla\varphi_i)\mathrm{p}.$$

Note that the supports of $G^{(i)}$, $i = 1, 2$, are contained in the bounded domain $D_1 \subset D$. Now we shall proceed by using induction on $m \in \mathbb{N}_0$.

Step 1: The case $m = 0$. It is clear that $G^{(1)} \in L^p(D)^d$ and $G^{(2)} \in L^p(\mathbb{R}^d)^d$. By applying Proposition 2.3 to $(u^{(1)}, \mathrm{p}^{(1)})$ and Proposition 2.1 to $(u^{(2)}, \mathrm{p}^{(2)})$ we conclude that there exists a constant $C := C(p, \Omega)$ such that

$$\begin{aligned}
\|\nabla^2 u^{(1)}\|_{L^p(D)} + \|\nabla\mathrm{p}^{(1)}\|_{L^p(D)} &\le C \left(\|g\|_{L^p(\Omega)} + \|G^{(1)}\|_{L^p(D)}\right), \\
\|\nabla^2 u^{(2)}\|_{L^p(\mathbb{R}^d)} + \|\nabla\mathrm{p}^{(2)}\|_{L^p(\mathbb{R}^d)} &\le C \left(\|g\|_{L^p(\Omega)} + \|G^{(2)}\|_{L^p(D)}\right).
\end{aligned} \tag{2.10}$$

Next we have to estimate the "error terms" $G^{(i)}$, $i = 1, 2$. Fix $\gamma \in (\frac{1}{p}, 1)$. Since $\mathrm{p} \in L_0^p(D)$ we can apply Lemma 2.9 to get

$$\|\mathrm{p}\|_{L^p(D)} \le C \left(\|\nabla u\|_{W_p^\gamma(D)} + \|g\|_{L^p(D)}\right).$$

Since $W_p^{\gamma,p}(D)^d = B_{p,p}^\gamma(D)^d$ is a real interpolation space we obtain

$$\|\mathrm{p}\|_{L^p(D)} \le C \left(\|\nabla^2 u\|_{L^p(D)}^{\frac{1}{\gamma}} \|\nabla u\|_{L^p(D)}^{1-\frac{1}{\gamma}} + \|\nabla u\|_{L^p(D)} + \|g\|_{L^p(D)}\right),$$

see also Remark 1.8. By using the Gagliardo-Nirenberg inequality, see Proposition 1.17, with $j = 1$, $p = r = q$, $m = 2$, $\theta = 1/2$ we have

$$\|\nabla u\|_{L^p(D)} \le C\|\nabla^2 u\|_{L^p(D)}^{\frac{1}{2}} \|u\|_{L^p(D)}^{\frac{1}{2}}. \tag{2.11}$$

By Young's inequality we conclude that

$$\|\mathrm{p}\|_{L^p(D)} \le C\left(\varepsilon\|\nabla^2 u\|_{L^p(D)} + \varepsilon^{-1}\|u\|_{L^p(D)} + \|g\|_{L^p(D)}\right) \tag{2.12}$$

for arbitrary $\varepsilon > 0$ and some constant $C := C(p, \Omega) > 0$. Moreover, by applying Proposition 1.87 we see that $\Delta\mathbb{B}_{D_1} \in \mathscr{L}(W_0^{1,p}(D_1), L^p(D_1)^d)$ and therefore

$$\|(\nabla\varphi_i) \cdot \nabla u\|_{L^p(D)} + \|(\Delta\varphi_i)u\|_{L^p(D)} + \|\Delta\mathbb{B}_{D_1}((\nabla\varphi_i) \cdot u)\|_{L^p(D)} \le C\|u\|_{W^{1,p}(D)}$$

for a constant $C > 0$. Applying again the Gagliardo-Nirenberg inequality (2.11) and Young's inequality we see that

$$\|u\|_{W^{1,p}(D)} \le C\left(\varepsilon\|\nabla^2 u\|_{L^p(D)} + \varepsilon^{-1}\|u\|_{L^p(D)}\right) \tag{2.13}$$

for arbitrary $\varepsilon > 0$ and some constant $C := C(p, \Omega) > 0$. Summing up, we obtain

$$\|G^{(i)}\|_{L^p(D)} \le C\left(\varepsilon\|\nabla^2 u\|_{L^p(D)} + \varepsilon^{-1}\|u\|_{L^p(D)} + \|g\|_{L^p(D)}\right)$$

for $\varepsilon > 0$ and some constant $C := C(p, \Omega) > 0$. By choosing $\varepsilon > 0$ sufficiently small, estimate (2.7) for $m = 0$ follows from the estimates in (2.10).

45

Step 2: Induction step. Let us assume that Proposition 2.7 holds for $m \in \mathbb{N}_0$. Then we know in particular that $u \in W^{m+2,p}_{\mathrm{loc}}(\Omega)$ and $\mathrm{p} \in W^{m+1,p}_{\mathrm{loc}}(\Omega)$. Hence we can conclude that $G^{(1)} \in W^{m+1,p}(D)^d$ and $G^{(2)} \in W^{m+1,p}(\mathbb{R}^d)^d$. By applying Proposition 2.3 to $(u^{(1)}, \mathrm{p}^{(1)})$ and Proposition 2.1 to $(u^{(2)}, \mathrm{p}^{(2)})$ we obtain

$$\|\nabla^2 u^{(1)}\|_{W^{m+1,p}(D)} + \|\nabla \mathrm{p}^{(1)}\|_{W^{m+1,p}(D)} \leq C \left(\|g\|_{W^{m+1,p}(\Omega)} + \|G^{(1)}\|_{W^{m+1,p}(D)} \right),$$

$$\|\nabla^2 u^{(2)}\|_{W^{m+1,p}(\mathbb{R}^d)} + \|\nabla \mathrm{p}^{(2)}\|_{W^{m+1,p}(\mathbb{R}^d)} \leq C \left(\|g\|_{W^{m+1,p}(\Omega)} + \|G^{(2)}\|_{W^{m+1,p}(D)} \right), \tag{2.14}$$

for a constant $C := C(p, \Omega)$. We shall now estimate $G^{(i)}$, $i = 1, 2$, in the $W^{m+1,p}(D)^d$-norm. By applying Poincaré's inequality and the induction hypothesis we have

$$\|\mathrm{p}\|_{W^{m+1,p}(D)} \leq C \|\nabla \mathrm{p}\|_{W^{m,p}(D)} \leq C \left(\|g\|_{W^{m,p}(\Omega)} + \|u\|_{L^p(D)} \right).$$

Moreover, by Proposition 1.87 we know that $\Delta \mathbb{B}_{D_1} \in \mathscr{L}(W^{m+2,p}_0(D_1), W^{m+1,p}_0(D_1)^d)$ and therefore

$$\|(\nabla \varphi_i) \cdot \nabla u\|_{W^{m+1,p}(D)} + \|(\Delta \varphi_i) u\|_{W^{m+1,p}(D)} + \|\Delta \mathbb{B}_{D_1}((\nabla \varphi_i) \cdot u)\|_{W^{m+1,p}(D)} \leq C \|u\|_{W^{m+2,p}(D)}.$$

By using the induction hypothesis and the Gagliardo-Nirenberg inequality (2.11) as in Step 1 we conclude that

$$\|u\|_{W^{m+2,p}(D)} \leq \|\nabla^2 u\|_{W^{m,p}(D)} + \|\nabla u\|_{W^{m,p}(D)} + \|u\|_{W^{m,p}(D)}$$

$$\leq C \left(\|g\|_{W^{m,p}(\Omega)} + \|u\|_{L^p(D)} \right).$$

Thus, we have

$$\|G^{(i)}\|_{W^{m+1,p}(D)} \leq C \left(\|g\|_{W^{m,p}(\Omega)} + \|u\|_{L^p(D)} \right)$$

for a constant $C := C(\Omega, p) > 0$. Estimate (2.7) for $m + 1$ is now a consequence of the estimate above and the estimates in (2.14).

Estimate (2.8) follows directly from (2.7) by applying Poincaré's inequality on D, which is possible since $D := \Omega \cap B_{R+4}$ and $u|_{\partial \Omega} = 0$. $\qquad \square$

Remark 2.10. If $p < \frac{d}{2}$, then instead of estimate (2.7) we even have

$$\|\nabla^2 u\|_{W^{k,p}(\Omega)} + \|\nabla \mathrm{p}\|_{W^{k,p}(\Omega)} \leq C \|g\|_{W^{k,p}(\Omega)} \tag{2.15}$$

for all $k \in [0, m]$ and a constant $C := C(p, k, \Omega) > 0$. See [MS90] for a proof.

2.2 The modified Stokes system with rotating effect

In this section we investigate the modified Stokes problem which is relevant in the study of fluid flows around a rotating obstacle. Let $M \in \mathbb{R}^{d \times d}$ be a skew-symmetric matrix, i.e.

$M^\top = -M$. Then we consider the stationary system

$$\begin{cases} -\Delta u - Mx \cdot \nabla u + Mu + \nabla \mathrm{p} &=\ g \quad \text{in } \Omega, \\ \operatorname{div} u &=\ 0 \quad \text{in } \Omega, \\ u &=\ 0 \quad \text{on } \partial\Omega, \end{cases} \qquad (2.16)$$

where $g \in L^p(\Omega)^d$, $1 < p < \infty$, is a given vector field and $\Omega \subset \mathbb{R}^d$, $d \geq 2$, is a domain specified later. The form of this stationary problem is motivated by the situation described in (14).

Problem (2.16) is (mainly) interesting and challenging in the situation of unbounded domains $\Omega \subset \mathbb{R}^d$, as in this case the term $Mx \cdot \nabla u$ has unbounded coefficients. In the 3-dimensional whole space situation $\mathbb{R}^3$ problem (2.16) was considered for the first time by Farwig, Hishida, Müller [FHM04]. They proved the following:

Proposition 2.11. *Let $1 < p < \infty$ and $g \in L^p(\mathbb{R}^3)^3$. Then there exists a solution $(u, \mathrm{p}) \in D^{2,p}(\mathbb{R}^3)^3 \times D^{1,p}(\mathbb{R}^3)$ to (2.16) for $\Omega := \mathbb{R}^3$. Moreover, there exists a constant $C := C(p) > 0$ such that*

$$\|\nabla^2 u\|_{L^p(\mathbb{R}^3)} + \|Mx \cdot \nabla u - Mu\|_{L^p(\mathbb{R}^3)} + \|\nabla \mathrm{p}\|_{L^p(\mathbb{R}^3)} \leq C\|g\|_{L^p(\mathbb{R}^3)}.$$

The proof of Farwig, Hishida and Müller makes heavily use of methods from Fourier analysis. Recently, a simpler proof was given by Galdi and Kyed in [GK11]. There the stationary problem (2.16) is transformed to a classical non-stationary Stokes problem and then maximal regularity results for the classical Stokes system in $\mathbb{R}^d$ are used. A crucial point in their proof is that the orthogonal transformation e^{tM} leaves the domain $\mathbb{R}^3$ invariant.

In this thesis, or more precisely in Chapter 5, we need an elliptic estimate similar to the one stated in Proposition 2.11, however not in the whole space $\mathbb{R}^3$, but in the setting of balls $B_R \subset \mathbb{R}^d$ for $R > 1$. Here it is important that the constant does not depend on the actual radius $R > 1$. This is a crucial point and by far not trivial, as the size of the coefficients of $Mx \cdot \nabla u$ grows linearly in R. Such a uniform estimate does not seem to be known in the literature so far. However, it turns out that such an estimate can be obtained by using a strategy similar to the one used in [GK11]. This is possible since the transformation e^{tM} is orthogonal and therefore leaves the ball B_R invariant, too. So, the problem reduces to a classical non-stationary Stokes problem in B_R. Then by applying the maximal regularity results from Proposition 1.76 and by using a scaling argument we obtain the following result:

Theorem 2.12. *Let $R > 1$ and $1 < p < \infty$. Let $(u, \mathrm{p}) \in W^{2,p}(B_R)^d \times W^{1,p}(B_R)$ be a solution to (2.16) for $\Omega := B_R$. Then there exists some constant $C := C(p) > 0$, independent of $R > 1$, such that*

$$\|\nabla^2 u\|_{L^p(B_R)} + \|Mx \cdot \nabla u - Mu\|_{L^p(B_R)} + \|\nabla \mathrm{p}\|_{L^p(B_R)} \leq C\|g\|_{L^p(B_R)}.$$

Note that in order to prove this result we cannot use directly scaling arguments, as done in

Proposition 2.6. This is due to the fact that the terms on the left hand side in (2.16) show different behaviour under scaling, as they are of different order.

Proof. Since M is skew-symmetric, e^{tM} is orthogonal for all $t \in \mathbb{R}$. For $(t, x) \in \mathbb{R} \times B_R$ set

$$v(t, x) := e^{tM} u(e^{-tM} x), \qquad \mathrm{q}(t, x) := \mathrm{p}(e^{-tM} x), \qquad G(t, x) := e^{tM} g(e^{-tM} x).$$

Note that for the definitions of v, q and g it is essential that $e^{-tM} x \in B_R$ for $x \in B_R$. Now let $T_0 > 0$ and set $J := (0, T_0)$. It is easy to check that

$$(v, \mathrm{q}) \in W^{1,p}(J; L^p_\sigma(B_R)) \cap L^p(J; \mathcal{D}(A_{B_R})) \times L^p(J; \widehat{W}^{1,p}(B_R))$$

holds and that (v, q) solves the non-stationary Stokes problem

$$\begin{cases} v_t - \Delta v + \nabla \mathrm{q} &= G \quad \text{in } J \times B_R, \\ \operatorname{div} v &= 0 \quad \text{in } J \times B_R, \\ v &= 0 \quad \text{on } J \times \partial B_R, \\ v|_{t=0} &= u \quad \text{in } B_R. \end{cases} \qquad (2.17)$$

Here $(A_{B_R}, \mathcal{D}(A_{B_R}))$ denotes the Stokes operator on $L^p_\sigma(B_R)$.

By using the maximal regularity results stated in Proposition 1.76 and Remarks 1.77 we obtain a solution

$$(v^{(1)}, \mathrm{q}^{(1)}) \in W^{1,p}(J; L^p_\sigma(B_R)) \cap L^p(J; \mathcal{D}(A_{B_R})) \times L^p(J; \widehat{W}^{1,p}(B_R))$$

to the Stokes problem

$$\begin{cases} v_t^{(1)} - \Delta v^{(1)} + \nabla \mathrm{q}^{(1)} &= \mathbb{P}_{B_R} G \quad \text{in } J \times B_R, \\ \operatorname{div} v^{(1)} &= 0 \quad \text{in } J \times B_R, \\ v^{(1)} &= 0 \quad \text{on } J \times \partial B_R, \\ v^{(1)}|_{t=0} &= 0 \quad \text{in } B_R, \end{cases}$$

and there exists some constant $C > 0$ independent of $T_0 > 0$ such that

$$\|A_{B_R} v^{(1)}\|_{L^p(J; L^p_\sigma(B_R))} + \|\nabla \mathrm{q}^{(1)}\|_{L^p(J; L^p(B_R))} \le C \|\mathbb{P}_{B_R} G\|_{L^p(J; L^p_\sigma(B_R))}. \qquad (2.18)$$

By applying the elliptic estimate in Proposition 2.6 we conclude that

$$\|\nabla^2 v^{(1)}\|_{L^p(J; L^p(B_R))} + \|\nabla \mathrm{q}^{(1)}\|_{L^p(J; L^p_\sigma(B_R))} \le C \|\mathbb{P}_{B_R} G\|_{L^p(J; L^p_\sigma(B_R))}, \qquad (2.19)$$

for a constant $C > 0$ independent of $T_0 > 0$. Note that it will be important later on that $C > 0$ is indeed independent of $T_0 > 0$. So far it is however not clear yet whether the maximal regularity constant $C > 0$ depends on the actual value of $R > 1$ or not. To show that C is indeed independent of $R > 1$ we use a scaling argument. For $(t, x) \in J \times B_R$ set

$$\tau := \frac{t}{R^2} \qquad \text{and} \qquad y := \frac{x}{R}.$$

Moreover, we introduce functions

$$w(\tau, y) := v^{(1)}(t, x), \qquad \pi(\tau, y) := R q^{(1)}(t, x), \qquad \widetilde{G}(\tau, y) := R^2 \mathbb{P}_{B_R} G(t, x),$$

and we set $\widetilde{J} := \left(0, \frac{T_0}{R^2}\right)$. Then by a straightforward computation one sees that the pair of functions

$$(w, \pi) \in W^{1,p}(\widetilde{J}; L_\sigma^p(B_1)) \cap L^p(\widetilde{J}; \mathcal{D}(A_{B_1})) \times L^p(\widetilde{J}; \widehat{W}^{1,p}(B_1))$$

solves the Stokes problem

$$\begin{cases} w_t - \Delta w + \nabla \pi &= \widetilde{G} \quad \text{in } \widetilde{J} \times B_1, \\ \operatorname{div} w &= 0 \quad \text{in } \widetilde{J} \times B_1, \\ w &= 0 \quad \text{on } \widetilde{J} \times \partial B_1, \\ w|_{t=0} &= 0 \quad \text{in } B_1, \end{cases}$$

in the ball B_1. Similarly as in the proof of Proposition 2.6, we see that

$$\|\nabla^2 v^{(1)}\|_{L^p(J; L^p(B_R))}^p = \frac{R^{d+2}}{R^{2p}} \|\nabla^2 w\|_{L^p(\widetilde{J}; L^p(B_1))}^p$$

$$\leq C \frac{R^{d+2}}{R^{2p}} \|\widetilde{G}\|_{L^p(\widetilde{J}; L^p(B_1))}^p = C \|\mathbb{P}_{B_R} G\|_{L^p(J; L^p(B_R))}^p$$

for a constant $C := C(p) > 0$. Here $C > 0$ is essentially the maximal regularity constant on $L^p(\widetilde{J}, L_\sigma^p(B_1))$. Here $\widetilde{J} = \left(0, \frac{T_0}{R^2}\right)$ depends on R, but by Proposition 1.76 the constant $C > 0$ is independent of $\frac{T_0}{R^2}$. This shows that the constant $C > 0$ in (2.19) is indeed independent of the actual value of $R > 1$. Next, set

$$v^{(2)}(t) := T_{B_R}(t) u, \qquad \nabla q^{(2)}(t) := (\operatorname{Id} - \mathbb{P}_{B_R}) \Delta v^{(2)}(t), \qquad \nabla q^{(3)}(t) := (\operatorname{Id} - \mathbb{P}_{B_R}) G(t),$$

where $\{T_{B_R}(t)\}_{t \geq 0}$ denotes the Stokes semigroup on $L_\sigma^p(B_R)$. The unique solvability of (2.17), see Proposition 1.76 and Remarks 1.77, implies that

$$v = v^{(1)} + v^{(2)} \qquad \text{and} \qquad \nabla q = \nabla q^{(1)} + \nabla q^{(2)} + \nabla q^{(3)}.$$

Fix $1 < r < p$. Using the maximal regularity estimate (2.19) and the decay estimates for the Stokes semigroup in Proposition 1.82, we obtain

$$(T_0 - 1) \|\nabla^2 u\|_{L^p(B_R)}^p \leq \int_1^{T_0} \int_{B_R} |\nabla^2 v(t, x)|^p \, dx \, dt$$

$$\leq C_0 \left(\|\nabla^2 v^{(1)}\|_{L^p(J; L^p(B_R))}^p + \int_1^{T_0} \|\nabla^2 T_{B_R}(t) u\|_{L^p(B_R)}^p \, dt \right)$$

$$\leq C_1 \|\mathbb{P}_{B_R} G\|_{L^p(J; L^p(B_R))}^p + C_2(R) \int_1^{T_0} t^{-\frac{d}{2}\left(\frac{1}{r} - \frac{1}{p}\right)p - p} \|u\|_{L^p(B_R)}^p \, dt$$

for constants $C_0 := C_0(p), C_1 := C_1(p) > 0$ independent of R and a constant $C_2(R) > 0$ which depends on $R > 1$. Since the norm of the Helmholtz projection $\mathbb{P}_{B_R}$ can be estimated independently of the actual value of $R > 1$, see Proposition 1.65, and since

$$\|G\|^p_{L^p(J;L^p(B_R))} \leq T_0 \|g\|^p_{L^p(B_R)},$$

we see that

$$(T_0 - 1)\|\nabla^2 u\|^p_{L^p(B_R)} \leq C_1 T_0 \|g\|^p_{L^p(B_R)} + C_2(R)\left(T_0^{-\frac{d}{2}\left(\frac{1}{r}-\frac{1}{p}\right)p-p+1} - 1\right)\|u\|^p_{L^p(B_R)} \qquad (2.20)$$

for a constant $C_1 := C_1(p) > 0$ independent of R and a constant $C_2(R) > 0$ which depends on $R > 1$. We divide both sides by $T_0 - 1$ and then we let[2] $T_0 \to \infty$. Then the second term on the right-hand side of (2.20) vanishes and we conclude that

$$\|\nabla^2 u\|_{L^p(B_R)} \leq C\|g\|_{L^p(B_R)}$$

for some constant $C := C(p) > 0$ independent of R. Analogously, we obtain

$$(T_0 - 1)\|\nabla p\|^p_{L^p(B_R)}$$

$$= \int_1^{T_0} \int_{B_R} |\nabla q(t,x)|^p \, dx \, dt$$

$$\leq C_0 \left(\|\nabla q^{(1)}\|^p_{L^p(J;L^p(B_R))} + \int_1^{T_0} \|\Delta T_{B_R,p}(t)u\|^p_{L^p(B_R)} \, dt + \|G\|^p_{L^p(J;L^p(B_R))}\right)$$

$$\leq C_1 \|G\|^p_{L^p(J;L^p(B_R))} + C_2(R) \int_1^{T_0} t^{-\frac{d}{2}\left(\frac{1}{r}-\frac{1}{p}\right)p-p} \|u\|^p_{L^p(B_R)} \, dt$$

$$\leq C_1 T_0 \|g\|^p_{L^p(B_R)} + C_2(R)\left(T_0^{-\frac{d}{2}\left(\frac{1}{r}-\frac{1}{p}\right)p-p+1} - 1\right)\|u\|^p_{L^p(B_R)}$$

for constants $C_0 := C_0(p), C_1 := C_1(p) > 0$ independent of R and a constant $C_2(R) > 0$. Thus, as above, we conclude that

$$\|\nabla p\|_{L^p(B_R)} \leq C\|g\|_{L^p(B_R)}$$

for some constant $C := C(p) > 0$ independent of R.

Finally, from equation (2.16) it follows that

$$\|Mx \cdot \nabla u - Mu\|_{L^p(B_R)} \leq \|\Delta u\|_{L^p(B_R)} + \|\nabla p\|_{L^p(B_R)} \leq C\|g\|_{L^p(B_R)}$$

for a constant $C := C(p) > 0$ independent of R. This concludes the proof. $\qquad \square$

---

[2]For that it is important that all constants are also independent of $T_0 > 0$.

3 Decay estimates for the Stokes semigroup in exterior domains

In order to obtain a global mild solution to the Navier-Stokes equations for small data by Kato's iteration method, see [Kat84], it is necessary to have decay estimates for the Stokes semigroup at hand. In this chapter such estimates are collected and proved in the case of exterior domains $\Omega \subset \mathbb{R}^d$, $d \geq 2$. The whole space case $\mathbb{R}^d$ and the case of bounded domains were already presented in Proposition 1.79 and Proposition 1.82, respectively.

3.1 Statement of the L^p-estimates

Let $\{T_{\Omega,p}(t)\}_{t \geq 0}$ be the Stokes semigroup on $L^p_\sigma(\Omega)$, $1 < p < \infty$, and $A_{\Omega,p}$ denotes the Stokes operator. The main objective of this chapter is to give a rather elementary and self-contained proof of the following well-known result, see [BS87, Iwa89, MS97, DS99a].

Theorem 3.1. *Let $\Omega \subset \mathbb{R}^d$, $d \geq 2$, be an exterior domain of class C^4. Then the following assertions hold:*

(a) *Let $1 < p \leq q < \infty$. Then there exists a constant $C := C(p, q, \Omega) > 0$ such that for all $f \in L^p_\sigma(\Omega)$*

$$\|T_{\Omega,p}(t)f\|_{L^q(\Omega)} \leq Ct^{-\frac{d}{2}\left(\frac{1}{p} - \frac{1}{q}\right)} \|f\|_{L^p(\Omega)}, \qquad t > 0.$$

(b) *Let $d \geq 3$ and $1 < p < \infty$. Then there exists a constant $C := C(p, \Omega) > 0$ such that for all $f \in L^p_\sigma(\Omega)$*

$$\|T_{\Omega,p}(t)f\|_{L^\infty(\Omega)} \leq Ct^{-\frac{d}{2p}} \|f\|_{L^p(\Omega)}, \qquad t > 0.$$

(c) *Let $1 < p \leq d$. Then there exists a constant $C := C(p, \Omega) > 0$ such that for all $f \in L^p_\sigma(\Omega)$*

$$\|\nabla T_{\Omega,p}(t)f\|_{L^p(\Omega)} \leq Ct^{-\frac{1}{2}} \|f\|_{L^p(\Omega)}, \qquad t > 0.$$

(d) *Let $d \geq 3$ and $d < p < \infty$. Then there exists a constant $C := C(p, \Omega) > 0$ such that for all $f \in L^p_\sigma(\Omega)$*

$$\|\nabla T_{\Omega,p}(t)f\|_{L^p(\Omega)} \leq Ct^{-\frac{d}{2p}} \|f\|_{L^p(\Omega)}, \qquad t > 1.$$

(e) *Let $d \geq 3$ and $1 < p < \infty$. Then there exists a constant $C := C(p, \Omega) > 0$ such that for all $f \in L^p_\sigma(\Omega)$*

$$\|A_{\Omega,p}T_{\Omega,p}(t)f\|_{L^p(\Omega)} \leq Ct^{-1} \|f\|_{L^p(\Omega)}, \qquad t > 0.$$

Remarks 3.2. 1. For $0 < t < T$ and $0 < T < \infty$ fixed these estimates were already stated in Proposition 1.78.

2. Assertion (a) for $p = q$ states that $\{T_{\Omega,p}(t)\}_{t \geq 0}$ is uniformly bounded on $L^p_\sigma(\Omega)$.

3. Assertion (e) states that $\{T_{\Omega,p}(t)\}_{t \geq 0}$ is bounded analytic; cf. Proposition 1.35.

4. In [MS97] it is shown that the estimate in assertion (d) is sharp. This means that such an estimate does not hold if we replace $t^{-\frac{d}{2p}}$ by $t^{-\frac{d}{2p}-\varepsilon}$ for any $\varepsilon > 0$.

For $p = q = 2$ the assertions (a) and (e) easily follow from abstract theory since the Stokes operator in $L^2_\sigma(\Omega)$ is induced by a symmetric form, see Proposition 1.68. The gradient estimate in (c) for $p = 2$ then follows easily from (e). However, for $p \neq 2$ the uniform boundedness of the Stokes semigroup on exterior domains remained open for some time and was first proved by Borchers and Sohr [BS87] for $d \geq 3$. In fact, Borchers and Sohr even show that the Stokes semigroup is bounded analytic which directly yields assertion (e). The analogous results for $d = 2$ were later proved by Borchers and Varnhorn [BV93]. Note that our theorem does not include the case $d = 2$ in assertion (e), as this case does not follow by the techniques used here. The L^p-L^q-estimates in assertion (a) and the gradient estimate in assertion (c) were first proved by Iwashita [Iwa89] for $d \geq 3$. The case $d = 2$ was proved by Dan and Shibata [DS99a]. All estimates included in our theorem were also proved by Maremonti and Solonnikov [MS97]. The L^∞-estimate in (b) does also hold for $d = 2$. This was proved by Dan and Shibata [DS99b]. However, in this case the techniques used here fail. Note that the case $d = 2$ is in general technically more involved as the cases $d \geq 3$.

As it was already said above, the estimates in the case $p = 2$ are rather easy to show, as they follow quite directly from the results in Section 1.4.1 and by using interpolation techniques. Therefore we shall first consider the case $p = 2$ in Section 3.2 and then based on these L^2-estimates we give a full proof of Theorem 3.1 in Section 3.3. Somehow the "philosophy" and the basic ideas of our proof follow the approach and ideas of Maremonti and Solonnikov [MS97]. However, our proof is adopted to the semigroup framework introduced in Section 1.2.1 and Section 1.4 and it also incorporates ideas of other approaches, see e.g. [DKS98] and [ES05]. We believe that the proof is rather elementary, as we mainly use L^2-theory, interpolation techniques, duality arguments and a suitable cut-off procedure.

3.2 Global L^2-estimates

In this section we prove global decay estimates for the Stokes semigroup in the Hilbert space case $L^2_\sigma(\Omega)$. These estimates mainly follow from Proposition 1.68.

Proposition 3.3. *Let* $\Omega \subset \mathbb{R}^d$, $d \geq 2$, *be an exterior domain of class* C^2. *Then there exists a constant* $C := C(\Omega) > 0$ *such that for all* $f \in L^2_\sigma(\Omega)$

(i) $\|T_{\Omega,2}(t)f\|_{L^2(\Omega)} \leq \|f\|_{L^2(\Omega)}, \quad t > 0,$

(ii) $\|\nabla T_{\Omega,2}(t)f\|_{L^2(\Omega)} \leq Ct^{-\frac{1}{2}}\|f\|_{L^2(\Omega)}, \quad t > 0,$

(iii) $\|A_{\Omega,2}T_{\Omega,2}(t)f\|_{L^2(\Omega)} \leq Ct^{-1}\|f\|_{L^2(\Omega)}, \quad t > 0,$

(iv) $\|\nabla^2 T_{\Omega,2}(t)f\|_{L^2(\Omega)} \leq Ct^{-\frac{1}{2}}\|f\|_{L^2(\Omega)}, \quad t > 1.$

Proof. Assertions (i) and (iii) follow directly from Proposition 1.68. By combining (i) and (iii) we see that

$$\|\nabla T_{\Omega,2}(t)f\|_{L^2(\Omega)}^2 = \langle \nabla T_{\Omega,2}(t)f, \nabla T_{\Omega,2}(t)f \rangle_{L^2(\Omega)} = \langle A_{\Omega,2}T_{\Omega,2}(t)f, T_{\Omega,2}f \rangle_{L^2(\Omega)}$$
$$\leq \|A_{\Omega,2}T_{\Omega,2}(t)f\|_{L^2(\Omega)}\|T_{\Omega,2}(t)f\|_{L^2(\Omega)} \leq Ct^{-1}\|f\|_{L^2(\Omega)}^2, \qquad t > 0.$$

This proves assertion (ii). The elliptic estimate in Proposition 2.7 yields

$$\|\nabla^2 T_{\Omega,2}(t)f\|_{L^2(\Omega)} \leq C\Big(\|A_{\Omega,p}T_{\Omega,2}(t)f\|_{L^2(\Omega)} + \|\nabla T_{\Omega,2}(t)f\|_{L^2(\Omega)}\Big), \qquad t > 0$$

for a constant $C := C(\Omega) > 0$. Thus, estimate (iv) follows from (ii) and (iii). $\square$

Remarks 3.4. 1. For $0 < t \leq 1$ we have

$$\|\nabla^2 T_{\Omega,2}(t)f\|_{L^2(\Omega)} \leq Ct^{-1}\|f\|_{L^2(\Omega)}$$

for a constant $C := C(\Omega) > 0$.

2. If $d \geq 5$, then by using the estimate in Remark 2.10 we conclude that instead of (iv) we have

$$\|\nabla^2 T_{\Omega,2}(t)f\|_{L^2(\Omega)} \leq Ct^{-1}\|f\|_{L^2(\Omega)}, \qquad t > 0$$

for a constant $C := C(\Omega) > 0$.

Next we combine Proposition 3.3 with standard interpolation techniques and duality arguments to show the following L^p-L^2-estimates; cf. [MS97, Lemma 5.3].

Lemma 3.5. *Let* $1 < p \leq 2$, $m \in \mathbb{N}_0$, *and let* $\Omega \subset \mathbb{R}^d$, $d \geq 2$, *be an exterior domain of class* C^{m+2}. *Then there exists a constant* $C := C(p,m,\Omega) > 0$ *such that for all* $f \in L^p_\sigma(\Omega)$

(i) $\|T_{\Omega,p}(t)f\|_{H^m(\Omega)} \leq Ct^{-\alpha(p)}\|f\|_{L^p(\Omega)}$, $t > 1$,

(ii) $\|\nabla T_{\Omega,p}(t)f\|_{H^m(\Omega)} \leq Ct^{-\alpha(p)-\frac{1}{2}}\|f\|_{L^p(\Omega)}$, $t > 1$,

(iii) $\|A_{\Omega,p}T_{\Omega,p}(t)f\|_{H^m(\Omega)} \leq Ct^{-\alpha(p)-1}\|f\|_{L^p(\Omega)}$, $t > 1$,

(iv) $\|\nabla^2 T_{\Omega,p}(t)f\|_{H^m(\Omega)} \leq Ct^{-\alpha(p)-\frac{1}{2}}\|f\|_{L^p(\Omega)}$, $t > 1$,

where $\alpha(p)$ *is given by*

$$\alpha(p) := \begin{cases} \frac{d}{2}\left(\frac{1}{p} - \frac{1}{2}\right), & \text{if } \frac{2d}{d+2} < p \leq 2, \\ \frac{1}{2}, & \text{if } 1 < p \leq \frac{2d}{d+2}. \end{cases} \tag{3.1}$$

Remark 3.6. For $m = 0$ this lemma has basically the same assertions as [MS97, Lemma 5.3]. However, Lemma 3.5 provides also estimates in the H^m-norm for $m \in \mathbb{N}$. In the proof of Theorem 3.1 it is essential that we can estimate higher order norms.

Proof of Lemma 3.5. We prove the lemma by using induction on $m \in \mathbb{N}_0$.

Step 1: The case $m = 0$. The case $p = 2$ is covered by Proposition 3.3. So next let us consider the case $1 < p < 2$ if $d = 2$ and $\frac{2d}{d+2} \leq p < 2$ if $d \geq 3$. Let $2 < p' < \infty$ be such that $\frac{1}{p} + \frac{1}{p'} = 1$. Take $g \in C_{c,\sigma}^{\infty}(\Omega)$ such that $\|g\|_{L^2(\Omega)} = 1$. Then

$$|\langle T_{\Omega,p}(t)f, g \rangle_{\Omega}| = |\langle f, T_{\Omega,p'}(t)g \rangle_{\Omega}| \leq \|f\|_{L^p(\Omega)} \|T_{\Omega,p'}(t)g\|_{L^{p'}(\Omega)}, \qquad t > 0. \qquad (3.2)$$

By applying the Gagliardo-Nirenberg inequality, see Proposition 1.17, with $\theta := d\left(\frac{1}{p} - \frac{1}{2}\right)$ we obtain

$$\|T_{\Omega,p'}(t)g\|_{L^{p'}(\Omega)} \leq C \|\nabla T_{\Omega,2}(t)g\|_{L^2(\Omega)}^{\theta} \|T_{\Omega,2}(t)g\|_{L^2(\Omega)}^{1-\theta}, \qquad t > 0.$$

Note that $\theta \in (0,1)$ if $d = 2$ and $\theta \in (0,1]$ if $d \geq 3$. Moreover, note that we used the consistency of the Stokes semigroup here, see Remark 1.75. Now Proposition 3.3 yields

$$\|T_{\Omega,p'}(t)g\|_{L^{p'}(\Omega)} \leq Ct^{-\frac{d}{2}\left(\frac{1}{p} - \frac{1}{2}\right)} \|g\|_{L^2(\Omega)}, \qquad t > 0.$$

Since $g \in C_{c,\sigma}^{\infty}(\Omega)$ with $\|g\|_{L^2(\Omega)} = 1$ is arbitrary, relation (3.2) implies

$$\|T_{\Omega,p}(t)f\|_{L^2(\Omega)} \leq Ct^{-\frac{d}{2}\left(\frac{1}{p} - \frac{1}{2}\right)} \|f\|_{L^p(\Omega)}, \qquad t > 0.$$

This proves estimate (i). Next we apply the semigroup property, Proposition 3.3 (ii) and estimate (i) to see that

$$\|\nabla T_{\Omega,p}(t)f\|_{L^2(\Omega)} \leq C\left(\frac{t}{2}\right)^{-\frac{1}{2}} \left\|T_{\Omega,p}\left(\frac{t}{2}\right)f\right\|_{L^2(\Omega)} \leq Ct^{-\frac{d}{2}\left(\frac{1}{p} - \frac{1}{2}\right) - \frac{1}{2}} \|f\|_{L^p(\Omega)}, \qquad t > 0.$$

Similarly, estimates (iii) and (iv) follow from (i) and Proposition 3.3 by applying the semigroup property. Next we consider the case $1 < p < \frac{2d}{d+2}$. We set $q := \frac{2d}{d+2}$. By using the semigroup property, the estimates above and the local estimates in Proposition 1.78 we obtain

$$\|T_{\Omega,p}(t)f\|_{L^2(\Omega)} \leq C\left(t - \frac{1}{2}\right)^{-\alpha(q)} \left\|T_{\Omega,p}\left(\frac{1}{2}\right)f\right\|_{L^q(\Omega)} \leq Ct^{-\frac{1}{2}} \|f\|_{L^p(\Omega)}, \qquad t > 1.$$

Estimates (ii)–(iv) follow by applying again the semigroup property.

Step 2: Induction step. Let us assume that Lemma 3.5 holds for $m \in \mathbb{N}_0$. If $m = 0, 1$, then the semigroup property and the local estimates in Proposition 1.78 imply

$$\|T_{\Omega,p}(t)f\|_{H^{m+1}(\Omega)} \leq C\left\|T_{\Omega,p}\left(t - \frac{1}{2}\right)f\right\|_{L^2(\Omega)}, \qquad t > 1,$$

and

$$\|A_{\Omega,p}T_{\Omega,p}(t)f\|_{H^{m+1}(\Omega)} = \|T_{\Omega,p}(t)A_{\Omega,p}f\|_{H^{m+1}(\Omega)} \leq C\left\|A_{\Omega,p}T_{\Omega,p}\left(t - \frac{1}{2}\right)f\right\|_{L^2(\Omega)}, \qquad t > 1.$$

Estimates (i) and (iii) for $m + 1$ now follow from Step 1 if $m = 0, 1$. Next, we assume that $m \geq 2$. Applying the elliptic estimate in Proposition 2.7 we conclude that

$$\|T_{\Omega,p}(t)f\|_{H^{m+1}(\Omega)} \leq \|\nabla^2 T_{\Omega,p}(t)f\|_{H^{m-1}(\Omega)} + \|T_{\Omega,p}(t)f\|_{H^1(\Omega)}$$
$$\leq C\Big(\|A_{\Omega,p}T_{\Omega,p}(t)f\|_{H^{m-1}(\Omega)} + \|T_{\Omega,p}(t)f\|_{H^1(\Omega)}\Big), \qquad t > 0.$$

Hence, estimate (i) for $m + 1$ follows from the induction hypothesis. Similarly, we see that

$$\|A_{\Omega,p}T_{\Omega,p}(t)f\|_{H^{m+1}(\Omega)}$$
$$\leq \|\nabla^2 T_{\Omega,p}(t)A_{\Omega,p}f\|_{H^{m-1}(\Omega)} + \|A_{\Omega,p}T_{\Omega,p}(t)f\|_{H^1(\Omega)}$$
$$\leq C\left(\left\|A_{\Omega,p}T_{\Omega,p}\left(t - \tfrac{1}{2}\right)A_{\Omega,p}T_{\Omega,p}\left(\tfrac{1}{2}\right)f\right\|_{H^{m-1}(\Omega)} + \|A_{\Omega,p}T_{\Omega,p}(t)f\|_{H^1(\Omega)}\right), \quad t > 1.$$

Thus, estimate (iii) for $m + 1$ follows from the induction hypothesis and the local estimates in Proposition 1.78. Next, we apply the elliptic estimate in Proposition 2.7 to conclude that

$$\|\nabla^2 T_{\Omega,p}(t)f\|_{H^{m+1}(\Omega)} \leq C\Big(\|A_{\Omega,p}T_{\Omega,p}(t)f\|_{H^{m+1}(\Omega)} + \|\nabla T_{\Omega,p}(t)f\|_{L^2(\Omega)}\Big), \qquad t > 0.$$

Estimate (iv) for $m + 1$ follows now from (iii) and from Step 1. Finally, estimate (ii) for $m + 1$ easily follows from the relation

$$\|\nabla T_{\Omega,p}(t)f\|_{H^{m+1}(\Omega)} \leq \|\nabla^2 T_{\Omega,p}(t)f\|_{H^m(\Omega)} + \|\nabla T_{\Omega,p}(t)f\|_{L^2(\Omega)}, \qquad t > 0$$

and what has been proved above. This concludes the proof. $\qquad\square$

3.3 Proof of the L^p-estimates

Now we give the proof of Theorem 3.1. In the following let $\Omega \subset \mathbb{R}^d$, $d \geq 2$, be an exterior domain of class C^4. We need this regularity of the boundary as we want to apply Lemma 3.5 for $m = 2$.

First of all, note that it suffices to consider $t > 1$ only since for $0 < t \leq 1$ the estimates are clear by Proposition 1.78. Therefore, for $f \in L^p_\sigma(\Omega)$, $1 < p < \infty$, set

$$b := T_{\Omega,p}(1)f \qquad \text{and} \qquad u(t) := T_{\Omega,p}(t)b = T_{\Omega,p}(t+1)f.$$

We are interested only in the asymptotic behaviour of $\{T_{\Omega,p}(t)\}_{t \geq 0}$, i.e. the behaviour as $t \to \infty$. It is clear that for large t the behaviour of $T_{\Omega,p}(t)f$ follows from the behavior of $u(t) = T_{\Omega,p}(t+1)f$. Therefore, we consider $u(t)$ instead of $T_{\Omega,p}(t)f$. This (technical) trick is also used e.g. in [DKS98], and it has two technical advantages: First of all, Remark 1.36 yields that $b \in \mathcal{D}(A^n_{\Omega,p})$ for all $n \in \mathbb{N}$. Moreover, by the local L^p-L^q-estimates for the Stokes semigroup, see Proposition 1.78, we conclude that $b \in L^q_\sigma(\Omega)$ for all $p \leq q \leq \infty$. Thus, the new initial value b is more regular than the function $f \in L^p_\sigma(\Omega)$. Second, if we consider $u(t) = T_{\Omega,p}(t+1)f$ instead of $T_{\Omega,p}(t)f$ we avoid integrability problems at $t = 0$.

If we choose $p \in C(\mathbb{R}_+; \widehat{W}^{1,p}(\Omega))$ such that $\nabla p = (\mathrm{Id} - \mathbb{P}_{\Omega,p})T_{\Omega,p}(t)b$, then the pair of functions (u, p) satisfies the homogeneous Stokes problem

$$\begin{cases} u_t - \Delta u + \nabla p = 0 & \text{in } \mathbb{R}_+ \times \Omega, \\ \operatorname{div} u = 0 & \text{in } \mathbb{R}_+ \times \Omega, \\ u = 0 & \text{on } \mathbb{R}_+ \times \partial\Omega, \\ u|_{t=0} = b & \text{in } \Omega. \end{cases} \tag{3.3}$$

In the following, we shall frequently use the consistency of the Stokes semigroups on $L^p_\sigma(\Omega)$, $1 < p < \infty$, and the fact that $T^*_{\Omega,p}(t) = T_{\Omega,p'}(t)$ with $\frac{1}{p} + \frac{1}{p'} = 1$, see Remark 1.75.

The general idea is to use a cut-off procedure, as introduced in Section 1.6.1, to decompose the problem into a problem in the whole space $\mathbb{R}^d$ and to a problem in a bounded domain close to the boundary $\partial\Omega$. Then we may use the decay estimates for the Stokes semigroup in $\mathbb{R}^d$ and in bounded domains, see Proposition 1.79 and Proposition 1.82, respectively.

Choose $R > 0$ sufficiently large such that $\mathcal{O} \subset B_R$. Next, set

$$D := \Omega \cap B_{R+4},$$
$$D_1 := \{x \in \Omega : R < |x| < R + 3\}.$$

and choose a cut-off function $\varphi_1 \in C^\infty_c(\mathbb{R}^d)$ such that $0 \leq \varphi_1 \leq 1$ and

$$\varphi_1(x) := \begin{cases} 1, & |x| \leq R + 1, \\ 0, & |x| \geq R + 2. \end{cases}$$

Set $\varphi_2 := (1 - \varphi_1)$.

Remark 3.7. Without loss of generality, we may assume that the pressure p is chosen such that $p(t) \in L^p_0(D)$ for all $t \in \mathbb{R}_+$, see Definition 2.8.

For $t \geq 0$ we introduce functions $u^{(1)}(t) \in W^{2,p}(D)^d \cap W^{1,p}_0(D)^d \cap L^p_\sigma(D)$ and $u^{(2)}(t) \in W^{2,p}(\mathbb{R}^d)^d \cap L^p_\sigma(\mathbb{R}^d)$ by setting

$$\begin{aligned} u^{(i)}(t) &:= \varphi_i u(t) - \mathbb{B}_{D_1}((\nabla\varphi_i) \cdot u(t)), \\ p^{(i)}(t) &:= \varphi_i p(t), \end{aligned} \tag{3.4}$$

where $i = 1, 2$. It is clear that $u = u^{(1)} + u^{(2)}$ and $p = p^{(1)} + p^{(2)}$. A short calculation shows that $u^{(1)}$ solves the inhomogeneous Stokes problem

$$\begin{cases} u^{(1)}_t - \Delta u^{(1)} + \nabla p^{(1)} = F^{(1)} & \text{in } \mathbb{R}_+ \times D, \\ \operatorname{div} u^{(1)} = 0 & \text{in } \mathbb{R}_+ \times D, \\ u^{(1)} = 0 & \text{on } \mathbb{R}_+ \times \partial D, \\ u^{(1)}|_{t=0} = b^{(1)} & \text{in } D, \end{cases}$$

in the bounded domain D, and $u^{(2)}$ solves the inhomogeneous Stokes problem

$$
\begin{cases}
u_t^{(2)} - \Delta u^{(2)} + \nabla p^{(2)} &=\ F^{(2)} \quad \text{in } \mathbb{R}_+ \times \mathbb{R}^d, \\
\operatorname{div} u^{(2)} &=\ 0 \quad \text{in } \mathbb{R}_+ \times \mathbb{R}^d, \\
u^{(2)}|_{t=0} &=\ b^{(2)} \quad \text{in } \mathbb{R}^d,
\end{cases}
$$

in the whole space $\mathbb{R}^d$. Here the "error terms" $F^{(i)}$, $i = 1, 2$, are given by

$$
F^{(i)}(t) = -2(\nabla \varphi_i) \cdot \nabla u(t) - (\Delta \varphi_i) u(t) - (\partial_t - \Delta) \mathbb{B}_{D_1}((\nabla \varphi_i) \cdot u(t)) + (\nabla \varphi_i) p(t),
$$

and the initial values $b^{(i)}$, $i = 1, 2$, are given by

$$
b^{(i)} = \varphi_i b - \mathbb{B}_{D_1}((\nabla \varphi_i) \cdot b).
$$

Remark 3.8. By applying Proposition 1.87 and Proposition 1.78 we see that for $i = 1, 2$, $m = 0, 1$ and $1 < p \le q < \infty$ or $1 < p < q \le \infty$ there exists a constant $C := C(p, q, m) > 0$ such that

$$
\|b^{(i)}\|_{W^{2m,p}(\Omega)} \le C \|b\|_{W^{2m,p}(\Omega)} \le C \|f\|_{L^p(\Omega)},
$$

and

$$
\|b^{(i)}\|_{L^q(\Omega)} \le C \|b\|_{L^q(\Omega)} \le C \|f\|_{L^p(\Omega)}.
$$

Moreover, note that the supports of $F^{(i)}(t)$, $i = 1, 2$, are contained in $D_1 \subset D$ for all $t \in \mathbb{R}_+$.

In the following $\{T_{D,p}(t)\}_{t>0}$ denotes the Stokes semigroup on $L^p_\sigma(D)$ and $\{T_{\mathbb{R}^d,p}(t)\}_{t\ge0}$ denotes the Stokes semigroup on $L^p_\sigma(\mathbb{R}^d)$. The Stokes operators on $L^p_\sigma(D)$ and $L^p_\sigma(\mathbb{R}^d)$ are denoted by $A_{D,p}$ and $A_{\mathbb{R}^d,p}$, respectively. Since $u^{(1)}$ and $u^{(2)}$ solve inhomogeneous Stokes problems we conclude that $u^{(1)}$ and $u^{(2)}$ are given via the variation of constants formula[1]:

$$
u^{(1)}(t) = T_{D,p}(t) b^{(1)} + \int_0^t T_{D,p}(t - s) \mathbb{P}_{D,p} F^{(1)}(s) \, ds, \tag{3.5}
$$

and

$$
u^{(2)}(t) = T_{\mathbb{R}^d,p}(t) b^{(2)} + \int_0^t T_{\mathbb{R}^d,p}(t - s) \mathbb{P}_{\mathbb{R}^d,p} F^{(2)}(s) \, ds. \tag{3.6}
$$

In the following we estimate the norms of the formulas (3.5) and (3.6). For that we use the decay estimates for $\{T_{D,p}(t)\}_{t\ge0}$ and $\{T_{\mathbb{R}^d,p}(t)\}_{t\ge0}$ stated in Proposition 1.82 and Proposition 1.79, respectively. In some sense, we reduced the problem of proving Theorem 3.1 to the problem of controlling the "error terms" $F^{(i)}$, $i = 1, 2$, in a suitable way. Dealing with the error terms has the big advantage that the supports of $F^{(i)}(t)$ are contained in $D_1 \subset D$ and therefore we have tools (e.g. Poincaré's inequality or the embedding $L^p(D) \hookrightarrow L^q(D)$ for $p > q$) at hand which are only applicable in bounded domains. This allows in some sense to reduce everything to the L^2-case studied in Section 3.2. For the proof of Theorem 3.1 the following lemma is a key ingredient.

---

[1]See the explanations at the end of Section 1.2.1.

Lemma 3.9. *Let $1 < p \leq 2$ and $m = 0, 1$. Then for $i = 1, 2$ there exists a constant $C := C(p, \Omega) > 0$ such that*

$$\|F^{(i)}(t)\|_{H^{2m}(\Omega)} \leq C(1 + t)^{-\alpha(p) - \frac{1}{2}} \|f\|_{L^p(\Omega)}, \qquad t > 0,$$

where $\alpha(p)$ is given in (3.1).

Proof. First of all let us note that $\|F^{(i)}(t)\|_{H^{2m}(\Omega)} = \|F^{(i)}(t)\|_{H^{2m}(D)}$ because the supports of $F^{(i)}(t)$ are contained in $D_1 \subset D$. Set

$$I_1^{(i)} + I_2^{(i)} + I_3^{(i)} + I_4^{(i)}$$
$$:= -2(\nabla \varphi) \cdot \nabla u(t) - (\Delta \varphi_i) u(t) - (\partial_t - \Delta) \mathbb{B}_{D_1}((\nabla \varphi_i) \cdot u(t)) + (\nabla \varphi_i) p(t).$$

In the following all constants may depend on Ω and p. Using Poincaré's inequality[2] on D implies that

$$\|I_1^{(i)}\|_{H^{2m}(D)} + \|I_2^{(i)}\|_{H^{2m}(D)} \leq C \|\nabla u\|_{H^{2m}(D)}. \tag{3.7}$$

Moreover, since $p(t) \in L_0^p(D)$ for all $t \in \mathbb{R}_+$, we may apply Poincaré's inequality for the pressure $p(t)$ on D and we use the fact $\nabla p = -u_t + \Delta u$ to obtain

$$\|p(t)\|_{L^2(D)} \leq C \|\nabla p(t)\|_{L^2(D)} \leq C \left(\|\nabla^2 u(t)\|_{L^2(D)} + \|u_t(t)\|_{L^2(D)} \right)$$

and

$$\|p(t)\|_{H^2(D)} \leq C \|\nabla p(t)\|_{H^1(D)} \leq C \left(\|\nabla^2 u(t)\|_{H^1(D)} + \|u_t(t)\|_{H^1(D)} \right).$$

Consequently, we conclude that

$$\|I_4^{(i)}\|_{H^{2m}(D)} \leq C \left(\|\nabla^2 u(t)\|_{H^{2m}(D)} + \|u_t(t)\|_{H^{2m}(D)} \right). \tag{3.8}$$

Proposition 1.87 implies $\Delta \mathbb{B}_{D_1} \in \mathscr{L}(H_0^{2m+1}(D_1), H_0^{2m}(D_1)^d)$. Hence, by applying again Poincaré's inequality, we obtain

$$\|\Delta \mathbb{B}_{D_1}((\nabla \varphi_i) \cdot u)\|_{H^{2m}(D)} \leq C \|u\|_{H^{2m+1}(D)} \leq C \|\nabla u\|_{H^{2m}(D)}.$$

Moreover, since $\partial_t \mathbb{B}_{D_1}((\nabla \varphi_i) \cdot u(t)) = \mathbb{B}_{D_1}((\nabla \varphi_i) \cdot u_t(t))$ and $\mathbb{B}_{D_1} \in \mathscr{L}(H_0^{2m}(D_1), H_0^{2m+1}(D_1)^d)$, we conclude

$$\|\partial_t \mathbb{B}_{D_1}((\nabla \varphi_i) \cdot u(t))\|_{H^{2m}(D)} \leq C \|u_t(t)\|_{H^{2m}(D)}.$$

This implies

$$\|I_3^{(i)}\|_{H^{2m}(D)} \leq C \left(\|\nabla u(t)\|_{H^{2m}(D)} + \|u_t(t)\|_{H^{2m}(D)} \right). \tag{3.9}$$

Combining (3.7), (3.8) and (3.9) implies

$$\|F^{(i)}(t)\|_{H^{2m}(\Omega)} \leq C \left(\|\nabla^2 u(t)\|_{H^{2m}(D)} + \|\nabla u(t)\|_{H^{2m}(D)} + \|u_t(t)\|_{H^{2m}(D)} \right), \qquad t > 0. \tag{3.10}$$

Since $u(t) = T_{\Omega,p}(t + 1)f$ and $u_t(t) = A_{\Omega,p} u(t) = A_{\Omega,p} T_{\Omega,p}(t + 1)f$, applying the estimates in Lemma 3.5 yields the assertion. $\qquad \square$

---

[2]Poincaré's inequality can be applied here on D since $D := \Omega \cap B_{R+4}$ and $u|_{\partial\Omega} = 0$.

Before we turn our attention to the actual proof of Theorem 3.1 we need some more preparations. In Proposition 1.79 and Proposition 1.82 we stated decay estimates for $\{T_{\mathbb{R}^d,p}(t)\}_{t \geq 0}$ and $\{T_{D,p}(t)\}_{t \geq 0}$, respectively. These estimates express a local behaviour of the Stokes semigroup at $t = 0$ and a global behaviour for $t \to \infty$. In particular, the norm estimates of first or second order derivates behave like $t^{-\frac{1}{2}}$ or t^{-1}, respectively, which expresses a singular behaviour at $t = 0$. Since in this chapter we are only interested in the behaviour as $t \to \infty$, we would like to remove this singularity for $t = 0$ in a suitable way. The prize we have to pay for this is that the estimates involve higher-order norms.

Lemma 3.10. *Let $1 < p < \infty$. Then there exist constants $\delta := \delta(p, D), C := C(p, D) > 0$ such that for all $f \in \mathcal{D}(A_{D,p})$*

(i) $\|\nabla T_{D,p}(t)f\|_{L^p(D)} \leq C(t+1)^{-\frac{1}{2}}e^{-\delta t}\|f\|_{W^{2,p}(D)}, \qquad t > 0,$

(ii) $\|\nabla^2 T_{D,p}(t)f\|_{L^p(D)} \leq C(t+1)^{-1}e^{-\delta t}\|f\|_{W^{2,p}(D)}, \qquad t > 0,$

(iii) $\|A_{D,p}T_{D,p}(t)f\|_{L^p(D)} \leq C(t+1)^{-1}e^{-\delta t}\|f\|_{W^{2,p}(D)}, \qquad t > 0.$

Proof. For $t > 1$ the assertions follow directly from Proposition 1.82, because $(t+1)/t$ is uniformly bounded for $t \in (1, \infty)$. So now let $0 < t \leq 1$. Since the Stokes operator $A_{D,p}$ commutes with the Stokes semigroup $\{T_{D,p}(t)\}_{t \geq 0}$ we obtain that

$$\|A_{D,p}T_{D,p}(t)f\|_{L^p(D)} = \|T_{D,p}(t)A_{D,p}f\|_{L^p(D)} \leq C\|A_{D,p}f\|_{L^p(D)} \leq C\|f\|_{W^{2,p}(D)}.$$

This yields estimate (iii). Next we use the elliptic estimate in Proposition 2.3 to conclude that

$$\|\nabla T_{D,p}(t)f\|_{L^p(D)} + \|\nabla^2 T_{D,p}(t)f\|_{L^p(D)} \leq \|T_{D,p}(t)f\|_{W^{2,p}(D)} \leq C\|A_{D,p}T_{D,p}(t)f\|_{L^p(D)}.$$

Consequently, estimates (i) and (ii) follow from estimate (iii). $\qquad\square$

Lemma 3.11. *Let $1 < p < \infty$. Then there exists a constant $C := C(p) > 0$ such that for all $f \in \mathcal{D}(A_{\mathbb{R}^d,p})$*

(i) $\|\nabla T_{\mathbb{R}^d,p}(t)f\|_{L^p(\mathbb{R}^d)} \leq C(t+1)^{-\frac{1}{2}}\|f\|_{W^{2,p}(\mathbb{R}^d)}, \qquad t > 0,$

(ii) $\|\nabla^2 T_{\mathbb{R}^d,p}(t)f\|_{L^p(\mathbb{R}^d)} \leq C(t+1)^{-1}\|f\|_{W^{2,p}(\mathbb{R}^d)}, \qquad t > 0,$

(iii) $\|A_{\mathbb{R}^d,p}T_{\mathbb{R}^d,p}(t)f\|_{L^p(\mathbb{R}^d)} \leq C(t+1)^{-1}\|f\|_{W^{2,p}(\mathbb{R}^d)}, \qquad t > 0.$

Proof. The assertions for $t > 1$ follow directly from the decay estimates in Proposition 1.79 and for $0 < t \leq 1$ we use Proposition 1.81. $\qquad\square$

In the following all appearing constants may depend on p, q and Ω. The proof of Theorem 3.1 is divided into 6 steps:

Step 1: Assertion (a). Let $1 < p \leq q \leq 2$. The other cases follow later by duality arguments. In the following, we consider the functions $u^{(1)}$ and $u^{(2)}$ separately. Let us start with $u^{(1)}$. The exponential decay of the semigroup $\{T_{D,p}(t)\}_{t\geq 0}$, see Proposition 1.82, yields that there exist constants $C, \delta > 0$ such that

$$\|u^{(1)}(t)\|_{L^q(D)} \leq C\left(e^{-\delta t}\|b^{(1)}\|_{L^q(D)} + \int_0^t e^{-\delta(t-s)}\|F^{(1)}(s)\|_{L^q(D)}\,\mathrm{d}s\right), \qquad t > 0.$$

By using Remark 3.8, the embedding $L^2(D) \hookrightarrow L^q(D)$ and the estimate for $F^{(1)}$ in Lemma 3.9 we obtain

$$\|u^{(1)}(t)\|_{L^q(D)}$$
$$\leq C\|f\|_{L^p(\Omega)}\left((1+t)^{-\frac{d}{2}\left(\frac{1}{p}-\frac{1}{q}\right)} + \int_0^t (1+t-s)^{-\frac{d}{2}\left(1-\frac{1}{q}\right)}(1+s)^{-\alpha(p)-\frac{1}{2}}\,\mathrm{d}s\right), \, t > 0, \qquad (3.11)$$

where $\alpha(p)$ is given in (3.1). In a similar way we can also deal with $u^{(2)}(t)$. However, the semigroup $\{T_{\mathbb{R}^d,p}(t)\}_{t\geq 0}$ has no exponential decay, but we use the decay estimates stated Proposition 1.79 and Proposition 1.80. Then we get

$$\|T_{\mathbb{R}^d,p}(t-s)\mathbb{P}_{\mathbb{R}^d,p}F^{(2)}(s)\|_{L^q(\mathbb{R}^d)} \leq C(t-s)^{-\frac{d}{2}\left(1-\frac{1}{q}\right)}\|F^{(2)}(s)\|_{L^1(D)}$$
$$\leq C(1+t-s)^{-\frac{d}{2}\left(1-\frac{1}{q}\right)}\|F^{(2)}(s)\|_{L^2(D)}, \qquad t-s > 1,$$

and

$$\|T_{\mathbb{R}^d,p}(t-s)\mathbb{P}_{\mathbb{R}^d,p}F^{(2)}(s)\|_{L^q(\mathbb{R}^d)} \leq C\|F^{(2)}(s)\|_{L^q(D)}$$
$$\leq C(1+t-s)^{-\frac{d}{2}\left(1-\frac{1}{q}\right)}\|F^{(2)}(s)\|_{L^2(D)}, \qquad 0 < t-s < 1.$$

Note that we have also applied the embeddigs $L^2(D) \hookrightarrow L^q(D) \hookrightarrow L^1(D)$. Similarly, by taking into account Remark 3.8, we see that

$$\|T_{\mathbb{R}^d,p}(t)b^{(2)}\|_{L^q(\mathbb{R}^d)} \leq C(1+t)^{-\frac{d}{2}\left(\frac{1}{p}-\frac{1}{q}\right)}\|f\|_{L^p(\Omega)}, \qquad t > 0.$$

Thus, by Lemma 3.9 we finally receive

$$\|u^{(2)}(t)\|_{L^q(\mathbb{R}^d)}$$
$$\leq C\|f\|_{L^p(\Omega)}\left((1+t)^{-\frac{d}{2}\left(\frac{1}{p}-\frac{1}{q}\right)} + \int_0^t (1+t-s)^{-\frac{d}{2}\left(1-\frac{1}{q}\right)}(1+s)^{-\alpha(p)-\frac{1}{2}}\,\mathrm{d}s\right), \, t > 0. \qquad (3.12)$$

Let us now evaluate the integrals on the right-hand sides of (3.11) and (3.12). For this purpose set

$$J_1(t) := \int_0^t (1+t-s)^{-\frac{d}{2}\left(1-\frac{1}{q}\right)}(1+s)^{-\alpha(p)-\frac{1}{2}}\,\mathrm{d}s, \qquad t > 0. \qquad (3.13)$$

Let $d = 2$. Then

$$J_1(t) \leq \left(1+\tfrac{t}{2}\right)^{-\left(1-\frac{1}{q}\right)}\int_0^{\frac{t}{2}}(1+s)^{-\frac{1}{p}}\,\mathrm{d}s + \left(1+\tfrac{t}{2}\right)^{-\frac{1}{p}}\int_{\frac{t}{2}}^t (1+t-s)^{-\left(1-\frac{1}{q}\right)}\,\mathrm{d}s$$
$$\leq C(1+t)^{-\left(\frac{1}{p}-\frac{1}{q}\right)}, \qquad t > 0.$$

Now let $d \geq 3$. Then

$$J_1(t) \leq \left(1 + \tfrac{t}{2}\right)^{-\frac{d}{2}\left(\frac{1}{p} - \frac{1}{q}\right)} \left(\int_0^{\frac{t}{2}} (1 + t - s)^{-\frac{d}{2}\left(1 - \frac{1}{p}\right)} (1 + s)^{-\alpha(p) - \frac{1}{2}} \, ds \right.$$

$$\left. + \int_{\frac{t}{2}}^t (1 + t - s)^{-\frac{d}{2}\left(1 - \frac{1}{q}\right)} (1 + s)^{-\alpha(p) - \frac{1}{2} + \frac{d}{2}\left(\frac{1}{p} - \frac{1}{q}\right)} \, ds \right), \quad t > 0.$$

By making the change of variable $r = t - s$ in the integral from $\frac{t}{2}$ to t and by using that $1 + t - s \geq 1 + s$ when $0 \leq s \leq \frac{t}{2}$ we obtain

$$J_1(t) \leq 2 \int_0^{\frac{t}{2}} (1 + s)^{-\frac{d}{2}\left(1 - \frac{1}{p}\right) - \alpha(p) - \frac{1}{2}} \, ds, \quad t > 0.$$

Since $\frac{d}{2}\left(1 - \frac{1}{p}\right) + \alpha(p) + \frac{1}{2} > 1$, it follows that $J_1(t)$ is uniformly bounded for $t \in \mathbb{R}_+$. Summing up we conclude that

$$\|T_{\Omega,p}(t)b\|_{L^q(\Omega)} = \|u(t)\|_{L^q(\Omega)} \leq \|u^{(1)}(t)\|_{L^q(D)} + \|u^{(2)}(t)\|_{L^q(\mathbb{R}^d)} \leq C(1 + t)^{-\frac{d}{2}\left(\frac{1}{p} - \frac{1}{q}\right)} \|f\|_{L^p(\Omega)}.$$

This proves assertion (a) for $1 < p \leq q \leq 2$. Next we consider $2 < p \leq q < \infty$. This case follows by duality arguments. Let $1 < q' \leq p' < 2$ be such that $\frac{1}{p} + \frac{1}{p'} = 1$ and $\frac{1}{q} + \frac{1}{q'} = 1$. Furthermore, we take some $g \in C_{c,\sigma}^\infty(\Omega)$ such that $\|g\|_{L^{q'}(\Omega)} = 1$. Then we obtain

$$|\langle T_{\Omega,p}(t)b, g\rangle_\Omega| = |\langle f, T_{\Omega,p'}(t+1)g\rangle_\Omega| \leq \|f\|_{L^p(\Omega)} \|T_{\Omega,p'}(t+1)g\|_{L^{p'}(\Omega)}$$

$$\leq C(1 + t)^{-\frac{d}{2}\left(\frac{1}{p} - \frac{1}{q}\right)} \|f\|_{L^p(\Omega)}, \quad t > 0.$$

Assertion (a) for $2 < p \leq q < \infty$ follows. Finally, we consider the case $1 < p \leq 2$ and $q > 2$. Again let $1 < q' < 2$ such that $\frac{1}{q} + \frac{1}{q'} = 1$ and let $g \in C_{c,\sigma}^\infty(\Omega)$ be such that $\|g\|_{L^{q'}(\Omega)} = 1$. Consequently, we see that

$$|\langle T_{\Omega,p}(t)b, g\rangle_\Omega| = \left|\left\langle T_{\Omega,p}\left(\tfrac{t+1}{2}\right) f, T_{\Omega,p'}\left(\tfrac{t+1}{2}\right) g\right\rangle_\Omega\right| \leq \left\|T_{\Omega,p}\left(\tfrac{t+1}{2}\right) f\right\|_{L^2(\Omega)} \left\|T_{\Omega,p'}\left(\tfrac{t+1}{2}\right) g\right\|_{L^2(\Omega)}$$

$$\leq C(1 + t)^{-\frac{d}{2}\left(\frac{1}{p} - \frac{1}{q}\right)} \|f\|_{L^p(\Omega)}, \quad t > 0.$$

This concludes the proof of assertion (a).

Step 2: Assertion (c) for $1 < p \leq 2$. The case $p = 2$ follows directly from Proposition 3.3. So we may assume that $1 < p < 2$. Similarly to (3.11) and (3.12) we obtain

$$\|\nabla u^{(1)}(t)\|_{L^p(D)} + \|\nabla u^{(2)}(t)\|_{L^p(\mathbb{R}^d)}$$

$$\leq C\|f\|_{L^p(\Omega)} \left((1 + t)^{-\frac{1}{2}} + \int_0^t (1 + t - s)^{-\frac{d}{2}\left(1 - \frac{1}{p}\right) - \frac{1}{2}} (1 + s)^{-\alpha(p) - \frac{1}{2}} \, ds \right), \, t > 0, \tag{3.14}$$

where $\alpha(p)$ is given in (3.1). Here we applied Lemma 3.10, Lemma 3.11 and the estimates for $F^{(i)}$, $i = 1, 2$, stated in Lemma 3.9. The integral

$$J_2(t) := \int_0^t (1 + t - s)^{-\frac{d}{2}\left(1 - \frac{1}{p}\right) - \frac{1}{2}} (1 + s)^{-\alpha(p) - \frac{1}{2}} \, ds, \quad t > 0$$

on the right-hand side of (3.14) is estimated analogously to $J_1(t)$ in (3.13):

$$J_2(t) \leq \left(1 + \tfrac{t}{2}\right)^{-\left(1-\frac{1}{p}\right)-\frac{1}{2}} \int_0^{\frac{t}{2}} (1+s)^{-\frac{1}{p}} \, ds + \left(1 + \tfrac{t}{2}\right)^{-\frac{1}{p}} \int_{\frac{t}{2}}^t (1+t-s)^{-\left(1-\frac{1}{p}\right)-\frac{1}{2}} \, ds$$

$$\leq C(1+t)^{-\frac{1}{2}}, \qquad t > 0,$$

if $d = 2$, and

$$J_2(t) \leq \left(1 + \tfrac{t}{2}\right)^{-\frac{1}{2}} \left(\int_0^{\frac{t}{2}} (1+t-s)^{-\frac{d}{2}\left(1-\frac{1}{p}\right)} (1+s)^{-\alpha(p)-\frac{1}{2}} , \, ds \right.$$

$$\left. + \int_{\frac{t}{2}}^t (1+t-s)^{-\frac{d}{2}\left(1-\frac{1}{p}\right)-\frac{1}{2}} (1+s)^{-\alpha(p)} \, ds \right)$$

$$\leq 2 \left(1 + \tfrac{t}{2}\right)^{-\frac{d}{2p}} \int_0^{\frac{t}{2}} (1+s)^{-\frac{d}{2}\left(1-\frac{1}{p}\right)-\frac{1}{2}-\alpha(p)} \, ds \leq C \left(1 + \tfrac{t}{2}\right)^{-\frac{1}{2}}, \qquad t > 0,$$

if $d \geq 3$. Assertion (c) for $1 < p \leq 2$ follows.

Step 3: Assertion (e). In this step we assume that $d \geq 3$. Note that $A_{\Omega,p} u(t) = u_t(t)$ and that $u_t(t) = u_t^{(1)} + u_t^{(2)}$. We start with the case $1 < p \leq \frac{2d}{d+2}$. The strategy is the same as in Step 1 and Step 2. First of all, note that

$$\|u_t^{(1)}(t)\|_{L^p(D)} \leq \|A_{D,p} u^{(1)}(t)\|_{L^p(D)} + \|F^{(1)}(t)\|_{L^p(D)}, \qquad t > 0$$

and

$$\|u_t^{(2)}(t)\|_{L^p(\mathbb{R}^d)} \leq \|A_{\mathbb{R}^d,p} u^{(2)}(t)\|_{L^p(\mathbb{R}^d)} + \|F^{(2)}(t)\|_{L^p(D)}, \qquad t > 0.$$

By using the embedding $L^2(D) \hookrightarrow L^p(D)$ and Lemma 3.9 we obtain

$$\|F^{(i)}(t)\|_{L^p(D)} \leq C \|F^{(i)}(t)\|_{L^2(D)} \leq C(1+t)^{-1} \|f\|_{L^p(\Omega)}, \qquad t > 0,$$

for $i = 1, 2$. Similarly to (3.14), we see that

$$\|A_{D,p} u^{(1)}(t)\|_{L^p(D)} + \|A_{\mathbb{R}^d,p} u^{(2)}(t)\|_{L^p(\mathbb{R}^d)}$$

$$\leq C \|f\|_{L^p(\Omega)} \left((1+t)^{-1} + \int_0^t (1+t-s)^{-\frac{d}{2}\left(1-\frac{1}{p}\right)-1} (1+s)^{-1} \, ds \right), \qquad t > 0. \tag{3.15}$$

The integral

$$J_3(t) := \int_0^t (1+t-s)^{-\frac{d}{2}\left(1-\frac{1}{p}\right)-1} (1+s)^{-1} \, ds, \qquad t > 0$$

on the right-hand side of (3.15) is estimated as follows:

$$J_3(t) \leq \left(1 + \tfrac{t}{2}\right)^{-1} \left(\int_0^{\frac{t}{2}} (1+t-s)^{-\frac{d}{2}\left(1-\frac{1}{p}\right)} (1+s)^{-1} \, ds + \int_{\frac{t}{2}}^t (1+t-s)^{-\frac{d}{2}\left(1-\frac{1}{p}\right)-1} ds \right)$$

$$\leq 2 \left(1 + \tfrac{t}{2}\right)^{-1} \int_0^{\frac{t}{2}} (1+s)^{-\frac{d}{2}\left(1-\frac{1}{p}\right)-1} ds \leq C \left(1 + \tfrac{t}{2}\right)^{-1}, \qquad t > 0.$$

Hence,

$$\|u_t(t)\|_{L^p(\Omega)} \le \|u_t^{(1)}(t)\|_{L^p(D)} + \|u_t^{(2)}(t)\|_{L^p(\mathbb{R}^d)} \le C(1+t)^{-1}\|f\|_{L^p(\Omega)}, \qquad t > 0.$$

This proves assertion (e) for $1 < p \le \frac{2d}{d+2}$. Next, consider the case $\frac{2d}{d-2} \le p < \infty$. Take $g \in C_{c,\sigma}^\infty(\Omega)$ with $\|g\|_{L^{p'}(\Omega)} = 1$, where $\frac{1}{p} + \frac{1}{p'} = 1$. Since $1 < p' < \frac{2d}{d+2}$, we obtain

$$|\langle A_{\Omega,p}T_{\Omega,p}(t)b, g\rangle_\Omega| = |\langle f, A_{\Omega,p'}T_{\Omega,p'}(t+1)g\rangle_\Omega| \le \|f\|_{L^p(\Omega)}\|A_{\Omega,p'}T_{\Omega,p'}(t+1)g\|_{L^{p'}(\Omega)}$$
$$\le C(1+t)^{-\frac{1}{2}}\|f\|_{L^p(\Omega)}, \qquad t > 0.$$

This yields the assertion for $\frac{2d}{d-2} \le p < \infty$. It only remains to prove the assertion for $\frac{2d}{d+2} < p < \frac{2d}{d-2}$. We set $p_0 = q_0 = \frac{2d}{d+2}$ and $p_1 = q_1 = \frac{2d}{d-2}$. So far we know that

$$\begin{aligned}
\|A_\Omega T_\Omega(t+1)\|_{\mathscr{L}(L^{p_0}(\Omega), L^{q_0}(\Omega))} &\le C(1+t)^{-1}, & t > 0 \\
\|A_\Omega T_\Omega(t+1)\|_{\mathscr{L}(L^{p_1}(\Omega), L^{q_1}(\Omega))} &\le C(1+t)^{-1}, & t > 0.
\end{aligned} \qquad (3.16)$$

By applying the Riesz-Thorin theorem, see Proposition 1.18, the remaining case directly follows from (3.16).

Step 4: Assertion (b). Let $d \ge 3$. Again we decompose the solution u to (3.3) appropriately. However, we do not use the decomposition introduced in (3.4), but here a different decomposition is more suitable. In the following we denote by $b_0 \in L_\sigma^p(\mathbb{R}^d)$ the zero extension of b to the whole $\mathbb{R}^d$. Then set

$$\begin{aligned}
u^{(3)}(t) &:= \varphi_2 T_{\mathbb{R}^d, p}(t)b_0 - \mathbb{B}_{D_1}((\nabla\varphi_2) \cdot T_{\mathbb{R}^d, p}(t)b_0), \\
u^{(4)}(t) &:= T_{\Omega, p}(t)b^{(3)},
\end{aligned} \qquad (3.17)$$

where $b^{(3)} := \varphi_1 b - \mathbb{B}_{D_1}((\nabla\varphi_1) \cdot b)$. Note that supp $b^{(3)} \subset D$. Moreover, let $u^{(5)}$, together with a suitable pressure q, be the solution to the inhomogeneous Stokes problem

$$\begin{cases}
u_t^{(5)} - \Delta u^{(5)} + \nabla q &= F^{(3)} & \text{in } \mathbb{R}_+ \times \Omega, \\
\operatorname{div} u^{(5)} &= 0 & \text{in } \mathbb{R}_+ \times \Omega, \\
u^{(5)} &= 0 & \text{on } \mathbb{R}_+ \times \partial\Omega, \\
u^{(5)}|_{t=0} &= 0 & \text{in } \Omega,
\end{cases}$$

where

$$F^{(3)} := 2(\nabla\varphi_2) \cdot \nabla T_{\mathbb{R}^d, p}(t)b_0 + (\Delta\varphi_2)T_{\mathbb{R}^d, p}(t)b_0 + (\partial_t - \Delta)\mathbb{B}_{D_1}((\nabla\varphi_2) \cdot T_{\mathbb{R}^d, p}(t)b_0).$$

Note that supp $F^{(3)} \subset D_1 \subset D$ for all $t \in \mathbb{R}_+$, and by the variation of constants formula we have

$$u^{(5)}(t) = \int_0^t T_{\Omega, p}(t-s)\mathbb{P}_{\Omega, p}F^{(3)}(s)\, \mathrm{d}s.$$

It is easy to see that $u = u^{(3)} + u^{(4)} + u^{(5)}$ holds. Such a decomposition was also used in [MS97, Section 7]. In the following we consider the terms $u^{(3)}$, $u^{(4)}$ and $u^{(5)}$ separately. For that we need the following decay estimate for the "error term" $F^{(3)}(t)$, cf. Lemma 3.9.

Lemma 3.12. *Let $1 < p < q \leq \infty$. Then there exists a constant $C := C(p, q, \Omega) > 0$ such that*

$$\|F^{(3)}(t)\|_{L^q(\Omega)} \leq C(1+t)^{-\frac{d}{2}\left(\frac{1}{p}-\frac{1}{q}\right)}\|f\|_{L^p(\Omega)}, \qquad t > 0.$$

Proof. Set

$$I_1 + I_2 + I_3 := 2(\nabla\varphi_2) \cdot \nabla T_{\mathbb{R}^d,p}(t)b_0 + (\Delta\varphi_2)T_{\mathbb{R}^d,p}(t)b_0 + (\partial_t - \Delta)\mathbb{B}_{D_1}((\nabla\varphi_2) \cdot T_{\mathbb{R}^d,p}(t)b_0).$$

By using the decay estimates for the semigroup $\{T_{\mathbb{R}^d,p}(t)\}_{t\geq 0}$ (see Proposition 1.79 and Lemma 3.11) and Remark 3.8 we see that

$$\|I_1\|_{L^q(\Omega)} + \|I_2\|_{L^q(\Omega)} \leq C(1+t)^{-\frac{d}{2}\left(\frac{1}{p}-\frac{1}{q}\right)}\|f\|_{L^p(\Omega)}, \qquad t > 0.$$

To estimate I_3, note that $\Delta\mathbb{B}_{D_1} \in \mathscr{L}(W_0^{1,q}(D_1), L^q(D_1)^d)$, $\mathbb{B}_{D_1} \in \mathscr{L}(L^q(D_1), W_0^{1,q}(D_1)^d)$ and that

$$\partial_t\mathbb{B}_{D_1}((\nabla\varphi_2) \cdot T_{\mathbb{R}^d,p}(t)b_0) = \mathbb{B}_{D_1}((\nabla\varphi_2) \cdot \partial_t T_{\mathbb{R}^d,p}(t)b_0) = \mathbb{B}_{D_1}((\nabla\varphi_2) \cdot A_{\mathbb{R}^d,p}T_{\mathbb{R}^d,p}(t)b_0).$$

Hence,

$$\|I_3\|_{L^q(\Omega)} \leq C\left(\|A_{\mathbb{R}^d,p}T_{\mathbb{R}^d,p}(t)b_0\|_{L^q(D)} + \|T_{\mathbb{R}^d,p}(t)b_0\|_{W^{1,q}(D)}\right)$$
$$\leq C(1+t)^{-\frac{d}{2}\left(\frac{1}{p}-\frac{1}{q}\right)}\|f\|_{L^p(\Omega)}, \qquad t > 0.$$

This concludes the proof. $\qquad\square$

Next, we shall estimate the L^∞-norm of $u^{(3)}$. We use the L^∞-estimates for the semigroup $\{T_{\mathbb{R}^d,p}(t)\}_{t\geq 0}$ stated in Proposition 1.79 and Proposition 1.80 to obtain

$$\|T_{\mathbb{R}^d,p}(t)b_0\|_{L^\infty(\mathbb{R}^d)} \leq C(1+t)^{-\frac{d}{2p}}\|b\|_{L^p(\Omega)}, \qquad t > 1,$$

and

$$\|T_{\mathbb{R}^d,p}(t)b_0\|_{L^\infty(\mathbb{R}^d)} \leq C\|b\|_{L^\infty(\Omega)} \leq C(1+t)^{-\frac{d}{2p}}\|b\|_{L^\infty(\Omega)}, \qquad 0 < t \leq 1.$$

Hence, using Remark 3.8, we conclude that

$$\|u^{(3)}(t)\|_{L^\infty(\Omega)} \leq C\left((1+t)^{-\frac{d}{2p}}\|f\|_{L^p(\Omega)} + \|\mathbb{B}_{D_1}((\nabla\varphi_2) \cdot T_{\mathbb{R}^d,p}(t)b_0)\|_{L^\infty(D_1)}\right), \qquad t > 0.$$

Further, by the Sobolev embedding $W^{1,q}(D_1) \hookrightarrow L^\infty(D_1)$ for $d < q < \infty$ and the fact that $\mathbb{B}_{D_1} \in \mathscr{L}(L^q(D_1), W_0^{1,q}(D_1)^d)$ we have

$$\|\mathbb{B}_{D_1}((\nabla\varphi_2) \cdot T_{\mathbb{R}^d,p}(t)b_0)\|_{L^\infty(D_1)} \leq C\|\mathbb{B}_{D_1}((\nabla\varphi_2) \cdot T_{\mathbb{R}^d,p}(t)b_0)\|_{W^{1,q}(D_1)}$$
$$\leq C\|T_{\mathbb{R}^d,p}(t)b_0\|_{L^q(D_1)}, \qquad t > 0.$$

The embedding $L^\infty(D_1) \hookrightarrow L^q(D_1)$ and as above the L^∞-estimate for the whole space semigroup $\{T_{\mathbb{R}^d,p}(t)\}_{t\geq 0}$ imply

$$\|\mathbb{B}_{D_1}((\nabla\varphi) \cdot T_{\mathbb{R}^d,p}(t)b_0)\|_{L^\infty(D_1)} \leq C\|T_{\mathbb{R}^d,p}(t)b_0\|_{L^\infty(\mathbb{R}^d)} \leq C(1+t)^{-\frac{d}{2p}}\|f\|_{L^p(\Omega)}, \qquad t > 0.$$

This yields

$$\|u^{(3)}(t)\|_{L^\infty(\Omega)} \le C(1+t)^{\frac{d}{2p}}\|f\|_{L^p(\Omega)}, \qquad t > 0. \qquad (3.18)$$

To deal with $u^{(4)}$ and $u^{(5)}$ we first prove the following lemma, cf. [MS97, Lemma 7.2].

Lemma 3.13. *Let $d \ge 3$, $q > d$ and $q > p > \frac{d}{2}$. Then, for $i = 4, 5$, there exists a constant $C := C(p, q, \Omega) > 0$ such that*

$$\|u^{(i)}(t)\|_{L^q(\Omega)} \le C(1+t)^{-\frac{d}{2p}}\|f\|_{L^p(\Omega)}, \qquad t > 0,$$

$$\|A_{\Omega,p}u^{(i)}(t)\|_{L^q(\Omega)} \le C(1+t)^{-\frac{d}{2p}}\|f\|_{L^p(\Omega)}, \qquad t > 0.$$

Proof. Let $\varepsilon > 0$ be sufficiently small and set

$$r := r(\varepsilon) := 1 + \varepsilon, \qquad \delta := \delta(\varepsilon) := \frac{d}{2}\frac{\varepsilon}{r}, \qquad \beta := \beta(\varepsilon) := \frac{d}{2}\left(\frac{1}{r} - \frac{1}{q}\right) = \frac{d}{2}\left(1 - \frac{1}{q}\right) - \delta.$$

Applying Theorem 3.1 (a) yields

$$\|u^{(4)}(t)\|_{L^q(\Omega)} \le C(1+t)^{-\beta}\|b^{(3)}\|_{L^r(D)}, \qquad t > 1,$$

and

$$\|u^{(4)}(t)\|_{L^q(\Omega)} \le C\|b^{(3)}\|_{L^q(D)} \le C(1+t)^{-\beta}\|b^{(3)}\|_{L^q(D)}, \qquad 0 < t \le 1.$$

The embedding $L^q(D) \hookrightarrow L^r(D)$ implies

$$\|u^{(4)}(t)\|_{L^q(\Omega)} \le C(1+t)^{-\beta}\|b^{(3)}\|_{L^q(\Omega)}, \qquad t > 0.$$

Similarly as in Remark 3.8, we see that

$$\|b^{(3)}\|_{L^q(\Omega)} \le C\|f\|_{L^p(\Omega)}.$$

Now, by using $1 - \frac{1}{q} > 1 - \frac{1}{d} - \frac{d-1}{d} \ge \frac{2}{d} > \frac{1}{p}$ and by making $\varepsilon > 0$ sufficiently small we conclude that

$$\|u^{(4)}(t)\|_{L^q(\Omega)} \le C(1+t)^{-\frac{d}{2p}}\|f\|_{L^p(\Omega)}, \qquad t > 0.$$

Analogously, by Theorem 3.1 (w) we have

$$\|A_{\Omega,p}u^{(4)}(t)\|_{L^q(\Omega)} \le C(1+t)^{-\frac{d}{2p}}\|f\|_{L^p(\Omega)}, \qquad t > 0.$$

Applying Theorem 3.1 (a), the embedding $L^\infty(D) \hookrightarrow L^r(D)$ and Lemma 3.12 shows that

$$\|u^{(5)}(t)\|_{L^q(\Omega)} \le C \int_0^t (1+t-s)^{-\beta}\|F^{(3)}(s)\|_{L^r(D)}\, ds \le C \int_0^t (1+t-s)^{-\beta}\|F^{(3)}(s)\|_{L^\infty(D)}\, ds$$

$$\le C\|f\|_{L^p(\Omega)} \int_0^t (1+t-s)^{-\beta}(1+s)^{-\frac{d}{2p}}\, ds, \qquad t > 0.$$

The integral

$$J_4(t) := \int_0^t (1+t-s)^{-\frac{d}{2}\left(1-\frac{1}{q}\right)+\delta}(1+s)^{-\frac{d}{2p}}\,ds, \qquad t > 0$$

is estimated as follows:

$$J_4(t) \le \left(1+\tfrac{t}{2}\right)^{-\frac{d}{2p}} \left(\int_0^{\frac{t}{2}} (1+t-s)^{-\frac{d}{2}\left(1-\frac{1}{q}\right)+\delta+\frac{d}{2p}} \right.$$

$$\left. (1+s)^{-\frac{d}{2p}}\,ds + \int_{\frac{t}{2}}^t (1+t-s)^{-\frac{d}{2}\left(1-\frac{1}{q}\right)+\delta}\,ds \right)$$

$$\le 2\left(1+\tfrac{t}{2}\right)^{-\frac{d}{2p}} \int_0^{\frac{t}{2}} (1+s)^{-\frac{d}{2}\left(1-\frac{1}{q}\right)+\delta}\,ds \le C\left(1+\tfrac{t}{2}\right)^{-\frac{d}{2p}}, \qquad t > 0.$$

Here we used that $\frac{d}{2} - \frac{d}{2q} - \delta > 1$ if we make $\delta(\varepsilon) > 0$ sufficiently small (this can be achieved by making $\varepsilon > 0$ sufficiently small). This implies the desired estimate

$$\|u^{(5)}(t)\|_{L^q(\Omega)} \le C(1+t)^{-\frac{d}{2p}}\|f\|_{L^p(\Omega)}, \qquad t > 0.$$

Analogously, we get

$$\|A_{\Omega,p}u^{(5)}(t)\|_{L^q(\Omega)} \le C(1+t)^{-\frac{d}{2p}}\|f\|_{L^p(\Omega)}, \qquad t > 0.$$

This concludes the proof of the lemma. $\qquad\square$

Having Lemma 3.13 at hand, we proceed in the proof of assertion (b). We use the the Sobolev embedding $W^{1,q}(\Omega) \hookrightarrow L^\infty(\Omega)$ for $q > d$ to see that

$$\|u^{(4)}(t) + u^{(5)}(t)\|_{L^\infty(\Omega)} \le C\|u^{(4)}(t) + u^{(5)}(t)\|_{W^{1,q}(\Omega)}, \qquad t > 0.$$

Next, by applying the Gagliardo-Nirenberg inequality, see Proposition 1.17, with $j = 1$, $m = 2$, $p = r = q$, $\theta = 1/2$ and Young's inequality we obtain

$$\|u^{(4)}(t) + u^{(5)}(t)\|_{W^{1,q}(\Omega)} \le C\left(\|\nabla^2 u^{(4)}(t) + \nabla^2 u^{(5)}(t)\|_{L^q(\Omega)} + \|u^{(4)}(t) + u^{(5)}(t)\|_{L^q(\Omega)} \right).$$

Then the elliptic estimate in Proposition 2.7 yields

$$\|u^{(4)}(t) + u^{(5)}(t)\|_{L^\infty(\Omega)}$$
$$\le C\left(\|\nabla^2 u^{(4)}(t) + \nabla^2 u^{(5)}(t)\|_{L^q(\Omega)} + \|u^{(4)}(t) + u^{(5)}(t)\|_{L^q(\Omega)} \right)$$
$$\le C\left(\|A_{\Omega,p}u^{(4)}(t) + A_{\Omega,p}u^{(5)}(t)\|_{L^q(\Omega)} + \|u^{(4)}(t) + u^{(5)}(t)\|_{L^q(\Omega)} + \|F^{(3)}(t)\|_{L^\infty(D)} \right)$$

for any $q > d$. Note that we also used the embedding $L^\infty(D) \hookrightarrow L^q(D)$ here. As a consequence of Lemma 3.13 and Lemma 3.12 we obtain

$$\|u^{(4)}(t) + u^{(5)}(t)\|_{L^\infty(\Omega)} \le C(1+t)^{\frac{d}{2p}}\|f\|_{L^p(\Omega)}, \qquad t > 0, \tag{3.19}$$

if $p > d/2$. Combining (3.18) and (3.19) yields assertion (b) for $p > d/2$. Now let $p \le d/2$.

Fix some $q > d/2$. Then by what has already been proved and by Theorem 3.1 (a) we have

$$\|u(t)\|_{L^\infty(\Omega)} \leq C(1+t)^{-\frac{d}{2q}} \left\|u\left(\tfrac{t}{2}\right)\right\|_{L^q(\Omega)} \leq C(1+t)^{-\frac{d}{2p}} \|f\|_{L^p(\Omega)}, \qquad t > 0.$$

This concludes the proof of assertion (b).

Step 5: Assertion (c) for $2 < p \leq d$. To prove the remaining part of assertion (c) we may assume that $d \geq 3$. The case $d = 2$ was already completed in Step 2. We use of the L^∞-estimate in assertion (b) to show the following decay estimates for the "error terms" $F^{(i)}$, $i = 1, 2$, cf. Lemma 3.9.

Lemma 3.14. *Let $d \geq 3$, $1 < p < \infty$ and $m = 0, 1$. Then, for $i = 1, 2$, there exists a constant $C := C(p, \Omega) > 0$ such that*

$$\|F^{(i)}(t)\|_{W^{2m,p}(\Omega)} \leq C(1+t)^{-\frac{d}{2p}} \|f\|_{L^p(\Omega)}, \qquad t > 0.$$

Proof. Analogously to the proof of Lemma 3.9, see estimate (3.10), we can show that

$$\|F^{(i)}(t)\|_{W^{2m,p}(\Omega)} \leq C\Big(\|\nabla^2 u(t)\|_{W^{2m,p}(D)} + \|\nabla u(t)\|_{W^{2m,p}(D)} + \|A_{\Omega,p}u(t)\|_{W^{2m,p}(D)}\Big), \qquad t > 0$$

for $i = 1, 2$ and a constant $C > 0$. Assume that $1 < p \leq d/2$. Then by using the embedding $L^{d/2}(\Omega) \hookrightarrow L^p(\Omega)$, the local estimates in Proposition 1.78 and Theorem 3.1 (d) we get

$$\|A_{\Omega,p}u(t)\|_{W^{2m,p}(D)} = \|A_{\Omega,p}T_{\Omega,p}(t)b\|_{W^{2m,p}(D)} \leq C\|A_{\Omega,p}T_{\Omega,p}(t+1)f\|_{W^{2m,d/2}(D)}$$
$$\leq C\left\|A_{\Omega,p}T_{\Omega,p}\left(t+\tfrac{1}{2}\right)f\right\|_{L^{d/2}(\Omega)} \leq C(1+t)^{-\frac{d}{2p}} \|f\|_{L^p(\Omega)}, \qquad t > 0.$$

Similarly, by using the elliptic estimate in Proposition 2.7 and Theorem 3.1 (b), we obtain

$$\|\nabla^2 u(t)\|_{W^{2m,p}(D)} =\leq C\left(\|A_{\Omega,p}u(t)\|_{W^{2m,d}(\Omega)} + \|u(t)\|_{L^\infty(D)}\right) \leq C(1+t)^{-\frac{d}{2p}} \|f\|_{L^p(\Omega)}, \qquad t > 0.$$

We apply the Gagliardo-Nirenberg inequality, see Proposition 1.17, with $j = 1$, $m = 2$, $p = r = q$, $\theta = 1/2$ and use the embedding $L^\infty(D) \hookrightarrow L^p(D)$ to see that

$$\|\nabla u(t)\|_{L^p(D)} \leq C\|\nabla^2 u(t)\|_{L^p(D)}^{\frac{1}{2}} \|u(t)\|_{L^\infty(D)}^{\frac{1}{2}} \leq C(1+t)^{-\frac{d}{2p}} \|f\|_{L^p(\Omega)}, \qquad t > 0.$$

Moreover, similarly as above, we conclude that

$$\|\nabla u(t)\|_{W^{2,p}(D)} \leq C\Big(\|\nabla^2 u(t)\|_{W^{1,p}(D)} + \|\nabla u(t)\|_{L^p(D)}\Big)$$
$$\leq C\Big(\|A_{\Omega,p}u(t)\|_{W^{1,d}(\Omega)} + \|\nabla u(t)\|_{L^p(D)}\Big)$$
$$\leq C(1+t)^{-\frac{d}{2p}} \|f\|_{L^p(\Omega)}, \qquad t > 0.$$

Now the assertion for $1 < p \leq d/2$ follows.

If $d/2 < p < \infty$, then

$$\|A_{\Omega,p} u(t)\|_{W^{2m,p}(D)} \leq C \left\|A_{\Omega,p} u\left(t - \tfrac{1}{2}\right) f\right\|_{L^p(\Omega)} \leq C(1+t)^{-\frac{d}{2p}} \|f\|_{L^p(\Omega)}, \qquad t > 0$$

since $\frac{d}{2p} < 1$. Now we proceed as above and obtain

$$\|\nabla^2 u(t)\|_{W^{2m,p}(D)} \leq C(1+t)^{-\frac{d}{2p}} \|f\|_{L^p(\Omega)}, \qquad t > 0$$

and

$$\|\nabla u(t)\|_{W^{2m,p}(D)} \leq C(1+t)^{-\frac{d}{2p}} \|f\|_{L^p(\Omega)}, \qquad t > 0.$$

This concludes the proof. $\qquad\square$

Now, we proceed similarly as in Step 2: We apply Lemma 3.14 to see that

$$\|\nabla u^{(1)}(t)\|_{L^p(\Omega)} + \|\nabla u^{(2)}(t)\|_{L^p(\Omega)}$$

$$\leq C\|f\|_{L^p(\Omega)} \left((1+t)^{-\frac{1}{2}} + \int_0^t (1+t-s)^{-\frac{d}{2}\left(1-\frac{1}{p}\right)-\frac{1}{2}} (1+s)^{-\frac{d}{2p}} \, ds \right), \qquad t > 0$$

for all $1 < p < \infty$. Set

$$J_5(t) := \int_0^t (1+t-s)^{-\frac{d}{2}\left(1-\frac{1}{p}\right)-\frac{1}{2}} (1+s)^{-\frac{d}{2p}} \, ds, \qquad t > 0.$$

For $2 < p \leq d$ the integral $J_5(t)$ is estimated as follows:

$$J_5(t) \leq \left(1 + \tfrac{t}{2}\right)^{-\frac{1}{2}} \left(\int_0^{\frac{t}{2}} (1+t-s)^{-\frac{d}{2}\left(1-\frac{1}{p}\right)} (1+s)^{-\frac{d}{2p}} \, ds \right.$$

$$\left. + \int_{\frac{t}{2}}^t (1+t-s)^{-\frac{d}{2}\left(1-\frac{1}{p}\right)-\frac{1}{2}} (1+s)^{-\frac{d}{2p}+\frac{1}{2}} \, ds \right)$$

$$\leq 2 \left(1 + \tfrac{t}{2}\right)^{-\frac{1}{2}} \int_0^{\frac{t}{2}} (1+s)^{-\frac{d}{2}\left(1-\frac{1}{p}\right)-\frac{1}{2}-\frac{d}{2p}} \, ds \leq C \left(1 + \tfrac{t}{2}\right)^{-\frac{1}{2}}, \qquad t > 0.$$

This yields the assertion.

Step 6: Assertion (d). For $d < p < \infty$ the integral $J_5(t)$ is estimated as follows:

$$J_5(t) \leq \left(1 + \tfrac{t}{2}\right)^{-\frac{d}{2p}} \left(\int_0^{\frac{t}{2}} (1+t-s)^{-\frac{d}{2}\left(1-\frac{1}{p}\right)-\frac{1}{2}+\frac{d}{2p}} (1+s)^{-\frac{d}{2p}} \, ds \right.$$

$$\left. + \int_{\frac{t}{2}}^t (1+t-s)^{-\frac{d}{2}\left(1-\frac{1}{p}\right)-\frac{1}{2}} \right)$$

$$\leq 2 \left(1 + \tfrac{t}{2}\right)^{-\frac{d}{2p}} \int_0^{\frac{t}{2}} (1+s)^{-\frac{d}{2}\left(1-\frac{1}{p}\right)-\frac{1}{2}} \, ds \leq C \left(1 + \tfrac{t}{2}\right)^{-\frac{d}{2p}}, \qquad t > 0.$$

This concludes the proof of assertion (d) and also the proof of Theorem 3.1. $\qquad\square$

4 Navier-Stokes flows around a moving obstacle: The non-autonomous case

In this chapter we study Navier-Stokes flows in the exterior of a moving and rotating obstacle. Particular emphasis is placed on the fact that the motion of the obstacle is non-autonomous, i.e. it may depend on time.

In the following we briefly sketch two physically motivated scenarios that both lead to the same mathematical problem. Let $\mathcal{O} \subset \mathbb{R}^d$, $d \geq 2$, be a compact set, also referred to as the *obstacle*, with boundary $\Gamma := \partial\mathcal{O}$ of class C^2. The exterior of the obstacle is denoted by $\Omega := \mathbb{R}^d \setminus \mathcal{O}$. Assume that Ω is a domain (i.e. Ω is connected).

Certainly the physically realistic case is $d = 3$. However, mathematically our approach works more generally for $d \geq 2$ and therefore we shall consider the general case in this chapter.

Scenario 1: Time-dependent rotation and time-dependent outflow conditions

We assume that $\mathcal{O}$ rotates with a prescribed time-dependent angular velocity, represented[1] by the matrix-valued function $m \in C^1([0, \infty); \mathbb{R}^{d \times d})$. Assume that the matrix $m(t)$ is *skew symmetric* for all $t \geq 0$. By $Q \in C^1([0, \infty); \mathbb{R}^{d \times d})$ we denote the solution to the linear ordinary differential equation

$$\begin{cases} \partial_t Q(t) &= m(t)Q(t), \quad t \geq 0, \\ Q(0) &= \mathrm{Id}. \end{cases} \tag{4.1}$$

Since $m(t)$ is skew symmetric for all $t \geq 0$, it easily follows that the matrix $Q(t)$ is *orthogonal* for all $t \geq 0$. This matrix-valued function Q describes the rotation of $\mathcal{O}$. The obstacle at time $t \geq 0$ and its boundary at time $t \geq 0$ are represented by

$$\mathcal{O}(t) := \{y(t) = Q(t)x : x \in \mathcal{O}\} \qquad \text{and} \qquad \Gamma(t) := \{y(t) = Q(t)x : x \in \Gamma\},$$

respectively. The exterior of the rotated obstacle at time $t \geq 0$ is represented by

$$\Omega(t) := \mathbb{R}^d \setminus \mathcal{O}(t) = \{y(t) = Q(t)x : x \in \Omega\}$$

Moreover, we prescribe a velocity field for the fluid at space-infinity. For this purpose let $v_\infty \in C^1([0, \infty); \mathbb{R}^d)$. The motion of the fluid past the rotating obstacle is assumed to governed by the incompressible Navier-Stokes equations with the usual no-slip boundary

---

[1] In the physically realistic case $d = 3$ the obstacle is rotating with an angular velocity $\omega(t)$ and then $m(t)$ denotes the matrix that represents the linear map $x \mapsto \omega(t) \times x$.

condition on the time-dependent domain $\Omega(t)$:

$$\begin{cases} v_t - \Delta v + v \cdot \nabla v + \nabla q &= 0 \qquad \text{for } t \in \mathbb{R}_+, \, y \in \Omega(t), \\ \operatorname{div} v &= 0 \qquad \text{for } t \in \mathbb{R}_+, \, y \in \Omega(t), \\ v(t, y) &= m(t)y \qquad \text{for } t \in \mathbb{R}_+, \, y \in \Gamma(t), \\ \lim_{|y| \to \infty} v(t, y) &= v_\infty(t) \qquad \text{for } t \in \mathbb{R}_+, \\ v|_{t=0} &= u_0 \qquad \text{for } y \in \Omega. \end{cases} \tag{4.2}$$

Here u_0 is a given initial value, and v and q are the unknown velocity field and the pressure of the fluid, respectively.

The disadvantages of this description are the time-dependence of the domain $\Omega(t)$ and the fact that the equations do not fit into the L^p-setting due to the velocity condition at infinity. To overcome these difficulties it is reasonable to rewrite (4.2) as a new problem on the fixed exterior domain Ω by using a linear coordinate transformation. Set

$$x = Q(t)^\top y, \qquad w(t, x) = Q(t)^\top (v(t, y) - v_\infty(t)), \qquad \text{p}(t, x) = \text{q}(t, y),$$

see also [His99a, GHH06a]. If we set $M(t) := Q(t)^\top m(t) Q(t)$, and use equation (4.1), then a short calculation yields

$$\partial_t v(t, y) = Q(t) \big[\partial_t w(t, x) - M(t)x \cdot \nabla_x w(t, x) + M(t)w(t, x) \\ + Q(t)^\top \partial_t v_\infty(t) \big],$$

$$\Delta_y v(t, y) = Q(t) \Delta_x w(t, x),$$

$$\nabla_y \text{q}(t, y) = Q(t) \nabla_x \text{p}(t, x),$$

$$v \cdot \nabla_y v = Q(t) \left[w \cdot \nabla_x w \right] + v_\infty(t) \cdot \nabla w,$$

$$\operatorname{div} v(t, y) = \operatorname{div} w(t, x).$$

Thus we obtain the following new equations on the fixed domain Ω:

$$\begin{cases} \left. \begin{aligned} w_t - \Delta w - M(t)x \cdot \nabla w + M(t)w \\ + Q(t)^\top v_\infty(t) \cdot \nabla w \\ + Q(t)^\top \partial_t v_\infty(t) + w \cdot \nabla w + \nabla \text{p} \end{aligned} \right\} &= 0 \qquad & \text{in } \mathbb{R}_+ \times \Omega, \\ \operatorname{div} w &= 0 \qquad & \text{in } \mathbb{R}_+ \times \Omega, \\ w(t, x) &= M(t)x - Q(t)^\top v_\infty(t) \qquad & \text{on } \mathbb{R}_+ \times \Gamma, \\ \lim_{|x| \to \infty} w(t, x) &= 0 \qquad & \text{for } t \in \mathbb{R}_+, \\ w|_{t=0} &= u_0 \qquad & \text{in } \Omega. \end{cases} \tag{4.3}$$

Scenario 2: Time-dependent rotation and time-dependent translation

We assume that $\mathcal{O}$ rotates with a time-dependent angular velocity, again represented by the matrix-valued function $m \in C^1([0, \infty); \mathbb{R}^{d \times d})$, and is at the same time translating with translational velocity $-v_\infty$, where $v_\infty \in C^1([0, \infty); \mathbb{R}^d)$. By $Q \in C^1([0, \infty); \mathbb{R}^{d \times d})$ we denote again the solution to (4.1). With the help of Q the exterior of the rotated and translated obstacle at time $t \geq 0$ is represented by

$$\widetilde{\Omega}(t) := \left\{ y(t) = Q(t)x - \int_0^t v_\infty(s) \ ds : x \in \Omega \right\}.$$

The boundary at time $t \geq 0$ is denoted by $\widetilde{\Gamma}(t) := \partial\widetilde{\Omega}(t)$. As in Scenario 1, the motion of the fluid is governed by the incompressible Navier-Stokes equations on time-dependent domain $\widetilde{\Omega}(t)$ with the no-slip boundary condition:

$$\left\{ \begin{array}{rcll} v_t - \Delta v + v \cdot \nabla v + \nabla \mathsf{q} & = & 0 & \text{for } t \in \mathbb{R}_+,\, y \in \widetilde{\Omega}(t),\\[4pt] \operatorname{div} v & = & 0 & \text{for } t \in \mathbb{R}_+,\, y \in \widetilde{\Omega}(t),\\[4pt] v(t,y) & = & m(t)\left(y + \int_0^t v_\infty(s)\,ds \right) - v_\infty(t) & \text{for } t \in \mathbb{R}_+,\, y \in \partial\widetilde{\Gamma}(t),\\[4pt] \lim_{|y| \to \infty} v(t,y) & = & 0 & \text{for } t \in \mathbb{R}_+,\\[4pt] v|_{t=0} & = & u_0 & \text{for } y \in \Omega. \end{array} \right. \tag{4.4}$$

Here u_0 is again a given initial value and v and q are the unknown velocity field and the pressure of the fluid, respectively.

As above, we rewrite (4.4) as a new problem on the fixed reference domain Ω. Similarly as in Scenario 1, set

$$x = Q(t)^\top \left(y + \int_0^t v_\infty(s) \ ds \right), \qquad w(t,x) = Q(t)^\top v(t,y), \qquad p(t,x) = \mathsf{q}(t,y).$$

Thus, we obtain

$$\partial_t v(t,y) = Q(t)\big[\partial_t w(t,x) - M(t)x \cdot \nabla_x w(t,x) + M(t)w(t,x).$$
$$+\, Q(t)^\top v_\infty(t) \cdot \nabla_x w(t,x) \big],$$

$$\Delta_y v(t,y) = Q(t)\Delta_x w(t,x),$$

$$\nabla_y \mathsf{q}(t,y) = Q(t)\nabla_x p(t,x),$$

$$v \cdot \nabla_y v = Q(t)\left[w \cdot \nabla_x w \right],$$

$$\operatorname{div} v(t,y) = \operatorname{div} w(t,x).$$

Here, we set again $M(t) := Q(t)^\top m(t)Q(t)$. Consequently, we have the following new equations on the fixed domain Ω:

$$\left\{ \begin{array}{rcll} \left. \begin{array}{l} w_t - \Delta w - M(t)x \cdot \nabla w + M(t)w \\[2pt] +\, Q(t)^\top v_\infty(t) \cdot \nabla w + w \cdot \nabla w + \nabla p \end{array} \right\} & = & 0 & \text{in } \mathbb{R}_+ \times \Omega,\\[10pt] \operatorname{div} w & = & 0 & \text{in } \mathbb{R}_+ \times \Omega,\\[4pt] w(t,x) & = & M(t)x - Q(t)^\top v_\infty(t) & \text{on } \mathbb{R}_+ \times \partial\Omega,\\[4pt] \lim_{|x| \to \infty} w(t,x) & = & 0 & \text{for } t \in \mathbb{R}_+,\\[4pt] w|_{t=0} & = & u_0 & \text{in } \Omega. \end{array} \right. \tag{4.5}$$

Approach to equations (4.3) **and** (4.5)

To simplify notation set $c(t) := -Q(t)^\top v_\infty(t)$. Note that the term $Q(t)\partial_t v_\infty(t)$ in equation (4.3) is constant in space. Therefore, this term can be formally absorbed in the pressure term, i.e. we look for a pressure of the form $\tilde{p} = p + Q(t)\partial_t v_\infty(t) \cdot x$, where (w, p) solves the following system of equations:

$$
\left\{
\begin{aligned}
w_t - \Delta w - M(t)x \cdot \nabla w - c(t) \cdot \nabla w \\
+ M(t)w + w \cdot \nabla w + \nabla p
\end{aligned}
\right\}
\;=\; 0 \qquad \text{in } \mathbb{R}_+ \times \Omega,
$$
$$
\begin{aligned}
\operatorname{div} w &= 0 && \text{in } \mathbb{R}_+ \times \Omega, \\
w(t, x) &= M(t)x + c(t) && \text{on } \mathbb{R}_+ \times \partial\Omega, \\
\lim_{|x|\to\infty} w(t, x) &= 0 && \text{for } t \in \mathbb{R}_+, \\
w|_{t=0} &= u_0 && \text{in } \Omega.
\end{aligned}
$$
$$\tag{4.6}$$

Thus, (4.2) and (4.4) are both mathematically equivalent to (4.6). Note that (4.6) is a non-autonomous system of equations, i.e. the coefficients depend on the time parameter t. The main difficulty here lies in the unbounded and time-dependent drift term $M(t)x \cdot \nabla w$.

In the following we assume that $u_0 \in L^p(\Omega)^d$ with $\operatorname{div} u_0 = 0$, and moreover we impose the following *compatibility condition*[2] on u_0:

$$u_0 \cdot \nu = (M(0)x + c(0)) \cdot \nu, \qquad \text{on } \partial\Omega. \tag{4.7}$$

As outlined in Section 1.6, we construct a suitable boundary extension $b \in C^1([0,\infty); C^2(\mathbb{R}^d))$ such that $b(t) \in C_{c,\sigma}^\infty(\mathbb{R}^d)$ for all $t \in \mathbb{R}_+$. In this situation here the function b is given by

$$b(t, x) := \zeta(x)(M(t)x + c(t)) - \mathbb{B}_{\Omega_b}((\nabla\zeta) \cdot (M(t)x + c(t))), \tag{4.8}$$

where ζ is a smooth cut-off function which is supported only close to Γ, and Ω_b is a bounded domain, see Section 1.6 for details. In particular, there exists a $R_0 > 0$ such that $\operatorname{supp} b(t) \subset B_{R_0}$ for all $t \in \mathbb{R}_+$. By setting $w := u + b$ problem (4.6) is equivalent to

$$
\left\{
\begin{aligned}
u_t - \Delta u - M(t)x \cdot \nabla u + c(t) \cdot \nabla u + M(t)u \\
+ b \cdot \nabla u + u \cdot \nabla b + u \cdot \nabla u + \nabla p
\end{aligned}
\right\}
\;=\; \widetilde{F} \quad \text{in } \mathbb{R}_+ \times \Omega,
$$
$$
\begin{aligned}
\operatorname{div} u &= 0 && \text{in } \mathbb{R}_+ \times \Omega, \\
u(t, x) &= 0 && \text{on } \mathbb{R}_+ \times \partial\Omega, \\
\lim_{|x|\to\infty} u(t, x) &= 0 && \text{for } t \in \mathbb{R}_+, \\
u|_{t=0} &= a && \text{in } \Omega,
\end{aligned}
$$
$$\tag{4.9}$$

where $a := u_0 - b(0)$ and the inhomogeneous right-hand side $\widetilde{F}$ is given by

$$\widetilde{F} := \Delta b + (M(t)x + c(t)) \cdot \nabla b - M(t)b - b \cdot \nabla b - b_t. \tag{4.10}$$

Note that $\operatorname{div} a = 0$ holds. Furthermore, the compatibility condition (4.7) even ensures that $a \in L_\sigma^p(\Omega)$, see the characterization of $L_\sigma^p(\Omega)$ in (1.14).

---

[2]This condition ensures that the initial value is compatible with the boundary condition.

The analysis of the system (4.9) is based on the family of linear operators $\{\mathcal{L}(t)\}_{t\geq 0}$, where the operator $\mathcal{L}(t)$ is formally defined for smooth vector fields $u = (u_1, \ldots, u_d)$ by

$$\mathcal{L}(t)u(x) := \Delta u(x) + M(t)x \cdot \nabla u(x) + c(t) \cdot \nabla u(x) - M(t)u(x), \qquad t \geq 0. \tag{4.11}$$

A single operator of this form was considered in Section 1.5, see (1.18). In the course of this chapter the L^p-theory for the family of non-autonomous operators $\{\mathcal{L}(t)\}_{t\geq 0}$ is developed and the associated non-autonomous linear Cauchy problems are studied in the framework of evolution systems. Based on these linear results we then finally come back to the full non-linear problem (4.9). More precisely, in this chapter we proceed as follows:

1. First, we consider the linear model problem associated to (4.9) in the whole space $\mathbb{R}^d$, see Section 4.1.

2. Second, we study the linearized version of (4.9) in a bounded domain, see Section 4.2.

3. In Section 4.3, we combine the results from Step 1 and Step 2 to study the linearized version of (4.9) in an exterior domain.

4. Finally, in Section 4.4 we return to the full non-linear problem (4.9) and show local in time existence and uniqueness of a mild solution to (4.9), or equivalently to (4.6). Moreover, we obtain long time existence under smallness assumptions on the data. If we consider (4.6) not in an exterior domain Ω but instead in the whole space $\mathbb{R}^d$ we even obtain a global mild solution for small data.

The linear results in Sections 4.1 – 4.3 are interesting in their own right and can be considered as generalizations of the results in Section 1.5 from the autonomous to the non-autonomous case. Therefore we state all linear results as possible, and we might impose weaker and more general assumptions on the coefficients than needed in the statement of problem (4.9).

In this chapter we use the following notation: For $T > 0$ set

$$
\begin{aligned}
\Lambda_T &:= \{(t,s) \in \mathbb{R}^2 : 0 \leq s \leq t \leq T\}, \\
\Lambda &:= \{(t,s) \in \mathbb{R}^2 : 0 \leq s \leq t\}.
\end{aligned}
$$

4.1 The linear model problem in $\mathbb{R}^d$

In this section we study the linear model problem associated to (4.9) in the situation of the whole space $\mathbb{R}^d$, $d \geq 2$. The results obtained here are important ingredients for the analysis of the original system of equations (4.9). Let $1 < p < \infty$. For some initial time $s \geq 0$ and some initial value $f \in L^p_\sigma(\mathbb{R}^d)$ we consider the system of equations

$$
\begin{cases}
u_t - \Delta u - M(t)x \cdot \nabla u - c(t) \cdot \nabla u + M(t)u + \nabla \mathrm{p} &= 0 \quad \text{in } (s,\infty) \times \mathbb{R}^d, \\
\operatorname{div} u &= 0 \quad \text{in } (s,\infty) \times \mathbb{R}^d, \\
u|_{t=s} &= f \quad \text{in } \mathbb{R}^d.
\end{cases} \tag{4.12}
$$

Here $u : (s,\infty) \times \mathbb{R}^d \to \mathbb{R}^d$ and $\mathrm{p} : (s,\infty) \times \mathbb{R}^d \to \mathbb{R}$ are the unknown velocity field and the pressure of the fluid, respectively.

73

Assumption 4.1. In this section we impose the following conditions on the coefficients:

(i) $M \in C([0, \infty); \mathbb{R}^{d \times d})$,

(ii) $c \in C([0, \infty); \mathbb{R}^d)$.

We define a family of linear operators $\{L_{\mathbb{R}^d, p}(t)\}_{t \geq 0}$ on $L_\sigma^p(\Omega)$ by setting

$$
\begin{aligned}
\mathcal{D}(L_{\mathbb{R}^d, p}(t)) &:= \{u \in W^{2,p}(\mathbb{R}^d)^d \cap L_\sigma^p(\Omega) : M(t)x \cdot \nabla u \in L^p(\mathbb{R}^d)^d\}, \\
L_{\mathbb{R}^d, p}(t)u &:= \mathcal{L}(t)u,
\end{aligned} \qquad t \geq 0,
$$

where $\mathcal{L}(t)$ is defined in (4.11). See also Definition 1.83. Note that Lemma 1.86 yields indeed an operator on $L_\sigma^p(\mathbb{R}^d)$. Although the domain $\mathcal{D}(L_{\mathbb{R}^d, p}(t))$ depends on the time parameter $t \geq 0$, the space $C_{c,\sigma}^\infty(\mathbb{R}^d)$ is contained in $\mathcal{D}(L_{\mathbb{R}^d, p}(t))$ for all $t \geq 0$.

Remark 4.2. If the actual value of $1 < p < \infty$ is clear from the context or not important, we simply write $L_{\mathbb{R}^d}(t)$.

By applying the Helmholtz projection $\mathbb{P}_{\mathbb{R}^d}$ to (4.12) the pressure p gets eliminated and we may rewrite the equations as a non-autonomous abstract Cauchy problem on $L_\sigma^p(\mathbb{R}^d)$:

$$
\begin{cases}
u'(t) = L_{\mathbb{R}^d}(t)u(t), & t > s \geq 0, \\
u(s) = f.
\end{cases} \tag{4.13}
$$

Due to the fact that for fixed $s \geq 0$ the linear operator $L_{\mathbb{R}^d}(s)$ generates a C_0-semigroup, which is not analytic, see Proposition 1.85, the Cauchy problem (4.13) does not fit into the setting of parabolic evolution equations in the sense of Assumption 1.56. Therefore the abstract results for parabolic evolution systems, presented in Section 1.2.2, cannot be applied here. In fact, the present situation fits more into the framework of hyperbolic evolution systems in the sense of Assumption 1.54. However, the abstract results in the hyperbolic setting, see Proposition 1.58, do not imply any smoothing properties. For further purposes we do not simply want to discuss well-posedness of (4.13) but we also need to derive L^p-L^q-estimates and gradient estimates. Therefore we shall not check all the necessary assumptions for hyperbolic evolution systems but we follow a different approach which is based – similarly to the approach by Hieber and Sawada in [HS05] for the autonomous case – on an explicit solution formula. This formula is derived in Section 4.1.1. The explicit formula then allows to derive norm estimates, in particular L^p-L^q-estimates and gradient estimates, in Section 4.1.2. Some of the results in this section have been partially published in [Han11, GH11]. In particular, the explicit solution formula was derived in [GH11].

4.1.1 The solution formula

The purpose of this section is to show well-posedness of the Cauchy problem (4.13) based on an explicit solution formula. For this purpose we consider the system of equations

$$
\begin{cases}
u_t(t, x) - \mathcal{L}(t)u(t, x) = 0, & t > s, x \in \mathbb{R}^d, \\
u(s, x) = f(x), & x \in \mathbb{R}^d,
\end{cases} \tag{4.14}
$$

where $\mathcal{L}(t)$ is given in (4.11) and $f : \mathbb{R}^d \to \mathbb{R}^d$ is a "nice" given initial value.

Problems of this type in the scalar-valued case were considered by Da Prato and Lunardi [DPL07] and Geissert and Lunardi [GL08] based on an explicit solution formula. In order to derive a solution formula for (4.14) we transform the equations in a suitable way to a non-autonomous heat equations which then can be explicitly solved, e.g. by means of the Fourier transform.

In this section we always denote by $\{U(t,s)\}_{t,s\geq 0}$ the evolution system on $\mathbb{R}^d$ generated by the family of matrices $\{-M(t)\}_{t\geq 0}$, i.e.

$$\begin{cases} \partial_t U(t,s) &= -M(t)U(t,s), \qquad t,s \geq 0 \\ U(s,s) &= \text{Id}. \end{cases}$$

The existence and uniqueness of $\{U(t,s)\}_{t,s\geq 0}$ follows from the classical theory of linear ordinary differential equations. It is clear that the map $(t,s) \mapsto U(t,s)$ is continuous and by uniqueness we conclude that

$$U(t,s) = U(t,r)U(r,s), \qquad t,s,r \geq 0.$$

This relation implies that $U(t,s)$ is invertible and its inverse is given by $U(s,t)$. Furthermore, by using $\partial_s(U(t,s)U(s,t)) = 0$, it is easy to see that $\partial_s U(t,s) = U(t,s)M(s)$ holds.

Moreover, set

$$g(t,s) := \int_s^t U(s,r)c(r)\,\mathrm{d}r \quad \text{and} \quad Q_{t,s} := \int_s^t U(s,r)U^\top(s,r)\,\mathrm{d}r, \quad t \geq s \geq 0. \qquad (4.15)$$

Now we are in the position to prove existence of classical solutions to (4.14). We refer also to [DPL07, Proposition 2.1] for a similar result in the scalar-valued case.

Proposition 4.3. *Let $s \geq 0$ be fixed and $f \in \mathcal{S}(\mathbb{R}^d)^d$. Then problem (4.14) admits a unique classical solution*

$$u(\cdot,\cdot;s,f) : [s,\infty) \times \mathbb{R}^d \to \mathbb{R}^d$$

such that $u(\cdot,x;s,f) \in C^1([s,\infty))^d$ for all $x \in \mathbb{R}^d$ and $u(t,\cdot;s,f) \in \mathcal{S}(\mathbb{R}^d)^d$ for all $t \in [s,\infty)$. The solution $u(\cdot,\cdot;s,f)$ is given by

$$u(t,x;s,f) := (k(t,s,\cdot) * f)(U(s,t)x + g(s,t)), \qquad 0 \leq s < t, \ x \in \mathbb{R}^d, \qquad (4.16)$$

where

$$k(t,s,x) := \frac{1}{(4\pi)^{\frac{d}{2}}(\det Q_{t,s})^{\frac{1}{2}}}U(t,s)e^{-\frac{1}{4}Q_{t,s}^{-1}x\cdot x}, \qquad 0 \leq s < t, \ x \in \mathbb{R}^d, \qquad (4.17)$$

and $g(t,s)$ and $Q_{t,s}$ are defined in (4.15). Moreover, for $0 \leq s \leq r \leq t$ the relation

$$u(t,x;s,f) = u(t,x;r,u(r,x;s,f)) \qquad (4.18)$$

holds.

Some words about the convolution in (4.16) seem to be in order. Here the kernel $k(t, s, \cdot)$ is matrix-valued and by the notation $k(t, s, \cdot) * f$ we mean

$$k(t, s, \cdot) * f = \int_{\mathbb{R}^d} k(t, s, x - y) f(y) \, \mathrm{d}y,$$

where $k(t, s, x - y) f(y)$ is understood as a multiplication between a matrix and a vector.

Proof. The idea is to do a coordinate transformation in order to eliminate the unbounded drift and the zero order term of the operator $\mathcal{L}(t)$. For this purpose, set

$$z = U(s, t)x + g(t, s) \qquad \text{and} \qquad w(t, z) = U(s, t)u(t, U(t, s)z - U(t, s)g(t, s)).$$

Thus, we have

$$u(t, x) = U(t, s)w(t, U(s, t)x + g(t, s)), \tag{4.19}$$

or component-wise

$$u_i(t, x) = \sum_{j=1}^{d} U_{ij}(t, s) w_j(t, U(s, t)x + g(t, s))$$

for $i = 1, \ldots, d$. We now compute

$$\partial_t u(t, x) = \partial_t U(t, s) w(t, U(s, t)x + g(t, s)) + U(t, s) \partial_t w(t, U(s, t)x + g(t, s))$$
$$+ U(t, s) \Big((\partial_t U(s, t)x + \partial_t g(t, s)) \cdot \nabla_z w_i(t, z) \Big)_{i=1}^{d}$$
$$= -M(t)U(t, s)w(t, z) + U(t, s)\partial_t w(t, z)$$
$$+ U(t, s) \Big((U(s, t)M(t)x + U(s, t)c(t)) \cdot \nabla_z w_i(t, z) \Big)_{i=1}^{d}.$$

Moreover, for $i = 1, \ldots, d$, we have

$$\nabla_x u_i(t, x) = \sum_{j=1}^{d} U_{ij}(t, s) U^\top(s, t) \nabla_z w_j(t, z),$$

$$\nabla_x^2 u_i(t, x) = \sum_{j=1}^{d} U_{ij}(t, s) U^\top(s, t) \nabla_z^2 w_j(t, z) U(s, t).$$

In particular, we obtain

$$(M(t)x + c(t)) \cdot \nabla_x u_i(t, x) = \sum_{j=1}^{d} U_{ij}(t, s) \big[(M(t)x + c(t)) \cdot U^\top(s, t) \nabla_z w_j(t, z) \big]$$

$$= \sum_{j=1}^{d} U_{ij}(t, s) \big[(U(s, t)M(t)x + U(s, t)c(t)) \cdot \nabla_z w_j(t, z) \big]$$

for $i = 1, \ldots, d$, and hence

$$(M(t)x + c(t)) \cdot \nabla_x u(t, x) = U(t, s) \Big((U(s, t)M(t)x + U(s, t)c(t)) \cdot \nabla_z w_i(t, z) \Big)_{i=1}^{d}.$$

Thus, we conclude that u solves (4.14) if and only if for every $i = 1, \ldots, d$ the function $w_i : [s, \infty) \times \mathbb{R}^d \to \mathbb{R}$ solves the non-autonomous scalar heat equation

$$\begin{cases} \partial_t w_i(t, z) &= \operatorname{tr}\left(U(s,t)U^\top(s,t)\nabla_z^2 w_i(t,z)\right), & t > s, z \in \mathbb{R}^d, \\ w_i(s, z) &= f_i(z), & z \in \mathbb{R}^d. \end{cases} \tag{4.20}$$

The coefficients in the equation above are independent of the space variable z. By taking the Fourier transform on both sides of (4.20) we obtain the linear ordinary differential equation

$$\begin{cases} \partial_t \widehat{w}_i(t, \xi) &= -\left(U(s,t)U^\top(s,t)\xi \cdot \xi\right) \widehat{w}_i(t,\xi), & t > s, \xi \in \mathbb{R}^d, \\ \widehat{w}_i(s, \xi) &= \widehat{f}_i(\xi), & \xi \in \mathbb{R}^d, \end{cases} \tag{4.21}$$

which is uniquely solved by

$$\widehat{w}_i(t, \xi) = e^{-Q_{t,s}\xi \cdot \xi} \widehat{f}_i(\xi),$$

where $Q_{t,s}$ is defined in (4.15). By taking the inverse Fourier transform we obtain

$$w_i(t, z) = \frac{1}{(4\pi)^{\frac{d}{2}} (\det Q_{t,s})^{\frac{1}{2}}} \int_{\mathbb{R}^d} f_i(z-y) e^{-\frac{1}{4}Q_{t,s}^{-1} y \cdot y} \, \mathrm{d}y.$$

The relation (4.19) yields that the solution to (4.14) is indeed given by (4.16). Note that $w_i(t, \cdot) \in \mathcal{S}(\mathbb{R}^d)$ for all $t \geq s$ and therefore by (4.19) it is clear that $u(t, \cdot; s, f) \in \mathcal{S}(\mathbb{R}^d)^d$ for all $t \geq s$. The uniqueness of the solution to the linear ordinary differential equation (4.21) in Fourier space implies that $u(\cdot, \cdot; s, f)$ is the unique solution to (4.14) with the property $u(t, \cdot; s, f) \in \mathcal{S}(\mathbb{R}^d)^d$ for all $t \geq s$. The uniqueness directly implies the relation (4.18). $\qquad \square$

Next, we briefly discuss the inhomogeneous problem

$$\begin{cases} u_t(t, x) - \mathcal{L}(t)u(t, x) &= F(t, x), & t > s, x \in \mathbb{R}^d, \\ u(s, x) &= f(x), & x \in \mathbb{R}^d, \end{cases} \tag{4.22}$$

where $f : \mathbb{R}^d \to \mathbb{R}^d$ and $F : [s, \infty) \times \mathbb{R}^d \to \mathbb{R}^d$ are "nice" given functions. The solution to the inhomogeneous problem is given via the variation of constants formula.

Proposition 4.4. *Let $s \geq 0$ be fixed and let $f \in \mathcal{S}(\mathbb{R}^d)^d$ and $F : [s, \infty) \times \mathbb{R}^d \to \mathbb{R}^d$ be such that $F(t, \cdot) \in \mathcal{S}(\mathbb{R}^d)^d$ for all $t \in [s, \infty)$ and $F(\cdot, x) \in C([s, \infty); \mathbb{R}^d)$ for all $x \in \mathbb{R}^d$. Then problem (4.22) admits a unique classical solution*

$$u(\cdot, \cdot; s, f, F) : [s, \infty) \times \mathbb{R}^d \to \mathbb{R}^d$$

such that $u(\cdot, x; s, f, F) \in C^1([s, \infty))^d$ for all $x \in \mathbb{R}^d$ and $u(t, \cdot; s, f, F) \in \mathcal{S}(\mathbb{R}^d)^d$ for all $t \in [s, \infty)$. The solution u is given by

$$\begin{aligned} u(t, x; s, f, F) :=&(k(t, s, \cdot) * f)(U(s,t)x + g(t, s)) \\ &+ \int_s^t (k(t, r, \cdot) * F(r))(U(r,t)x + g(t, r)) \, \mathrm{d}r, \end{aligned} \tag{4.23}$$

for $0 \leq s < t$, $x \in \mathbb{R}^d$, where $k(t, s, \cdot)$ and $g(t, s)$ are defined in (4.17) and (4.15), respectively.

Proof. The proof is very similar to the one in Proposition 4.3 to solve the homogeneous problem. Set again

$$z = U(s,t)x + g(t,s), \qquad w(t,z) = U(s,t)u(t, U(t,s)z - U(t,s)g(t,s)),$$

and

$$G(t,z) = U(s,t)F(t, U(t,s)z - U(t,s)g(t,s)).$$

Thus, we have

$$u(t,x) = U(t,s)w(t, U(s,t)x + g(t,s)). \tag{4.24}$$

Then by the same computation as in the proof of Proposition 4.3, we see that u solves (4.22) if and only if for every $i = 1, \ldots, d$, the function $w_i : [s, \infty) \times \mathbb{R}^d \to \mathbb{R}$ solves

$$\begin{cases} \partial_t w_i(t,z) &= \operatorname{tr}\left[U(s,t)U^\top(s,t)\nabla_z^2 w_i(t,z)\right] + G_i(t,z), & t > s, z \in \mathbb{R}^d, \\ w_i(s,z) &= f_i(z), & z \in \mathbb{R}^d. \end{cases}$$

Taking the Fourier transform on both sides of the equation above implies

$$\begin{cases} \partial_t \widehat{w}_i(t,\xi) &= -\left(U(s,t)U^\top(s,t)\xi \cdot \xi\right)\widehat{w}_i(t,\xi) + \widehat{G}_i(t,\xi), & t > s, \xi \in \mathbb{R}^d, \\ \widehat{w}_i(s,\xi) &= \widehat{f}_i(\xi), & \xi \in \mathbb{R}^d, \end{cases}$$

which is uniquely solved by

$$\widehat{w}_i(t,\xi) = \mathrm{e}^{-Q_{t,s}\xi\cdot\xi}\widehat{f}_i(\xi) + \int_s^t \mathrm{e}^{-\int_r^t U(s,\tau)U^\top(s,\tau)\,\mathrm{d}\tau\,\xi\cdot\xi}\,\widehat{G}_i(r,\xi)\,\mathrm{d}r.$$

By applying the inverse Fourier transform and by the transformation above we conclude that the unique solution u to (4.22) is given by (4.23). □

The formula on the right-hand side of (4.16) serves as the starting point to show well-posedness of the non-autonomous abstract Cauchy problem (4.13). In particular, this formula allows to define an evolution system on $L_\sigma^p(\mathbb{R}^d)$.

Proposition 4.5. *Let $1 < p < \infty$. For $f \in L_\sigma^p(\mathbb{R}^d)$ and $(t,s) \in \Lambda$ set*

$$S_{\mathbb{R}^d,p}(t,s)f(x) := \begin{cases} f(x) & \text{if } t = s, \\ (k(t,s,\cdot) * f)(U(s,t)x + g(t,s)) & \text{if } t \neq s, \end{cases} \qquad x \in \mathbb{R}^d, \tag{4.25}$$

where $k(t,s,\cdot)$ and $g(t,s)$ are defined in (4.17) and (4.15), respectively. Then the two-parameter family $\{S_{\mathbb{R}^d,p}(t,s)\}_{(t,s)\in\Lambda}$ defines a strongly continuous evolution system on $L_\sigma^p(\Omega)$.

We frequently write $S_{\mathbb{R}^d}(t,s)$ instead of $S_{\mathbb{R}^d,p}(t,s)$ if the actual value of $1 < p < \infty$ is clear from the context or is not important.

Proof of Proposition 4.5. First, we show that for $f \in L_\sigma^p(\mathbb{R}^d)$ and $(t,s) \in \Lambda$ with $t \neq s$, the function $S_{\mathbb{R}^d}(t,s)f$ defined in (4.25) is indeed an L^p-function. By the change of variables

$\xi = U(s,t)x$ and by Young's inequality for convolutions we obtain

$$\|S_{\mathbb{R}^d}(t,s)f\|_{L^p(\mathbb{R}^d)} \leq |\det U(s,t)|^{\frac{1}{p}} \|k(t,s,\cdot)\|_{L^1(\mathbb{R}^d)} \|f\|_{L^p(\mathbb{R}^d)}, \quad 0 \leq s \leq t. \tag{4.26}$$

By (4.17) it is easy to see that $k(t,s,\cdot) \in L^1(\mathbb{R}^d)^{d \times d}$. Thus, $S_{\mathbb{R}^d}(t,s)f \in L^p(\mathbb{R}^d)^d$. Next, we show that $S_{\mathbb{R}^d}(t,s)f$ is even solenoidal. Let $f \in C_{c,\sigma}^\infty(\mathbb{R}^d)$. Then a direct computation shows

$$\text{div}\,[(k(t,s,\cdot) * f)(U(s,t)x + g(t,s))]$$

$$= \frac{1}{(4\pi)^{\frac{d}{2}}(\det Q_{t,s})^{\frac{1}{2}}} \sum_{i=1}^{d} \partial_i \left(\sum_{j=1}^{d} U_{ij}(t,s) \int_{\mathbb{R}^d} f_j(U(s,t)x + g(t,s) - y) e^{-\frac{1}{4}Q_{t,s}^{-1}y \cdot y}\, dy \right)$$

$$= \frac{1}{(4\pi)^{\frac{d}{2}}(\det Q_{t,s})^{\frac{1}{2}}} \sum_{i,j,l=1}^{d} U_{li}(s,t)U_{ij}(t,s) \int_{\mathbb{R}^d} \partial_l f_j(U(s,t)x + g(t,s) - y) e^{-\frac{1}{4}Q_{t,s}^{-1}y \cdot y}\, dy$$

$$= \frac{1}{(4\pi)^{\frac{d}{2}}(\det Q_{t,s})^{\frac{1}{2}}} \int_{\mathbb{R}^d} \text{div}\, f(U(s,t)x + g(t,s) - y) e^{-\frac{1}{4}Q_{t,s}^{-1}y \cdot y}\, dy$$

$$= 0.$$

Here we have used the fact $U(t,s)U(s,t) = \text{Id}$. Since $C_{c,\sigma}^\infty(\mathbb{R}^d)$ is dense in $L_\sigma^p(\mathbb{R}^d)$, it follows that $S_{\mathbb{R}^d}(t,s)f \in L_\sigma^p(\mathbb{R}^d)$ for all $f \in L_\sigma^p(\mathbb{R}^d)$. Moreover, estimate (4.26) shows that $S_{\mathbb{R}^d}(t,s) \in \mathscr{L}(L_\sigma^p(\mathbb{R}^d))$ for fixed $(t,s) \in \Lambda$.

Proposition 4.3 states that

$$S_{\mathbb{R}^d}(t,r)S_{\mathbb{R}^d}(r,s)f = S_{\mathbb{R}^d}(t,s)f, \quad 0 \leq s \leq r \leq t,$$

for all $f \in \mathcal{S}(\mathbb{R}^d)^d \cap L_\sigma^p(\mathbb{R}^d)$. Since $C_{c,\sigma}^\infty(\mathbb{R}^d) \subset \mathcal{S}(\mathbb{R}^d)^d \cap L_\sigma^p(\mathbb{R}^d)$ is dense in $L_\sigma^p(\mathbb{R}^d)$ this relation even holds for all $f \in L_\sigma^p(\mathbb{R}^d)$.

It only remains to show the strong continuity of the map $\Lambda \ni (t,s) \mapsto S_{\mathbb{R}^d,p}(t,s)$. To see that, we apply a change of variables to obtain

$$S_{\mathbb{R}^d}(t,s)f(x) = \frac{1}{(4\pi)^{\frac{d}{2}}} U(t,s) \int_{\mathbb{R}^d} f\left(U(s,t)x + g(t,s) - Q_{t,s}^{1/2}z\right) e^{-\frac{|z|^2}{4}}\, dz. \tag{4.27}$$

For $(t,s) \in \Lambda$ fixed, we pick a sequences $\{(t_n, s_n)\}_{n \in \mathbb{N}} \subset \Lambda$ such that $(t_n, s_n) \to (t,s)$ as $n \to \infty$. For every $f \in C_{c,\sigma}^\infty(\mathbb{R}^d)$ and every $x \in \mathbb{R}^d$ we have

$$f\left(U(s_n, t_n)x + g(t_n, s_n) - Q_{t_n,s_n}^{1/2}z\right) \to f\left(U(s,t)x + g(t,s) - Q_{t,s}^{1/2}z\right)$$

as $n \to \infty$. Then formula (4.27) and Lebegue's theorem yield $S_{\mathbb{R}^d}(t_n, s_n)f \to S_{\mathbb{R}^d}(t,s)f$ as $n \to \infty$ for every $f \in C_{c,\sigma}^\infty(\mathbb{R}^d)$. The denseness of $C_{c,\sigma}^\infty(\mathbb{R}^d)$ in $L_\sigma^p(\mathbb{R}^d)$ implies the strong continuity.

Thus, $\{S_{\mathbb{R}^d}(t,s)\}_{(t,s) \in \Lambda}$ is indeed a strongly continuous evolution system on $L_\sigma^p(\Omega)$. $\qquad\square$

In order to discuss well-posedness of the non-autonomous abstract Cauchy problem (4.13), in the sense of Definition 1.48, we introduce the spaces

$$\widetilde{Y}_{\mathbb{R}^d,p} := \left\{ f \in W^{1,p}(\mathbb{R}^d)^d \cap L^p_\sigma(\mathbb{R}^d) : |x| \nabla f_i(x) \in L^p(\mathbb{R}^d)^d \text{ for } i = 1, \ldots, d \right\},$$
$$Y_{\mathbb{R}^d,p} := W^{2,p}(\mathbb{R}^d)^d \cap \widetilde{Y}_{\mathbb{R}^d,p}.$$

It is clear that $Y_{\mathbb{R}^d,p} \subset \mathcal{D}(L_{\mathbb{R}^d,p}(t))$ holds for all $t \geq 0$. Moreover, if we equip $\widetilde{Y}_{\mathbb{R}^d,p}$ and $Y_{\mathbb{R}^d,p}$ with the norms $\|\cdot\|_{\widetilde{Y}_{\mathbb{R}^d,p}}$ and $\|\cdot\|_{Y_{\mathbb{R}^d,p}}$ defined by

$$\|f\|_{\widetilde{Y}_{\mathbb{R}^d,p}} := \|f\|_{W^{1,p}(\mathbb{R}^d)} + \sum_{i=1}^d \||x| \nabla f_i\|_{L^p(\mathbb{R}^d)}, \qquad f \in \widetilde{Y}_{\mathbb{R}^d,p},$$

$$\|f\|_{Y_{\mathbb{R}^d,p}} := \|f\|_{W^{2,p}(\mathbb{R}^d)} + \sum_{i=1}^d \||x| \nabla f_i\|_{L^p(\mathbb{R}^d)}, \qquad f \in Y_{\mathbb{R}^d,p},$$

respectively, then they become Banach spaces. It is easy to see that $C^\infty_{c,\sigma}(\mathbb{R}^d) \subset Y_{\mathbb{R}^d,p}$ and that $C^\infty_{c,\sigma}(\mathbb{R}^d)$ is even dense in $Y_{\mathbb{R}^d,p}$. In the following the space $Y_{\mathbb{R}^d,p}$ serves as a *regularity space* for the evolution system $\{S_{\mathbb{R}^d,p}(t,s)\}_{(t,s)\in\Lambda}$.

Now we are in position to state the following well-posedness result.

Theorem 4.6. *Let $1 < p < \infty$. The two-parameter family $\{S_{\mathbb{R}^d,p}(t,s)\}_{(t,s)\in\Lambda}$, defined in (4.25), is the unique strongly continuous evolution system on $L^p_\sigma(\mathbb{R}^d)$ with the following properties:*

(a) For all $(t,s) \in \Lambda$, the operator $S_{\mathbb{R}^d,p}(t,s)$ maps $Y_{\mathbb{R}^d,p}$ into $Y_{\mathbb{R}^d,p}$.

(b) For every $f \in Y_{\mathbb{R}^d,p}$ and $s \geq 0$, the map $t \mapsto S_{\mathbb{R}^d,p}(t,s)f$ is differentiable in (s,∞) and

$$\partial_t S_{\mathbb{R}^d,p}(t,s)f = L_{\mathbb{R}^d,p}(t)S_{\mathbb{R}^d,p}(t,s)f.$$

(c) For every $f \in Y_{\mathbb{R}^d,p}$ and $t > 0$, the map $s \mapsto S_{\mathbb{R}^d,p}(t,s)f$ is differentiable in $[0,t)$ and

$$\partial_s S_{\mathbb{R}^d,p}(t,s)f = -S_{\mathbb{R}^d,p}(t,s)L_{\mathbb{R}^d,p}(s)f.$$

Moreover, the non-autonomous abstract Cauchy problem (4.13) is well-posed on the regularity space $Y_{\mathbb{R}^d,p}$ (in the sense of Definition 1.48).

Proof. **Step 1: Invariance of the regularity space.** Fix $(t,s) \in \Lambda$ with $t \neq s$. Since $k(t,s,\cdot) \in C^\infty(\mathbb{R}^d)^{d\times d}$ we immediately obtain that $k(t,s,\cdot) * f \in C^\infty(\mathbb{R}^d)^d$ for all $f \in L^p_\sigma(\mathbb{R}^d)$. Now let $f \in Y_{\mathbb{R}^d,p}$ and set $u := S_{\mathbb{R}^d}(t,s)f$. Then, by using formula (4.25), we see that

$$\partial_j u_i(x) = \frac{1}{(4\pi)^{\frac{d}{2}} (\det Q_{t,s})^{\frac{1}{2}}} \sum_{l=1}^d U_{il}(t,s)$$

$$\int_{\mathbb{R}^d} [\nabla f_l(U(s,t)x + g(t,s) - y) \cdot U_{\cdot j}(s,t)] e^{-\frac{1}{4}Q_{t,s}^{-1}y\cdot y} \, dy$$

for all $i, j = 1, \ldots, d$. Next, take some $h \in L^{p'}(\mathbb{R}^d)$ with $\frac{1}{p} + \frac{1}{p'} = 1$. For notational simplicity, set

$$z(x) := U(s, t)x + g(t, s) \qquad \text{and} \qquad C_{t,s} := (4\pi)^{-\frac{d}{2}}(\det Q_{t,s})^{-\frac{1}{2}}.$$

Then we obtain

$$\left| \int_{\mathbb{R}^d} |x| \partial_j u_i(x) h(x) \, dx \right|$$

$$\leq C_{t,s} \sum_{l=1}^{d} |U_{il}(t, s)| \left| \int_{\mathbb{R}^d} |x| h(x) \int_{\mathbb{R}^d} [\nabla f_l(z(x) - y) \cdot U_{\cdot j}(s, t)] e^{-\frac{1}{4} Q_{t,s}^{-1} y \cdot y} \, dy \, dx \right|.$$

Now we use the fact $|x| = |U(t, s)(U(s, t)x + g(t, s) - y - g(t, s) + y)|$ and a simple change of variables to see that

$$\left| \int_{\mathbb{R}^d} |x| \partial_j u_i(x) h(x) \, dx \right|$$

$$\leq C_{t,s} |U(t, s)|^2 \sum_{l=1}^{d} \left(\left| \int_{\mathbb{R}^d} e^{-\frac{1}{4} Q_{t,s}^{-1} y \cdot y} \int_{\mathbb{R}^d} |z(x) - y| [\nabla f_l(z(x) - y) \cdot U_{\cdot j}(s, t)] h(x) \, dx \, dy \right| \right.$$

$$\left. + \left| \int_{\mathbb{R}^d} |y - g(t, s)| e^{-\frac{1}{4} Q_{t,s}^{-1} y \cdot y} \int_{\mathbb{R}^d} [\nabla f_l(z(x) - y) \cdot U_{\cdot j}(s, t)] h(x) \, dx \, dy \right| \right)$$

$$\leq C \|f\|_{\tilde{Y}_{\mathbb{R}^d, p}} \|h\|_{L^{p'}(\mathbb{R}^d)}$$

$$(4.28)$$

for all $i, j = 1, \ldots, d$ and a suitable constant $C > 0$. This shows that $u \in Y_{\mathbb{R}^d, p}$ and that $\{S_{\mathbb{R}^d, p}(t, s)\}_{(t,s) \in \Lambda}$ defined in (4.25) fulfills property (a).

Step 2: Differentiability with respect to t**.** By Proposition 4.3 it is clear that

$$\partial_t S_{\mathbb{R}^d}(t, s)f = L_{\mathbb{R}^d}(t) S_{\mathbb{R}^d}(t, s)f$$

holds for all $f \in C_{c,\sigma}^\infty(\mathbb{R}^d) \subset \mathcal{S}(\mathbb{R}^d)^d \cap L_\sigma^p(\mathbb{R}^d)$. Now let $f \in Y_{\mathbb{R}^d, p}$ and let $\{f_n\}_{n \in \mathbb{N}} \subset C_{c,\sigma}^\infty(\mathbb{R}^d)$ be such that $f_n \to f$ in $Y_{\mathbb{R}^d, p}$ as $n \to \infty$. Then we see that

$$\partial_t S_{\mathbb{R}^d}(t, s)f_n = L_{\mathbb{R}^d}(t) S_{\mathbb{R}^d}(t, s)f_n \to L_{\mathbb{R}^d}(t) S_{\mathbb{R}^d}(t, s)f$$

as $n \to \infty$. By standard arguments we conclude that $\partial_t S_{\mathbb{R}^d}(t, s)f = L_{\mathbb{R}^d}(t) S_{\mathbb{R}^d}(t, s)f$. Thus, $\{S_{\mathbb{R}^d, p}(t, s)\}_{(t,s) \in \Lambda}$ defined in (4.25) fulfills property (b).

Step 3: Differentiability with respect to s**.** For $f \in C_{c,\sigma}^\infty(\mathbb{R}^d)$ and $(t, s) \in \Lambda$ we set $u(t) := S_{\mathbb{R}^d}(t, s)f - f$ for . Then u is a classical solution to the inhomogeneous problem

$$\begin{cases} u'(t) - L_{\mathbb{R}^d}(t)u(t) &= L_{\mathbb{R}^d}(t)f, \quad t > s \geq 0, \\ u(s) &= 0. \end{cases}$$

Proposition 4.4 states that

$$S_{\mathbb{R}^d}(t,s)f - f = \int_s^t S_{\mathbb{R}^d}(t,r)L_{\mathbb{R}^d}(r)f\,\mathrm{d}r, \qquad (t,s)\in\Lambda.$$

Thus, we conclude that

$$\frac{S_{\mathbb{R}^d}(t,s+h)f - S_{\mathbb{R}^d}(t,s)f}{h} = \frac{1}{h}\int_{s+h}^s S_{\mathbb{R}^d}(t,r)L_{\mathbb{R}^d}(r)f\,\mathrm{d}r, \qquad 0\le s,s+h<t.$$

Assumption 4.1 implies that the map $r\mapsto L_{\mathbb{R}^d}(r)f$ is continuous in $[0,t)$ with values in $L_\sigma^p(\mathbb{R}^d)$. By using this together with the strong continuity of the evolution system $\{S_{\mathbb{R}^d}(t,s)\}_{(t,s)\in\Lambda}$ we even obtain that the map $r\mapsto S_{\mathbb{R}^d}(t,r)L_{\mathbb{R}^d}(r)f$ is continuous in $[0,t)$ with values in $L_\sigma^p(\mathbb{R}^d)$. Therefore the integral on the right-hand side converges to $-S_{\mathbb{R}^d}(t,s)L_{\mathbb{R}^d}(s)f$ as $h\to 0$, and we see that $\partial_s S_{\mathbb{R}^d}(t,s)f = -S_{\mathbb{R}^d}(t,s)L_{\mathbb{R}^d}(s)f$ holds for all $f\in C_{c,\sigma}^\infty(\mathbb{R}^d)$. The general case $f\in Y_{\mathbb{R}^d,p}$ follows now as in Step 2 by approximation. Thus, we see that $\{S_{\mathbb{R}^d,p}(t,s)\}_{(t,s)\in\Lambda}$ defined in (4.25) fulfills property (c).

Step 4: Uniqueness. Assume there exists another evolution system $\{V(t)\}_{(t,s)\in\Lambda}$ such that (a) – (c) holds. Then for $(t,s)\in\Lambda$ with $t\ne s$ and for $f\in Y_{\mathbb{R}^d,p}$ we set $u(r):= V(t,r)S_{\mathbb{R}^d}(r,s)f$ for $s\le r\le t$. Note that $u(\cdot)$ is uniformly continuous on $[s,t]$. By using the properties (a) – (c) we see that

$$\partial_r u(r) = -V(t,r)L_{\mathbb{R}^d}(r)S(r,s)f + V(t,r)L_{\mathbb{R}^d}(r)S(r,s)f = 0$$

for $r\in(s,t)$. Hence the map $r\mapsto u(r)$ is constant on (s,t) and by continuity we conclude that $u(s)=V(t,s)f = S_{\mathbb{R}^d}(t,s)f = u(t)$ holds.

The well-posedness of (4.13) follows directly from Proposition 1.53. $\qquad\square$

4.1.2 L^p-L^q-estimates

In this section we prove certain smoothing properties and norm estimates for the evolution system $\{S_{\mathbb{R}^d}(t,s)\}_{(t,s)\in\Lambda}$. In particular, we derive L^p-L^q-estimates and gradient estimates. Since the evolution system is not of parabolic type, such estimates do not follow from the general theory, cf. Proposition 1.59 and Proposition 1.60. However, in our case they can be shown directly by using the explicit formula.

For fixed $T>0$ set

$$M_T := \sup\{|U(t,s)| : t,s\in[0,T]\}.$$

We say the evolution system $\{U(t,s)\}_{t,s\ge 0}$ is *uniformly bounded* if M_T can be chosen independently of T, i.e. $M_T\le M_\infty$ for all $T>0$ and some constant $M_\infty>0$. Since the matrix $U(t,s)$ is invertible with inverse $U(s,t)$, we conclude that for all $0\le s\le t\le T$ it holds that

$$\frac{1}{M_T}\le|U(t,s)|\le M_T. \tag{4.29}$$

Remark 4.7. If $M(t)$ is skew-symmetric for all $t\ge 0$, i.e. $M(t)=-M^\top(t)$, then the matrices $U(t,s)$ are orthogonal for all $t,s\ge 0$ and therefore in this case $M_T=1$ for all $T\le\infty$.

Lemma 4.8. *Let $T > 0$. There exists a constant $C := C(d, M_T) > 0$ such that*

$$|Q_{t,s}^{-\frac{1}{2}}| \leq C(t-s)^{-\frac{1}{2}},$$

$$(\det Q_{t,s})^{\frac{1}{2}} \geq C(t-s)^{\frac{d}{2}}, \tag{4.30}$$

for all $0 \leq s < t < T$. If the evolution system $\{U(t,s)\}_{s,t\geq 0}$ is uniformly bounded, then $T = \infty$ is allowed.

Proof. Let $T > 0$ and $x \in \mathbb{R}^d$. The defnition of $Q_{t,s}$ in (4.15) and inequality (4.29) imply

$$Q_{t,s}x \cdot x = \int_s^t U(s,r)U^\top(s,r)x \cdot x \, dr = \int_s^t |U^\top(s,r)x|^2 \, dr \geq \frac{1}{M_T^2}(t-s)|x|^2.$$

Since $Q_{t,s}$ is symmetric and positive definite we conclude that

$$|Q_{t,s}^{-\frac{1}{2}}| \leq C(t-s)^{-\frac{1}{2}}$$

holds for all $0 \leq s < t < T$ and a constant $C := C(M_T) > 0$. It is also clear that we can allow $T = \infty$ if M_T can be chosen independently of T. To show the second estimate we first observe that

$$\det Q_{t,s}^{-1} \leq C_1 |Q_{t,s}^{-1}|^d$$

holds for a constant $C_1 := C_1(d) > 0$. By using the first estimate in (4.30) we obtain

$$\det Q_{t,s} = (\det Q_{t,s}^{-1})^{-1} \geq C_1 (|Q_{t,s}^{-1}|^d)^{-1} \geq C(t-s)^d,$$

for $C := C(d, M_T) > 0$. Again, $T = \infty$ is allowed if M_T is independent of T. $\qquad\square$

We are now in position to state the following main result.

Theorem 4.9. *Let $T > 0$, $1 < p < \infty$ and $p \leq q \leq \infty$. Moreover, let $\{S_{\mathbb{R}^d,p}(t,s)\}_{(t,s)\in\Lambda}$ be the evolution system from Theorem 4.6. There exists a constant $C := C(p,q,d,M_T) > 0$ such that for all $f \in L_\sigma^p(\mathbb{R}^d)$*

$$\|S_{\mathbb{R}^d,p}(t,s)f\|_{L^q(\mathbb{R}^d)} \leq C(t-s)^{-\frac{d}{2}\left(\frac{1}{p}-\frac{1}{q}\right)}\|f\|_{L^p(\mathbb{R}^d)}, \qquad 0 \leq s < t < T, \tag{4.31}$$

$$\|\nabla S_{\mathbb{R}^d,p}(t,s)f\|_{L^q(\mathbb{R}^d)} \leq C(t-s)^{-\frac{d}{2}\left(\frac{1}{p}-\frac{1}{q}\right)-\frac{1}{2}}\|f\|_{L_\sigma^p(\mathbb{R}^d)}, \qquad 0 \leq s < t < T, \tag{4.32}$$

hold. Moreover, there exists a constant $C := C(p,d,M_T) > 0$ such that

$$\|S_{\mathbb{R}^d,p}(t,s)f\|_{W^{k,p}(\mathbb{R}^d)} \leq C\|f\|_{W^{k,p}(\mathbb{R}^d)}, \qquad 0 \leq s < t < T, \tag{4.33}$$

for all $f \in W^{k,p}(\mathbb{R}^d)^d \cap L_\sigma^p(\mathbb{R}^d)$ and $k = 1, 2$,

$$\|S_{\mathbb{R}^d,p}(t,s)f\|_{W^{2,p}(\mathbb{R}^d)} \leq C(t-s)^{-\frac{1}{2}}\|f\|_{W^{1,p}(\mathbb{R}^d)}, \qquad 0 \leq s < t < T, \tag{4.34}$$

for all $f \in W^{1,p}(\mathbb{R}^d)^d \cap L_\sigma^p(\mathbb{R}^d)$ and

$$\|S_{\mathbb{R}^d,p}(t,s)f\|_{\widetilde{Y}_{\mathbb{R}^d,p}} \leq C\|f\|_{\widetilde{Y}_{\mathbb{R}^d,p}}, \qquad 0 \leq s < t < T, \tag{4.35}$$

for all $f \in \widetilde{Y}_{\mathbb{R}^d,p}$. If the evolution system $\{U(t,s)\}_{s,t\geq 0}$ is uniformly bounded, then $T = \infty$ is allowed.

Proof. Let $T > 0$. We start by showing (4.31). By the change of variables $\xi = U(s,t)x$ and by Young's inequality for convolutions we obtain

$$\|S_{\mathbb{R}^d}(t,s)f\|_{L^q(\mathbb{R}^d)} \leq |\det U(s,t)|^{\frac{1}{q}} \|k(t,s,\cdot)\|_{L^r(\mathbb{R}^d)} \|f\|_{L^p(\mathbb{R}^d)}, \quad 0 \leq s < t < T,$$

where $1 < r < \infty$ with $\frac{1}{p} + \frac{1}{r} = 1 + \frac{1}{q}$. Further, by the change of variables $y = Q_{t,s}^{1/2} z$ we obtain

$$
\begin{aligned}
\|k(t,s,\cdot)\|_{L^r(\mathbb{R}^d)}^r &\leq |U(t,s)|^r \int_{\mathbb{R}^d} \left| \frac{1}{(4\pi)^{\frac{d}{2}}} e^{-\frac{1}{4} Q_{t,s}^{-1} y \cdot y} \right|^r \, dy \\
&= |U(t,s)|^r \int_{\mathbb{R}^d} \left(\frac{1}{(4\pi)^{\frac{d}{2}}} e^{-\frac{|z|^2}{4}} \right)^r (\det Q_{t,s})^{\frac{1}{2}(1-r)} \, dz \\
&\leq C\, M_T^r (\det Q_{t,s})^{\frac{1-r}{2}}, \qquad\qquad\qquad 0 \leq s < t < T,
\end{aligned}
$$

$$(4.36)$$

for some constant $C > 0$. Now the first estimate in (4.30) yields (4.31). To prove the gradient estimate (4.32), we first observe that for $x \in \mathbb{R}^d$ we have

$$\nabla S_{\mathbb{R}^d}(t,s)f(x) = \int_{\mathbb{R}^d} f(U(s,t)x + g(t,s)) k(t,s,y) \left(U^\top(s,t) Q_{t,s}^{-1} y \right)^\top \, dy, \ 0 \leq s < t < T.$$

Again we use the change of variables $\xi = U(s,t)x$ and we apply Young's inequality for convolutions to obtain

$$\|\nabla S_{\mathbb{R}^d}(t,s)f\|_{L^q(\mathbb{R}^d)} \leq |\det U(s,t)|^{\frac{1}{q}} \left\| k(t,s,\cdot) \left(U^\top(s,t) Q_{t,s}^{-1} \cdot \right)^\top \right\|_{L^r(\mathbb{R}^d)} \|f\|_{L^p(\mathbb{R}^d)}$$

for $0 \leq s < t < T$, where $1 < r < \infty$ with $\frac{1}{p} + \frac{1}{r} = 1 + \frac{1}{q}$. As above, by the change of variables $y = Q_{t,s}^{1/2} z$, we obtain

$$
\begin{aligned}
&\left\| k(t,s,\cdot) \left(U^\top(s,t) Q_{t,s}^{-1} \cdot \right)^\top \right\|_{L^r(\mathbb{R}^d)}^r \\
&\leq |U(t,s)|^r \int_{\mathbb{R}^d} \left| \frac{1}{(4\pi)^{\frac{d}{2}}} e^{-\frac{1}{4} Q_{t,s}^{-1} y \cdot y} \left(U^\top(s,t) Q_{t,s}^{-1} y \right)^\top \right|^r \, dy \\
&= |U(t,s)|^r \int_{\mathbb{R}^d} \left(\frac{1}{(4\pi)^{\frac{d}{2}}} e^{-\frac{|z|^2}{4}} \left| U^*(s,t) Q_{t,s}^{-\frac{1}{2}} z \right| \right)^r (\det Q_{t,s})^{\frac{1}{2}(1-r)} \, dz \\
&\leq |U(t,s)|^{2r} |Q_{t,s}^{-\frac{1}{2}}|^r \int_{\mathbb{R}^d} \left(\frac{1}{(4\pi)^{\frac{d}{2}}} e^{-\frac{|z|^2}{4}} |z| \right)^r (\det Q_{t,s})^{\frac{1}{2}(1-r)} \, dz \\
&\leq C\, M_T^{2r} |Q_{t,s}^{-\frac{1}{2}}|^r (\det Q_{t,s})^{\frac{1-r}{2}}, \qquad\qquad 0 \leq s < t < T.
\end{aligned}
$$

Now the estimates in (4.30) yield (4.32). To prove (4.33), note that for $f \in W^{k,p}(\mathbb{R}^d) \cap L^p_\sigma(\mathbb{R}^d)$,

$k = 1, 2$, we have

$$\nabla^k S_{\mathbb{R}^d}(t,s)f(x) = \int\limits_{\mathbb{R}^d} \nabla^k f(U(s,t)x + g(t,s))k(t,s,y)U^k(s,t)\,\mathrm{d}y, \quad 0 \leq s < t < T, \ x \in \mathbb{R}^d.$$

Again by the change of variables $\xi = U(s,t)x$ and by Young's inequality for convolutions we obtain

$$\|\nabla^k S_{\mathbb{R}^d}(t,s)f\|_{L^p(\mathbb{R}^d)} \leq |U(s,t)|^k |\det U(s,t)|^{\frac{1}{p}} \|k(t,s,\cdot)\|_{L^1(\mathbb{R}^d)} \|\nabla^k f\|_{L^p(\mathbb{R}^d)}, \quad 0 \leq s < t < T.$$

By the calculation (4.36) from above we see that

$$\|k(t,s,\cdot)\|_{L^1(\mathbb{R}^d)} \leq C\,M_T, \qquad\qquad 0 \leq s < t < T.$$

Hence, estimate (4.33) follows. Estimate (4.34) follows similarly by using the relation

$$\nabla^2 S_{\mathbb{R}^d}(t,s)f(x) = \int\limits_{\mathbb{R}^d} \nabla f(U(s,t)x + g(t,s))k(t,s,y)\left(U^\top(s,t)Q_{t,s}^{-1}y\right)^\top U(s,t)\,\mathrm{d}y$$

for $0 \leq s < t < T$ and $x \in \mathbb{R}^d$. Now it only remains to prove estimate (4.35). Let $f \in Y_{\mathbb{R}^d,p}$. By using the calculations in (4.28) we see that

$$\left|\int\limits_{\mathbb{R}^d} |x|\partial_k S_{\mathbb{R}^d}(t,s)f(x)\cdot h(x)\,\mathrm{d}x\right| \leq CM_T^2\|f\|_{\tilde{Y}_{\mathbb{R}^d,p}}\|h\|_{L^{p'}(\mathbb{R}^d)}, \qquad 0 \leq s < t < T,$$

holds for all $k = 1, \ldots, d$ and all $h \in L^{p'}(\mathbb{R}^d)^d$ with $\frac{1}{p} + \frac{1}{p'} = 1$. Hence this estimate together with (4.33) yields the assertion. $\qquad\square$

To conclude this section let us examine the behaviour of $\{S_{\mathbb{R}^d}(t,s)\}_{(t,s)\in\Lambda}$ near $t = s$.

Proposition 4.10. *Let $1 < p < q < \infty$ and $f \in L_\sigma^p(\mathbb{R}^d)$. Then the following holds:*

$$(t-s)^{\frac{d}{2}\left(\frac{1}{p}-\frac{1}{q}\right)}\|S_{\mathbb{R}^d,p}(t,s)f\|_{L^q(\mathbb{R}^d)} \to 0 \quad as \quad t \to s,$$

$$(t-s)^{\frac{1}{2}}\|\nabla S_{\mathbb{R}^d,p}(t,s)f\|_{L^p(\mathbb{R}^d)} \to 0 \quad as \quad t \to s.$$

Proof. Let $0 < t - s \leq 1$ and take a sequence $\{f_n\}_{n\in\mathbb{N}} \subset C_{c,\sigma}^\infty(\mathbb{R}^d) \subset L_\sigma^p(\mathbb{R}^d) \cap L_\sigma^q(\mathbb{R}^d)$ such that $f_n \to f$ in $L^p(\mathbb{R}^d)$ as $n \to \infty$. Let $\varepsilon > 0$. The triangle inequality and the L^p-L^q-estimates (4.31) imply that there exist constants $C_1, C_2 > 0$ such that

$$(t-s)^{\frac{d}{2}\left(\frac{1}{p}-\frac{1}{q}\right)}\|S_{\mathbb{R}^d}(t,s)f\|_{L^q(\mathbb{R}^d)}$$

$$\leq (t-s)^{\frac{d}{2}\left(\frac{1}{p}-\frac{1}{q}\right)}\|S_{\mathbb{R}^d}(t,s)f - S_{\mathbb{R}^d}(t,s)f_n\|_{L^q(\mathbb{R}^d)} + (t-s)^{\frac{d}{2}\left(\frac{1}{p}-\frac{1}{q}\right)}\|S_{\mathbb{R}^d}(t,s)f_n\|_{L^q(\mathbb{R}^d)}$$

$$\leq C_1\|f - f_n\|_{L^p(\mathbb{R}^d)} + C_2(t-s)^{\frac{d}{2}\left(\frac{1}{p}-\frac{1}{q}\right)}\|f_n\|_{L^q(\mathbb{R}^d)}.$$

Now we fix $n \in \mathbb{N}$ sufficiently large such that

$$\|f - f_n\|_{L^p(\mathbb{R}^d)} \leq \frac{\varepsilon}{2C_1}.$$

Then we choose $t - s$ sufficiently small such that

$$(t - s)^{\frac{d}{2}\left(\frac{1}{p} - \frac{1}{q}\right)} \leq \frac{\varepsilon}{2C_2 \|f_n\|_{L^q(\Omega)}}.$$

Then we see that

$$(t - s)^{\frac{d}{2}\left(\frac{1}{p} - \frac{1}{q}\right)} \|S_{\mathbb{R}^d}(t, s)f\|_{L^q(\mathbb{R}^d)} \leq \varepsilon$$

if $t - s$ is sufficiently small, and therefore the first assertion follows.

Similarly, by using (4.32) and (4.33) we obtain

$$(t - s)^{\frac{1}{2}} \|\nabla S_{\mathbb{R}^d}(t, s)f\|_{L^p(\mathbb{R}^d)}$$
$$\leq (t - s)^{\frac{1}{2}} \|\nabla S_{\mathbb{R}^d}(t, s)f - \nabla S_{\mathbb{R}^d}(t, s)f_n\|_{L^p(\mathbb{R}^d)} + (t - s)^{\frac{1}{2}} \|\nabla S_{\mathbb{R}^d}(t, s)f_n\|_{L^p(\mathbb{R}^d)}$$
$$\leq C_1 \|f - f_n\|_{L^p(\mathbb{R}^d)} + C_2 (t - s)^{\frac{1}{2}} \|f_n\|_{W^{1,p}(\mathbb{R}^d)}.$$

Now the second assertion follows by arguing as above. $\qquad\square$

4.2 The linear model problem in bounded domains

In this section we study the linear model problem associated to (4.9) in the situation of bounded domains. Let $\Omega \subset \mathbb{R}^d$, $d \geq 2$, be a bounded domain of class C^2 and let $1 < p < \infty$. For some initial time $s \geq 0$ and an initial value $f \in L^p_\sigma(\Omega)$ we consider the system of equations

$$\begin{cases} \left.\begin{aligned} u_t - \Delta u - M(t)x \cdot \nabla u - c(t) \cdot \nabla u \\ + M(t)u + b(t) \cdot \nabla u + u \cdot \nabla b(t) + \nabla \mathrm{p} \end{aligned}\right\} &= 0 &\text{in } (s, \infty) \times \Omega, \\ \operatorname{div} u &= 0 &\text{in } (s, \infty) \times \Omega, \\ u &= 0 &\text{on } (s, \infty) \times \partial\Omega, \\ u|_{t=s} &= f &\text{in } \Omega. \end{cases} \tag{4.37}$$

Here $u : (s, \infty) \times \Omega \to \mathbb{R}^d$ and $\mathrm{p} : (s, \infty) \times \Omega \to \mathbb{R}$ are the unknown velocity field and the pressure of the fluid, respectively.

Assumption 4.11. In this section we impose the following conditions on the coefficients:

(i) $M \in C^{0,\alpha}([0, T]; \mathbb{R}^{d \times d})$ for every $T > 0$ and some $\alpha \in (0, 1)$.

(ii) $c \in C^{0,\alpha}([0, T]; \mathbb{R}^d)$ for every $T > 0$ and some $\alpha \in (0, 1)$.

(iii) $b \in C^{0,\alpha}([0, T]; C^1(\overline{\Omega})^d)$ for every $T > 0$ and some $\alpha \in (0, 1)$. For simplicity, we interpret b as a function from $[0, \infty) \times \Omega$ to $\mathbb{R}^d$ and we write $b(t, x)$ instead of $b(t)(x)$.

In order to analyse the system (4.37) we consider the classical Stokes operator $(A_{\Omega,p}, \mathcal{D}(A_{\Omega,p}))$ on $L^p_\sigma(\Omega)$ and two families of perturbing operators $\{\mathcal{B}_1(t)\}_{t \geq 0}$ and $\{\mathcal{B}_2(t)\}_{t \geq 0}$, formally defined on smooth vector fields $u = (u_1, ..., u_d)$ by

$$\mathcal{B}_1(t)u(x) := M(t)x \cdot \nabla u(x) + c(t) \cdot \nabla u(x) - M(t)u(x), \qquad x \in \Omega,\ t \geq 0,$$

and

$$\mathcal{B}_2(t)u(x) := -b(t,x) \cdot \nabla u(x) - u(x) \cdot \nabla b(t,x), \qquad x \in \Omega,\ t \geq 0.$$

In contrast to the situation in Section 4.1 the coefficients of the term $M(t)x \cdot \nabla$ are bounded over the bounded domain Ω. Thus $\mathcal{B}_1(t)$ and $\mathcal{B}_2(t)$ can be considered as "small" perturbations of the Stokes operator. Therefore, for $t \geq 0$ we define an operator $\tilde{L}_{\Omega,p}(t)$ on $L^p(\Omega)^d$ by setting

$$\begin{aligned}
\mathcal{D}(\tilde{L}_{\Omega,p}(t)) &:= \mathcal{D}(\Delta_{\Omega,p}) := W^{2,p}(\Omega)^d \cap W_0^{1,p}(\Omega)^d, \\
\tilde{L}_{\Omega,p}(t)u &:= \Delta u + \mathcal{B}_1(t)u + \mathcal{B}_2(t)u,
\end{aligned} \qquad t \geq 0,$$

and an operator $L_{\Omega,p}(t)$ on $L_\sigma^p(\Omega)$ by setting

$$\begin{aligned}
\mathcal{D}(L_{\Omega,p}(t)) &:= \mathcal{D}(A_{\Omega,p}) := \mathcal{D}(\Delta_{\Omega,p}) \cap L_\sigma^p(\Omega), \\
L_{\Omega,p}(t)u &:= A_{\Omega,p} + \mathbb{P}_{\Omega,p}\mathcal{B}_1(t)u + \mathbb{P}_{\Omega,p}\mathcal{B}_2(t)u,
\end{aligned} \qquad t \geq 0.$$

Remark 4.12. If it is clear from the context what the actual value of $1 < p < \infty$ is, or if it is not important, we simply write $\tilde{L}_\Omega(t)$ and $L_\Omega(t)$.

The classical perturbation theory for the Stokes operator yields the following result.

Proposition 4.13. *Let $1 < p < \infty$, $T > 0$ and $s \in [0,T]$. Then the linear operator $(L_{\Omega,p}(s), \mathcal{D}(A_{\Omega,p}))$ generates an analytic semigroup on $L_\sigma^p(\Omega)$ which is denoted by $\{S_s(t)\}_{t \geq 0}$. Moreover there exist constants $\omega := \omega(p, \Omega, T) \in \mathbb{R}$ and $C := C(p, \Omega, T) > 0$ such that*

$$\|S_s(t)f\|_{L^p(\Omega)} \leq Ce^{\omega t}\|f\|_{L^p(\Omega)}, \qquad t > 0, \tag{4.38}$$

for all $f \in L_\sigma^p(\Omega)$ and all $s \in [0,T]$.

Proof. Let $T > 0$ be arbitrary but fixed and $s \in [0,T]$. Certainly, the operator $B := \mathbb{P}_\Omega \mathcal{B}_1(s) + \mathbb{P}_\Omega \mathcal{B}_2(s)$ is well-defined on $\mathcal{D}(A_{\Omega,p})$. In the following let $u \in \mathcal{D}(A_{\Omega,p})$. Then, by using the Gagliardo-Nirenberg inequality, see Proposition 1.17, for $j = 1$, $m = 2$, $p = q = r$, $\theta = 1/2$, Young's inequality and Proposition 2.3, we have

$$\begin{aligned}
\|Bu\|_{L^p(\Omega)} &\leq C_1 \left(\|\nabla u\|_{L^p(\Omega)} + \|u\|_{L^p(\Omega)} \right) \leq \varepsilon \|\nabla^2 u\|_{L^p(\Omega)} + C_2 \|u\|_{L^p(\Omega)} \\
&\leq \varepsilon C_3 \|A_\Omega u\|_{L^p(\Omega)} + C_2 \|u\|_{L^p(\Omega)}
\end{aligned}$$

for arbitrary $\varepsilon > 0$ and constants $C_1 := C_1(p, \Omega, T)$, $C_2 := C_2(\varepsilon, p, \Omega, T)$, $C_3 := C_3(p, \Omega) > 0$. The perturbation result for analytic semigroups, see Proposition 1.37, yields the existence of the semigroup $\{S_s(t)\}_{t \geq 0}$. Estimate (4.38) follows by having a closer look at the proof of the perturbation result, see e.g. the proof of [ABHN01, Theorem 3.7.23], and by noting that the constant C_2 above only depends on $T > 0$ and not directly on $s \in [0,T]$. $\qquad\square$

By applying the Helmholtz projection $\mathbb{P}_\Omega$ to (4.37) the pressure p gets eliminated and we may rewrite the equations as a non-autonomous abstract Cauchy problem on $L^p_\sigma(\Omega)$:

$$\begin{cases} u'(t) & = & L_\Omega(t)u(t), \quad t > s \geq 0, \\ u(s) & = & f. \end{cases} \tag{4.39}$$

Proposition 4.13 indicates that this problem fits into the framework of parabolic evolution equations in the sense of Assumption 1.56. Therefore, we can apply the abstract results presented in Section 1.2.2. This is done in Section 4.2.1. In Section 4.2.2 we return to the original problem (4.37) and derive decay estimates for the pressure term.

4.2.1 Well-posedness and L^p-L^q-estimates

Proposition 4.13 states that for every fixed $s \geq 0$, the operator $(L_{\Omega,p}(s), \mathcal{D}(A_{\Omega,p}))$ generates an analytic semigroup on $L^p_\sigma(\Omega)$. Furthermore, Assumption 4.11 implies that $t \mapsto L_{\Omega,p}(t)$ belongs to $C^{0,\alpha}([0,T]; \mathscr{L}(\mathcal{D}(A_{\Omega,p}), L^p_\sigma(\Omega)))$ for all $T > 0$ and some $\alpha \in (0,1)$. Thus, the following result follows from the theory of parabolic evolution systems, see Proposition 1.59.

Proposition 4.14. *Let $1 < p < \infty$ and $\Omega \subset \mathbb{R}^d$ be a bounded domain of class C^2. Then there exists a unique evolution system $\{S_{\Omega,p}(t,s)\}_{(t,s)\in\Lambda}$ on $L^p_\sigma(\Omega)$ with the following properties:*

(a) *For $(t,s) \in \Lambda$ with $t \neq s$, the operator $S_{\Omega,p}(t,s)$ maps $L^p_\sigma(\Omega)$ into $\mathcal{D}(A_{\Omega,p})$.*

(b) *For every $s \geq 0$, the map $t \mapsto S_{\Omega,p}(t,s)$ is differentiable in (s,∞) with values in $\mathscr{L}(L^p_\sigma(\Omega))$ and*

$$\partial_t S_{\Omega,p}(t,s) = L_{\Omega,p}(t)S_{\Omega,p}(t,s).$$

(c) *For every $f \in \mathcal{D}(A_{\Omega,p})$ and $t > 0$, the map $s \mapsto S_{\Omega,p}(t,s)f$ is differentiable in $[0,t)$ and*

$$\partial_s S_{\Omega,p}(t,s)f = -S_{\Omega,p}(t,s)L_{\Omega,p}(s)f.$$

(d) *Let $T > 0$. Then there exists a constant $C := C(p,\Omega,T) > 0$ such that*

$$\|S_{\Omega,p}(t,s)f\|_{L^p(\Omega)} \leq C\|f\|_{L^p(\Omega)},$$

and

$$\|L_{\Omega,p}(t)S_{\Omega,p}(t,s)f\|_{L^p(\Omega)} \leq C(t-s)^{-1}\|f\|_{L^p(\Omega)}$$

for all $f \in L^p_\sigma(\Omega)$ and all $(t,s) \in \Lambda_T$ with $t \neq s$.

Moreover the non-autonomous abstract Cauchy problem (4.39) is well-posed on $L^p_\sigma(\Omega)$ (in the sense of Definition 1.48).

In the following, we simply write $S_\Omega(t,s)$ instead of $S_{\Omega,p}(t,s)$ if the actual value of $1 < p < \infty$ is clear from the context, or is not important.

The following estimates follow basically from Proposition 4.14 (d) and Proposition 1.60.

Proposition 4.15. *Let $T > 0$ and $1 < p \leq q < \infty$. Moreover, let $\{S_{\Omega,p}(t,s)\}_{(t,s)\in\Lambda}$ be the evolution system from Proposition 4.14. Then there exists a constant $C := C(p,q,\Omega,T) > 0$ such that for every $f \in L^p_\sigma(\Omega)$*

$$\|S_{\Omega,p}(t,s)f\|_{L^q(\Omega)} \leq C(t-s)^{-\frac{d}{2}\left(\frac{1}{p}-\frac{1}{q}\right)}\|f\|_{L^p(\Omega)}, \qquad 0 \leq s < t \leq T \qquad (4.40)$$

$$\|\nabla S_{\Omega,p}(t,s)f\|_{L^q(\Omega)} \leq C(t-s)^{-\frac{d}{2}\left(\frac{1}{p}-\frac{1}{q}\right)-\frac{1}{2}}\|f\|_{L^p(\Omega)}, \qquad 0 \leq s < t \leq T, \qquad (4.41)$$

$$\|\nabla^2 S_{\Omega,p}(t,s)f\|_{L^q(\Omega)} \leq C(t-s)^{-\frac{d}{2}\left(\frac{1}{p}-\frac{1}{q}\right)-1}\|f\|_{L^p(\Omega)}, \qquad 0 \leq s < t \leq T, \qquad (4.42)$$

hold. Moreover, there exists a constant $C := C(p,\Omega,T) > 0$ such that

$$\|S_{\Omega,p}(t,s)f\|_{W^{k,p}(\Omega)} \leq C\|f\|_{W^{k,p}(\Omega)}, \quad 0 \leq s \leq t \leq T, \qquad (4.43)$$

for all $f \in W^{k,p}(\Omega)^d \cap W_0^{1,p}(\Omega)^d \cap L^p_\sigma(\Omega)$ and $k = 1,2$, and

$$\|S_{\Omega,p}(t,s)f\|_{W^{2,p}(\Omega)} \leq C(t-s)^{-\frac{1}{2}}\|f\|_{W^{1,p}(\Omega)}, \quad 0 \leq s < t \leq T, \qquad (4.44)$$

for all $f \in W_0^{1,p}(\Omega)^d \cap L^p_\sigma(\Omega)$.

Proof. From Proposition 4.14 (d) and Lemma 1.57 it follows that

$$\|\nabla^2 S_\Omega(t,s)f\|_{L^p(\Omega)} \leq C(t-s)^{-1}\|f\|_{L^p(\Omega)}, \qquad 0 \leq s < t \leq T, \qquad (4.45)$$

holds for all $f \in L^p_\sigma(\Omega)$. This proves (4.42) for $p = q$. Estimate (4.40) for $p = q$ is clear by Proposition 4.14 (d). Now let us assume that $1 < p < q < \infty$ such that $0 < 1/p - 1/q \leq 2/d$. In this case we have $0 < d/2\,(1/p - 1/q) \leq 1$, and hence (4.40) follows from (4.45) by applying the Gagliardo-Nirenberg inequality, see Proposition 1.17, with $j = 0$, $m = 2$, $r = p$ and $\theta = d/2\,(1/p - 1/q)$. Now assume that $1 < p < q < \infty$ such that $2/d < 1/p - 1/q \leq 4/d$. Set $1/r = 1/q + 2d$. Then $1 < p < r < q < \infty$, $1/r - 1/q = 2/d$ and $0 < 1/p - 1/r \leq 2/d$. So, by using the algebraic property of evolution systems and the first step we obtain

$$
\begin{aligned}
\|S_\Omega(t,s)f\|_{L^q(\Omega)} &= \left\|S_\Omega\left(t,s+\tfrac{t-s}{2}\right)S_\Omega\left(s+\tfrac{t-s}{2},s\right)f\right\|_{L^q(\Omega)} \\
&\leq C\left(\frac{t-s}{2}\right)^{-\frac{d}{2}\left(\frac{1}{r}-\frac{1}{q}\right)}\left\|S_\Omega\left(s+\tfrac{t-s}{2},s\right)f\right\|_{L^r(\Omega)} \\
&\leq C\left(\frac{t-s}{2}\right)^{-\frac{d}{2}\left(\frac{1}{r}-\frac{1}{q}\right)}\left(\frac{t-s}{2}\right)^{-\frac{d}{2}\left(\frac{1}{p}-\frac{1}{r}\right)}\|f\|_{L^p(\Omega)} \\
&\leq C(t-s)^{-\frac{d}{2}\left(\frac{1}{p}-\frac{1}{q}\right)}\|f\|_{L^p(\Omega)}, \qquad 0 \leq s < t \leq T.
\end{aligned}
$$

By iterating this argument[3] we obtain (4.40) for all $1 < p < q < \infty$. Estimate (4.41) for $p = q$ follows from (4.45), Proposition 4.14 (d) and the Gagliardo-Nirenberg inequality, see

---

[3]The number of iteration steps depends only on the dimension d.

89

Proposition 1.17, with $j = 1$, $m = 2$, $r = p = q$ and $\theta = 1/2$. Moreover, we obtain

$$
\begin{aligned}
\|\nabla S_\Omega(t,s)f\|_{L^q(\Omega)} &= \left\|\nabla S_\Omega\left(t, s + \tfrac{t-s}{2}\right) S_\Omega\left(s + \tfrac{t-s}{2}, s\right) f\right\|_{L^q(\Omega)} \\
&\leq C\left(\frac{t-s}{2}\right)^{-\frac{1}{2}} \left\|S_\Omega\left(s + \tfrac{t-s}{2}, s\right) f\right\|_{L^q(\Omega)} \\
&\leq C(t-s)^{-\frac{d}{2}\left(\frac{1}{p}-\frac{1}{q}\right)-\frac{1}{2}}\|f\|_{L^p(\Omega)}, \qquad 0 \leq s < t \leq T.
\end{aligned}
$$

This proves (4.41) for $1 < p < q < \infty$. Analogously, we prove (4.42) for $1 < p < q < \infty$. Estimate (4.43) for $k = 2$ follows directly from the first estimate in Proposition 1.60. Next, let us prove (4.43) for $k = 1$. For $s \geq 0$ fixed and $\omega > 0$ sufficiently large we conclude[4] that

$$
\begin{aligned}
\mathcal{D}((-L_{\Omega,p}(s) + \omega)^{\frac{1}{2}}) = \mathcal{D}((-A_{\Omega,p})^{\frac{1}{2}}) &= [L_\sigma^p(\Omega), W^{2,p}(\Omega)^d \cap W_0^{1,p}(\Omega)^d \cap L_\sigma^p(\Omega)]_{\frac{1}{2}} \\
&= [L^p(\Omega), W^{2,p}(\Omega)^d \cap W_0^{1,p}(\Omega)^d]_{\frac{1}{2}} \cap L_\sigma^p(\Omega)
\end{aligned}
$$

holds with equivalent norms, see [Tri95, Section 1.15.3] and [Gig85]. Note that for fixed $s \geq 0$ the operator $(-L_{\Omega,p}(s) + \omega)^{1/2}$ commutes[5] with the semigroup $\{S_s(t)\}_{t \geq 0}$. By (4.38) we conclude that there exists a constant $C := C(p, \Omega, T) > 0$ such that

$$
\|S_s(t)f\|_{\mathcal{D}((-A_{\Omega,p})^{1/2})} \leq C\|f\|_{\mathcal{D}((-A_{\Omega,p})^{1/2})}, \qquad 0 \leq s, t \leq T.
$$

The last estimate in Proposition 1.60 implies that there exists a $C := C(p, \Omega, T) > 0$ such that

$$
\|S_\Omega(t,s)f\|_{\mathcal{D}((-A_{\Omega,p})^{1/2})} \leq C\|f\|_{\mathcal{D}((-A_{\Omega,p})^{1/2})}, \qquad 0 \leq s \leq t \leq T \tag{4.46}
$$

for all $f \in \mathcal{D}((-A_{\Omega,p})^{\frac{1}{2}})$. Moreover, it is clear that

$$
\mathcal{D}((-A_{\Omega,p})^{\frac{1}{2}}) = [L_\sigma^p(\Omega), W^{2,p}(\Omega)^d \cap W_0^{1,p}(\Omega)^d \cap L_\sigma^p(\Omega)]_{\frac{1}{2}} \hookrightarrow W^{1,p}(\Omega)^d
$$

and by [Tri95, Section 4.3.2] we obtain

$$
W_0^{1,p}(\Omega)^d \cap L_\sigma^p(\Omega) \hookrightarrow [L^p(\Omega), W^{2,p}(\Omega)^d \cap W_0^{1,p}(\Omega)^d]_{\frac{1}{2}} \cap L_\sigma^p(\Omega) = \mathcal{D}((-A_{\Omega,p})^{\frac{1}{2}}).
$$

Thus, estimate (4.43) with $k = 1$ follows from (4.46). Now it only remains to prove (4.44). By applying the second estimate in Proposition 1.60 we obtain

$$
\|S_\Omega(t,s)f\|_{W^{2,p}(\Omega)} \leq C(t-s)^{\frac{1}{2}}\|f\|_{(L_\sigma^p(\Omega), \mathcal{D}(A_{\Omega,p}))_{1/2,\infty}}, \qquad 0 \leq s < t \leq T
$$

for all $f \in (L_\sigma^p(\Omega), \mathcal{D}(A_{\Omega,p}))_{1/2,\infty}$. Now, by using the embedding (see [Ama95, Section I.2.5])

$$
[L_\sigma^p(\Omega), \mathcal{D}(A_{\Omega,p})]_{\frac{1}{2}} \hookrightarrow (L_\sigma^p(\Omega), \mathcal{D}(A_{\Omega,p}))_{\frac{1}{2},\infty}
$$

the assertion follows. □

---

[4]For the definition of the fractional powers $(-L_{\Omega,p}(s) + \omega)^{\frac{1}{2}}$ and $(-A_{\Omega,p})^{\frac{1}{2}}$ we refer e.g. to [DHP03, Chapter 2]. The characterization of fractional power domains by complex interpolation spaces, which is used here, holds for operators belonging to the class $\mathcal{BIP}$ or for the smaller class of operators admitting a bounded $\mathcal{H}^\infty$-calculus (we refer e.g. to [DHP03, Chapter 2] for definitions). By [NS03] the operator $-A_\Omega$ has a bounded $\mathcal{H}^\infty$-calculus. By simple perturbation arguments this carries over to the operator $-L_{\Omega,p}(s) + \omega$.

[5]It is clear that $-L_{\Omega,p}(s) + \omega$ commutes with $\{S_s(t)\}_{t \geq 0}$. Hence, by abstract arguments from the theory of functional calculus, see e.g. [Haa06, Theorem 2.3.3], the same holds for $(-L_{\Omega,p}(s) + \omega)^{\frac{1}{2}}$.

4.2.2 Pressure estimates

After having solved the abstract Cauchy problem (4.39), we return now to the original problem

$$
\begin{cases}
\left. \begin{aligned}
u_t - \Delta u - M(t)x \cdot \nabla u - c(t) \cdot \nabla u \\
+ M(t)u + b(t) \cdot \nabla u + u \cdot \nabla b(t) + \nabla \mathrm{p}
\end{aligned} \right\} &= 0 \quad \text{in } (s, \infty) \times \Omega, \\
\operatorname{div} u &= 0 \quad \text{in } (s, \infty) \times \Omega, \\
u &= 0 \quad \text{on } (s, \infty) \times \partial\Omega, \\
u|_{t=s} &= f \quad \text{in } \Omega,
\end{cases}
\tag{4.47}
$$

for $s \geq 0$ and $f \in L^p_\sigma(\Omega)$. Proposition 4.14 states that this problem is solved by (u, p), where $u(t) := S_\Omega(t, s)f$ and the pressure $\mathrm{p} : (s, \infty) \times \Omega \to \mathbb{R}^d$ is defined in such a way that

$$
\nabla \mathrm{p}(t) = (\mathrm{Id} - \mathbb{P}_\Omega)\tilde{L}_\Omega(t)u(t). \tag{4.48}
$$

In the case of a bounded domain Ω we can always assume that

$$
\mathrm{p}(t) \in L^p_0(\Omega) := \left\{ \mathrm{q} \in L^p(\Omega) : \int_\Omega \mathrm{q} \, dx = 0 \right\}
$$

for all $t \in (s, \infty)$, see the explanations below Definition 2.8 for details. The solution (u, p) with $\mathrm{p}(t) \in L^p_0(\Omega)$ for all $t \in (s, \infty)$ is unique.

By Proposition 4.14 we know that

$$
u \in C^1((s, \infty); L^p_\sigma(\Omega)) \cap C((s, \infty); \mathcal{D}(A_{\Omega, p})) \cap C([s, \infty); L^p_\sigma(\Omega)).
$$

So in particular we have $\mathrm{p} \in C((s, \infty); \widehat{W}^{1,p}(\Omega))$. Then by applying Poincaré's inequality, which is possible since $\mathrm{p}(t) \in L^p_0(\Omega)$, we conclude that we even have

$$
\mathrm{p} \in C((s, \infty); W^{1,p}(\Omega)).
$$

The purpose of this section is to show the following decay estimates for the pressure term p which are needed later in Section 4.3.

Proposition 4.16. *Let $1 < p < \infty$, $s \geq 0$ and $\Omega \subset \mathbb{R}^d$ be a bounded domain with boundary of class C^2. Let (u, p) be the unique solution of (4.47) with $\mathrm{p}(t) \in L^p_0(\Omega)$ for all $t \in (s, \infty)$ and let $\gamma \in (1 + \frac{1}{p}, 2)$. Then for every $T > s$ there exists a constant $C := C(p, \gamma, \Omega, T) > 0$ such that*

$$
\|\mathrm{p}(t)\|_{L^p(\Omega)} \leq C(t - s)^{-\frac{\gamma}{2}} \|f\|_{L^p(\Omega)}, \qquad s < t \leq T.
$$

Moreover, if $f \in W^{1,p}_0(\Omega)^d \cap L^p_\sigma(\Omega)$, then for every $T > s$ there exists a constant $C := C(p, \Omega, T) > 0$ such that

$$
\|\nabla \mathrm{p}(t)\|_{L^p(\Omega)} \leq C(t - s)^{-\frac{1}{2}} \|f\|_{W^{1,p}(\Omega)}, \qquad s < t \leq T.
$$

Since p(t) $\in L_0^p(\Omega)$ we may apply Poincaré's inequality and relation (4.48) to obtain

$$\|p(t)\|_{L^p(\Omega)} \leq C\|u(t)\|_{W^{2,p}(\Omega)}, \qquad s < t \leq T.$$

Then applying Proposition 4.15 yields

$$\|p(t)\|_{L^p(\Omega)} \leq C(t-s)^{-1}\|f\|_{L^p(\Omega)}, \qquad s < t \leq T.$$

However, for later purposes this estimate is not suitable as $(t-s)^{-1}$ is not integrable at $t = s$. Therefore, in the statement of Lemma 4.16 it is important that we have $\gamma < 2$. A similar estimate was proved in [GHH06a, Lemma 3.5] for the solution to the corresponding resolvent problem, see also [NS03] for the case of the Stokes operator. Our proof here is in the spirit of Lemma 2.9 and it follows again the ideas of Shibata and Shimada [SS07a] and Shibata [Shi08, Section 4].

Proof of Lemma 4.16. Let $1 < p' < \infty$ with $\frac{1}{p} + \frac{1}{p'} = 1$. Moreover, let $\varphi \in C_c^\infty(\Omega)$ and set $\tilde{\varphi} = \varphi - |\Omega|^{-1} \int_\Omega \varphi \, dx$. We consider the classical Neumann problem (see [ADN59])

$$\begin{cases} \Delta\psi &= \tilde{\varphi} \quad \text{in } \Omega, \\ \nu \cdot \nabla\psi &= 0 \quad \text{on } \partial\Omega. \end{cases}$$

It is well-known that there exits a unique solution $\psi \in W^{2,p'}(\Omega)$ to this Neumann problem which satisfies the estimate

$$\|\psi\|_{W^{2,p'}(\Omega)} \leq C\|\tilde{\varphi}\|_{L^{p'}(\Omega)} \leq 2C\|\varphi\|_{L^{p'}(\Omega)}$$

for some constant $C > 0$. Then, we obtain

$$\begin{aligned} \langle p(t), \varphi\rangle_\Omega &= \langle p(t), \tilde{\varphi}\rangle_\Omega = \langle p(t), \Delta\psi\rangle_\Omega = -\langle \nabla p(t), \nabla\psi\rangle_\Omega = -\langle (\text{Id} - \mathbb{P}_\Omega)\tilde{L}_\Omega(t)u(t), \nabla\psi\rangle_\Omega \\ &= -\langle \Delta u(t), \nabla\psi\rangle_\Omega - \langle (M(t)x + c(t)) \cdot \nabla u(t), \nabla\psi\rangle_\Omega + \langle M(t)u(t), \nabla\psi\rangle_\Omega \\ &\quad + \langle b(t) \cdot \nabla u(t), \nabla\psi\rangle_\Omega + \langle u(t) \cdot \nabla b(t), \nabla\psi\rangle_\Omega \\ &= -\langle \nu \cdot \nabla u(t), \nabla\psi\rangle_{\partial\Omega} + \langle \nabla u(t), \nabla^2\psi\rangle_\Omega - \langle (M(t)x + c(t)) \cdot \nabla u(t), \nabla\psi\rangle_\Omega \\ &\quad + \langle M(t)u(t), \nabla\psi\rangle_\Omega + \langle b(t) \cdot \nabla u(t), \nabla\psi\rangle_\Omega + \langle u(t) \cdot \nabla b(t), \nabla\psi\rangle_\Omega. \end{aligned}$$

By using the embedding $W_p^{1/p+\varepsilon}(\Omega) \hookrightarrow L^p(\partial\Omega)$ for $0 < \varepsilon \leq 1 - \frac{1}{p}$, see Proposition 1.10, we get

$$\begin{aligned} |\langle p(t), \varphi\rangle_\Omega| &\leq C \left(\|\nu \cdot \nabla u(t)\|_{L^p(\partial\Omega)} + \|\nabla u(t)\|_{L^p(\Omega)} + \|u(t)\|_{L^p(\Omega)} \right) \|\psi\|_{W^{2,p'}(\Omega)} \\ &\leq C\|u(t)\|_{W_p^{1+1/p+\varepsilon}(\Omega)}\|\varphi\|_{L^{p'}(\Omega)}. \end{aligned}$$

Now choose $\varepsilon := \varepsilon(\gamma)$ such that $\gamma = 1 + \frac{1}{p} + \varepsilon$. Note that $W_p^\gamma(\Omega)^d$ is the real interpolation space between $L^p(\Omega)^d$ and $W^{2,p}(\Omega)^d$ and thus we have

$$\|u(t)\|_{W_p^\gamma(\Omega)} \leq C\|u(t)\|_{L^p(\Omega)}^{1-\gamma/2}\|u(t)\|_{W^{2,p}(\Omega)}^{\gamma/2},$$

see Remark 1.8. Thus the first assertion follows from Proposition 4.15. The second estimate follows easily from relation (4.48) and estimate (4.44) in Proposition 4.15. $\qquad\square$

4.3 The linear model problem in exterior domains

In this section we study the linear model problem associated to (4.9) in the situation of exterior domains. Let $1 < p < \infty$ and let $\Omega \subset \mathbb{R}^d$, $d \geq 2$, be an exterior domain of class C^2. For some initial time $s \geq 0$ and an initial value $f \in L^p_\sigma(\Omega)$ we consider the system of equations

$$
\left\{
\begin{aligned}
\left.
\begin{aligned}
u_t - \Delta u - M(t)x \cdot \nabla u - c(t) \cdot \nabla u \\
+ M(t)u + b(t) \cdot \nabla u + u \cdot \nabla b(t) + \nabla \mathrm{p}
\end{aligned}
\right\} &= 0 \quad \text{in } (s, \infty) \times \Omega, \\
\operatorname{div} u &= 0 \quad \text{in } (s, \infty) \times \Omega, \\
u &= 0 \quad \text{on } (s, \infty) \times \partial\Omega, \\
u|_{t=s} &= f \quad \text{in } \Omega.
\end{aligned}
\right.
\tag{4.49}
$$

Here $u : (s, \infty) \times \Omega \to \mathbb{R}^d$ and $\mathrm{p} : (s, \infty) \times \Omega \to \mathbb{R}$ are the unknown velocity field and the pressure of the fluid, respectively.

Assumption 4.17. In this section we impose the following conditions on the coefficients:

(i) $M \in C^{0,\alpha}([0, T]; \mathbb{R}^{d \times d})$ for every $T > 0$ and some $\alpha \in (0, 1)$.

(ii) $c \in C^{0,\alpha}([0, T]; \mathbb{R}^d)$ for every $T > 0$ and some $\alpha \in (0, 1)$.

(iii) $b \in C^{0,\alpha}([0, T]; C^1(\overline{\Omega})^d)$ for every $T > 0$ and some $\alpha \in (0, 1)$. Moreover, we assume that $\operatorname{supp} b(t) \subset B_{R_0} \cap \overline{\Omega}$ for all $t \geq 0$ and some $R_0 > 0$. For simplicity, we interpret b as a function from $[0, \infty) \times \Omega$ to $\mathbb{R}^d$ and we write $b(t, x)$ instead of $b(t)(x)$.

We consider a family of linear operators $\{\tilde{L}_{\Omega,p}(t)\}_{t \geq 0}$ on $L^p(\Omega)^d$ which is defined by

$$
\begin{aligned}
\mathcal{D}(\tilde{L}_{\Omega,p}(t)) &:= \{u \in W^{2,p}(\Omega)^d \cap W_0^{1,p}(\Omega) : M(t)x \cdot \nabla u \in L^p(\Omega)^d\}, \\
\tilde{L}_{\Omega,p}(t)u &:= \mathcal{L}(t)u + \mathcal{B}_2(t)u,
\end{aligned}
\quad t \geq 0.
$$

Here $\mathcal{L}(t)$ is given in (4.11) and $\mathcal{B}_2(t)$ is a small perturbation of the main part $\mathcal{L}(t)$ which is formally defined on smooth vector fields $u := (u_1, \ldots, u_d)$ by

$$
\mathcal{B}_2(t)u(x) := -b(t, x) \cdot \nabla u(x) - u(x) \cdot \nabla b(t, x), \quad x \in \Omega, \ t \geq 0.
$$

As usual in the theory of the Navier-Stokes equtions we do not work directly on $L^p(\Omega)^d$ but on $L^p_\sigma(\Omega)$. Therefore we introduce a family of linear operators $\{L_{\Omega,p}(t)\}_{t \geq 0}$ on $L^p_\sigma(\Omega)$ by setting

$$
\begin{aligned}
\mathcal{D}(L_{\Omega,p}(t)) &:= \mathcal{D}(\tilde{L}_{\Omega,p}(t)) \cap L^p_\sigma(\Omega), \\
L_{\Omega,p}(t)u &:= \mathbb{P}_{\Omega,p}\tilde{L}_{\Omega,p},
\end{aligned}
\quad t \geq 0.
$$

Note that the domain $\mathcal{D}(L_{\Omega,p}(t))$ depends on the time parameter $t \geq 0$. To handle this difficulty we introduce the spaces

$$
\begin{aligned}
\tilde{Y}_{\Omega,p} &:= \left\{ f \in W_0^{1,p}(\Omega)^d \cap L^p_\sigma(\Omega) : |x| \nabla f_i(x) \in L^p(\Omega)^d \text{ for } i = 1, \ldots, d \right\}, \\
Y_{\Omega,p} &:= W^{2,p}(\Omega)^d \cap \tilde{Y}_{\Omega,p}.
\end{aligned}
$$

It is clear that $C^\infty_{c,\sigma}(\Omega) \subset Y_{\Omega,p} \subset \mathcal{D}(L_{\Omega,p}(t))$ holds for all $t \geq 0$. Moreover, if we equip $\widetilde{Y}_{\Omega,p}$ and $Y_{\Omega,p}$ with the norms $\|\cdot\|_{\widetilde{Y}_{\Omega,p}}$ and $\|\cdot\|_{Y_{\Omega,p}}$ defined by

$$\|f\|_{\widetilde{Y}_{\Omega,p}} := \|f\|_{W^{1,p}(\Omega)} + \sum_{i=1}^{d} \||x|\nabla f_i\|_{L^p(\Omega)}, \qquad f \in \widetilde{Y}_{\Omega,p},$$

$$\|f\|_{Y_{\Omega,p}} := \|f\|_{W^{2,p}(\Omega)} + \sum_{i=1}^{d} \||x|\nabla f_i\|_{L^p(\Omega)}, \qquad f \in Y_{\Omega,p},$$

respectively, then they become Banach spaces.

Remark 4.18. When it is clear from the context what the actual value of $1 < p < \infty$ is, or if it is not important, then we simply write $\tilde{L}_\Omega(t)$ and $L_\Omega(t)$.

By applying the Helmholtz projection $\mathbb{P}_\Omega$ to (4.49) the pressure p gets eliminated and we may rewrite the equations as a non-autonomous abstract Cauchy problem on $L^p_\sigma(\Omega)$:

$$\begin{cases} u'(t) &= L_\Omega(t)u(t), \quad t > s \geq 0, \\ u(s) &= f. \end{cases} \tag{4.50}$$

Here the situation is similar to the one in Section 4.1. We cannot use the theory of parabolic evolution equations in the sense of Assumption 1.56, as for fixed $s \geq 0$ the C_0-semigroup generated by $L_\Omega(s)$ is not analytic, see Proposition 1.85. Therefore the approach in this section is as follows: We use a localization (cut-off) procedure to reduce the exterior domain problem to two problems: One on $\mathbb{R}^d$ and one in some bounded domain close to the boundary $\partial\Omega$. In these cases we can apply the results from Section 4.1 and Section 4.2, respectively. This localization technique also allows to prove L^p-L^q-estimates and gradient estimates in Section 4.3.2.

4.3.1 Construction of the evolution system

The aim of this section is to prove the following well-posedness result.

Theorem 4.19. *Let $1 < p < \infty$, and let $\Omega \subset \mathbb{R}^d$ be an exterior domain of class C^2. There exists a unique evolution system $\{S_{\Omega,p}(t,s)\}_{(t,s)\in\Lambda}$ on $L^p_\sigma(\Omega)$ with the following properties:*

(a) For $(t,s) \in \Lambda$, the operator $S_{\Omega,p}(t,s)$ maps $Y_{\Omega,p}$ into $Y_{\Omega,p}$.

(b) For every $f \in Y_{\Omega,p}$ and $s \geq 0$, the map $t \mapsto S_{\Omega,p}(t,s)f$ is differentiable in (s,∞) and

$$\partial_t S_{\Omega,p}(t,s)f = L_{\Omega,p}(t)S_{\Omega,p}(t,s)f.$$

(c) For every $f \in Y_{\Omega,p}$ and $t > 0$, the map $s \mapsto S_{\Omega,p}(t,s)f$ is differentiable in $[0,t)$ and

$$\partial_s S_{\Omega,p}(t,s)f = -S_{\Omega,p}(t,s)L_{\Omega,p}(s)f.$$

Moreover, the non-autonomous abstract Cauchy problem (4.50) is well-posed on the regularity space $Y_{\Omega,p}$ (in the sense of Definition 1.48).

In the following we simply write $S_\Omega(t, s)$ instead of $S_{\Omega,p}(t, s)$ if the actual value of $1 < p < \infty$ is clear from the context, or is not important.

As already mentioned above, the general idea is to derive the result for an exterior domain Ω from the corresponding results in the case of $\mathbb{R}^d$ and in the case of bounded domains by using some localization (cut-off) technique. A similar localization (cut-off) technique is briefly sketched in Section 1.6.1. However, the situtation here is a bit different compared to that situation. There we already have a solution to an exterior domain problem. In this section we first need to construct the exterior domain solution. This is done in the following.

Setting of the localization procedure: Set $\mathcal{O} := \mathbb{R}^d \setminus \Omega$ and let $R > R_0$ be such that $\mathcal{O} \subset B_R$ (here R_0 is the number from Assumption 4.17 (iii)). Then set

$$D := \Omega \cap B_{R+8},$$
$$D_1 := \Omega \cap B_{R+2},$$
$$D_2 := \{x \in \Omega : R + 2 < |x| < R + 5\},$$
$$D_3 := \{x \in \Omega : R + 5 < |x| < R + 8\}.$$

In the following $\{S_{\mathbb{R}^d}(t, s)\}_{(t,s)\in\Lambda}$ denotes the evolution system on $L^p_\sigma(\mathbb{R}^d)$ from Theorem 4.6 and $\{S_D(t, s)\}_{(t,s)\in\Lambda}$ denotes the evolution system on $L^p_\sigma(D)$ from Proposition 4.14. Next, choose cut-off functions $\varphi, \xi, \eta \in C^\infty(\mathbb{R}^d)$ such that $0 \leq \varphi, \xi, \eta \leq 1$ and

$$\varphi(x) := \begin{cases} 1, & |x| \geq R + 4, \\ 0, & |x| \leq R + 3, \end{cases} \qquad \xi(x) := \begin{cases} 1, & |x| \geq R + 1, \\ 0, & |x| \leq R, \end{cases}$$

and

$$\eta(x) := \begin{cases} 1, & |x| \leq R + 6, \\ 0, & |x| \geq R + 7. \end{cases}$$

For a given function $f \in L^p_\sigma(\Omega)$ we define functions $f_R \in L^p_\sigma(\mathbb{R}^d)$ and $f_D \in L^p_\sigma(D)$ by setting

$$f_R(x) := \begin{cases} \xi(x)f(x) - \mathbb{B}_{D_1}((\nabla\xi) \cdot f)(x), & x \in \Omega, \\ 0, & x \in \mathcal{O}, \end{cases} \tag{4.51}$$

and

$$f_D(x) := \eta(x)f(x) - \mathbb{B}_{D_3}((\nabla\eta) \cdot f)(x), \qquad x \in D. \tag{4.52}$$

Here $\mathbb{B}_{D_1}$ and $\mathbb{B}_{D_3}$ denote the Bogovskiĭ operator defined on the bounded domains D_1 and D_3, respectively. Note that $\mathbb{B}_{D_1}((\nabla\xi)\cdot f)$ and $\mathbb{B}_{D_3}((\nabla\eta)\cdot f)$ are interpreted as functions on $\mathbb{R}^d$ and D by extending these functions by zero outside of D_1 and D_3, respectively. Integration by parts and the divergence theorem yield

$$\int_{D_1} \nabla\xi(x) \cdot f(x) \, \mathrm{d}x = \int_{|x|=R+2} f(x) \cdot \nu \, \mathrm{d}\sigma - \int_{D_1} \xi(x)\mathrm{div}\, f(x) \, \mathrm{d}x = 0$$

since $f \in L^p_\sigma(\Omega)$. Analogously, we see that

$$\int_{D_3} \nabla\eta(x) \cdot f(x) \, \mathrm{d}x = 0.$$

Hence, Proposition 1.87 ensures that $f_R \in L^p_\sigma(\mathbb{R}^d)$ and $f_D \in L^p_\sigma(D)$ indeed hold. Moreover, by applying again Proposition 1.87 it is easy to see that if $f \in Y_{\Omega,p}$, then $f_R \in Y_{\mathbb{R}^d,p}$ and $f_D \in \mathcal{D}(A_{D,p})$.

In the following, all constants may depend on R and on the cut-off functions.

A candidate for the solution: Next we construct a candidate for the solution to the Cauchy problem (4.50) by combining the evolution systems $\{S_{\mathbb{R}^d}(t,s)\}_{(t,s)\in\Lambda}$ and $\{S_D(t,s)\}_{(t,s)\in\Lambda}$. For $(t,s) \in \Lambda$ and $f \in L^p_\sigma(\Omega)$, we define a function $W(t,s)f$ by setting

$$
\begin{aligned}
W(t,s)f := {}& \varphi S_{\mathbb{R}^d}(t,s)f_R + (1-\varphi)S_D(t,s)f_D \\
& - \mathbb{B}_{D_2}((\nabla\varphi) \cdot (S_{\mathbb{R}^d}(t,s)f_R - S_D(t,s)f_D)).
\end{aligned} \tag{4.53}
$$

Here $\mathbb{B}_{D_2}$ denotes the Bogovskiĭ operator defined on the bounded domain D_2, and the term $\mathbb{B}_{D_2}((\nabla\varphi)\cdot(S_{\mathbb{R}^d}(t,s)f_R - S_D(t,s)f_D))$ is considered as a function on Ω by extending it outside of D_2 by zero.

By Proposition 1.87 it is clear that $W(t,s)f \in L^p_\sigma(\Omega)$ and moreover, we conclude that $W(t,s) \in \mathscr{L}(L^p_\sigma(\Omega))$. If $f \in Y_{\Omega,p}$, then it is easily seen that $W(t,s)f \in W^{2,p}(\Omega)^d \cap W^{1,p}_0(\Omega)^d \cap L^p_\sigma(\Omega)$ holds. Moreover, a short calculation yields

$$
\begin{aligned}
\nabla W(t,s)f = {}& \varphi \nabla S_{\mathbb{R}^d}(t,s)f_R + (1-\varphi)\nabla S_D(t,s)f_D + \nabla\varphi \cdot (S_{\mathbb{R}^d}(t,s)f_R - S_D(t,s)f_D) \\
& - \nabla \mathbb{B}_{D_2}((\nabla\varphi) \cdot (S_{\mathbb{R}^d}(t,s)f_R - S_D(t,s)f_D)).
\end{aligned} \tag{4.54}
$$

By using this relation we conclude that $W(t,s)f \in Y_{\Omega,p}$ if $f \in Y_{\Omega,p}$. At this point, let us note that $\varphi\xi = \varphi$ and $(1-\varphi)\eta = (1-\varphi)$. Thus, we see that

$$
W(s,s)f = f, \qquad s \geq 0. \tag{4.55}
$$

For notational simplicity, set $u_R(t) := S_{\mathbb{R}^d}(t,s)f_R$, $u_D(t) := S_D(t,s)f_D$, and let p_D be the pressure associated[6] to u_D. As in Section 4.2.2 we may assume $\mathrm{p}_D(t) \in L^p_0(D)$ for all (s,∞).

A short calculation yields that for a function $f \in Y_{\Omega,p}$ we have

$$
\begin{aligned}
\Delta W(t,s)f = {}& \varphi\Delta u_R + (1-\varphi)\Delta u_D + 2(\nabla\varphi) \cdot (\nabla u_R - \nabla u_D) + \Delta\varphi\,(u_R - u_D) \\
& - \Delta \mathbb{B}_{D_2}(\nabla\varphi(u_R - u_D)).
\end{aligned} \tag{4.56}
$$

By using (4.54), (4.55) and (4.56) we see that $u(t) := W(t,s)f$ solves the inhomogeneous problem

$$
\begin{cases}
u_t - \tilde{L}_\Omega(t)u + \nabla q & = -F(t,s)f & \text{in } (s,\infty) \times \Omega, \\
\operatorname{div} u & = 0 & \text{in } (s,\infty) \times \Omega, \\
u & = 0 & \text{on } (s,\infty) \times \partial\Omega, \\
u|_{t=0} & = f & \text{in } \Omega,
\end{cases} \tag{4.57}
$$

---

[6]This means $\nabla \mathrm{p}_D(t) = (\mathrm{Id} - \mathbb{P}_D)\tilde{L}_D(t)u_D(t)$.

where $q := (1 - \varphi) p_D$ and the inhomogeneous right-hand side $F(t,s)f$ is given by

$$
\begin{aligned}
F(t,s)f := {} & 2(\nabla\varphi) \cdot (\nabla u_R - \nabla u_D) + (\Delta\varphi + (M(t)x + c(t)) \cdot (\nabla\varphi)) (u_R - u_D) \\
& + \mathbb{B}_{D_2}((\nabla\varphi) \cdot (\partial_t u_R - \partial_t u_D)) - \tilde{L}_D(t) \mathbb{B}_{D_2}((\nabla\varphi) \cdot (u_R - u_D)) \\
& - (\nabla\varphi) p_D, \qquad\qquad\qquad\qquad (t,s) \in \Lambda, t \neq s.
\end{aligned}
\tag{4.58}
$$

For that calculation we also used the fact that $\operatorname{supp} b(t) \subset B_{R_0}$ for all $t \in (s, \infty)$.

Certainly, the function $F(t,s)f$ in (4.58) is even well-defined for every $f \in L^p_\sigma(\Omega)$ and $(t,s) \in \Lambda$, $t \neq s$. Later we need properties of $F(t,s)f$ stated in the next lemma.

Lemma 4.20. *Let $1 < p < \infty$. Then for every $(t,s) \in \Lambda$ with $t \neq s$ we have $F(t,s) \in \mathscr{L}(L^p(\Omega)^d)$. Moreover, the map*

$$
\{(t,s) \in \Lambda : t \neq s\} \ni (t,s) \mapsto F(t,s)
$$

is strongly continuous. Let $T > 0$ be arbitrary but fixed and let $\gamma \in (1 + \frac{1}{p}, 2)$. Then there exists a constant $C := C(p, \gamma, \Omega, T) > 0$ such that

$$
\|F(t,s)f\|_{L^p(\Omega)} \leq C(t-s)^{-\frac{\gamma}{2}} \|f\|_{L^p(\Omega)}, \qquad 0 \leq s < t \leq T, \ f \in L^p_\sigma(\Omega),
$$

$$
\|F(t,s)f\|_{W^{1,p}(\Omega)} \leq C(t-s)^{-\frac{\gamma}{2}} \|f\|_{W^{1,p}(\Omega)}, \qquad 0 \leq s < t \leq T, \ f \in W^{1,p}_0(\Omega)^d \cap L^p_\sigma(\Omega),
$$

$$
\|F(t,s)f\|_{\tilde{Y}_{\Omega,p}} \leq C(t-s)^{-\frac{\gamma}{2}} \|f\|_{\tilde{Y}_{\Omega,p}}, \qquad 0 \leq s < t \leq T, \ f \in \tilde{Y}_{\Omega,p}.
$$

Proof. Set

$$
\begin{aligned}
I_1 + I_2 + I_3 + I_4 + I_5 := {} & 2(\nabla\varphi) \cdot (\nabla u_R - \nabla u_D) + (\Delta\varphi + (M(t)x + c(t)) \cdot (\nabla\varphi)) (u_R - u_D) \\
& + \mathbb{B}_{D_2}((\nabla\varphi) \cdot (\partial_t u_R - \partial_t u_D)) - \tilde{L}_D(t) \mathbb{B}_{D_2}((\nabla\varphi) \cdot (u_R - u_D)) \\
& - (\nabla\varphi) p_D.
\end{aligned}
$$

Let $T > 0$ and suppose that $(t,s) \in \Lambda_T$ with $t \neq s$. In the following all appearing constants may depend on p, γ, Ω and $T > 0$. Let us start with the norm estimates for I_1 and I_2. By using Theorem 4.9 and Proposition 4.15 we obtain

$$
\|I_1\|_{L^p(\Omega)} \leq C(t-s)^{-\frac{1}{2}} \|f\|_{L^p(\Omega)}, \qquad \|I_2\|_{L^p(\Omega)} \leq C\|f\|_{L^p(\Omega)}
$$

for $f \in L^p_\sigma(\Omega)$ and

$$
\|I_1\|_{W^{1,p}(\Omega)} \leq C(t-s)^{-\frac{1}{2}} \|f\|_{W^{1,p}(\Omega)}, \qquad \|I_2\|_{W^{1,p}(\Omega)} \leq C\|f\|_{W^{1,p}(\Omega)}
$$

for $f \in W^{1,p}_0(\Omega)^d \cap L^p_\sigma(\Omega)$. To estimate the norm of I_4 we first note that Proposition 1.87 implies that $\tilde{L}_D(t)\mathbb{B}_{D_2} \in \mathscr{L}(W^{1,p}_0(D_2), L^p(D_2)^d)$ and $\tilde{L}_D(t)\mathbb{B}_{D_2} \in \mathscr{L}(W^{2,p}_0(D_2), W^{1,p}_0(D_2)^d)$. Thus, again by using Theorem 4.9 and Proposition 4.15, we obtain

$$
\|I_4\|_{L^p(\Omega)} \leq C(t-s)^{-\frac{1}{2}} \|f\|_{L^p(\Omega)}
$$

for $f \in L^p_\sigma(\Omega)$ and

$$
\|I_4\|_{W^{1,p}(\Omega)} \leq C(t-s)^{-\frac{1}{2}} \|f\|_{W^{1,p}(\Omega)}
$$

for $f \in W_0^{1,p}(\Omega)^d \cap L_\sigma^p(\Omega)$. Next we consider the term I_3. We can write

$$\mathbb{B}_{D_2}((\nabla\varphi) \cdot (\partial_t u_R(t) - \partial_t u_D(t)))$$
$$= \mathbb{B}_{D_2}((\nabla\varphi) \cdot (\tilde{L}_{\mathbb{R}^d}(t)u_R(t))) - \mathbb{B}_{D_2}((\nabla\varphi) \cdot (\tilde{L}_D(t)u_D(t) - \nabla p_D)).$$

Now for a test function $\psi \in C_c^\infty(\mathbb{R}^d)$ we have

$$|\langle (\nabla\varphi) \cdot \tilde{L}_{\mathbb{R}^d}(t)u_R(t), \psi \rangle_{\mathbb{R}^d}| = |\langle \tilde{L}_{\mathbb{R}^d}(t)u_R(t), \psi(\nabla\varphi) \rangle_{\mathbb{R}^d}|$$
$$\leq |\langle \nabla u_R(t), \nabla (\psi(\nabla\varphi)) \rangle_{\mathbb{R}^d}| + |\langle (M(t)x + c(t)) \cdot \nabla u_R(t) - M(t)u_R(t), \psi(\nabla\varphi) \rangle_{\mathbb{R}^d}|$$
$$\leq C\|u_R(t)\|_{W^{1,p}(\mathbb{R}^d)}\|\psi\|_{W^{1,p'}(\mathbb{R}^d)} + C\|u_R(t)\|_{W^{1,p}(D_2)}\|\psi\|_{L^{p'}(D_2)}$$
$$\leq C\|u_R(t)\|_{W^{1,p}(\mathbb{R}^d)}\|\psi\|_{W^{1,p'}(\mathbb{R}^d)},$$

where $\frac{1}{p} + \frac{1}{p'} = 1$. This shows that

$$\|(\nabla\varphi) \cdot \tilde{L}_{\mathbb{R}^d}(t)u_R(t)\|_{W_0^{-1,p}(\mathbb{R}^d)} \leq C\|u_R(t)\|_{W^{1,p}(\mathbb{R}^d)}$$

holds. Analogously, since supp $\nabla\varphi \subset D_2$, we obtain

$$\|(\nabla\varphi) \cdot \tilde{L}_D(t)u_D(t)\|_{W_0^{-1,p}(D)} \leq C\left(\|u_D(t)\|_{W^{1,p}(D)} + \|p_D\|_{L^p(D)}\right).$$

By Proposition 1.87 we know that $\mathbb{B}_{D_2} \in \mathscr{L}(W_0^{-1,p}(D_2), L^p(D_2)^d)$. Thus, using this together with Theorem 4.9, Proposition 4.15 and Proposition 4.16 yields

$$\|I_3\|_{L^p(\Omega)} \leq C(t-s)^{-\frac{\gamma}{2}}\|f\|_{L^p(\Omega)}$$

for $f \in L_\sigma^p(\Omega)$. Moreover, by Proposition 1.87 we also know that $\mathbb{B}_{D_2} \in \mathscr{L}(L^p(D_2), W_0^{1,p}(D_2)^d)$. So again by Theorem 4.9, Proposition 4.15 and Proposition 4.16 we see that

$$\|I_3\|_{W^{1,p}(\Omega)} \leq C(t-s)^{-\frac{1}{2}}\|f\|_{W^{1,p}(\Omega)}$$

for $f \in W_0^{1,p}(\Omega)^d \cap L_\sigma^p(\Omega)$. Finally, by Proposition 4.16 we obtain that

$$\|I_5\|_{L^p(\Omega)} \leq C(t-s)^{-\frac{\gamma}{2}}\|f\|_{L^p(\Omega)}$$

for $f \in L_\sigma^p(\Omega)$ and

$$\|I_5\|_{W^{1,p}(\Omega)} \leq C(t-s)^{-\frac{\gamma}{2}}\|f\|_{W^{1,p}(\Omega)}$$

for $f \in W_0^{1,p}(\Omega)^d \cap L_\sigma^p(\Omega)$. This proves the first two estimates in Lemma 4.20, and it shows in particular that $F(t,s) \in \mathscr{L}(L^p(\Omega)^d)$ for $(t,s) \in \Lambda$ with $t \neq s$. The third estimate in Lemma 4.20 follows directly from the second, since supp $F(t,s) \subset D$ for all $(t,s) \in \Lambda$ with $t \neq s$.

The strong continuity of the map $\{(t,s) \in \Lambda : t \neq s\} \ni (t,s) \mapsto F(t,s)$ follows basically from the strong continuity of $\Lambda \ni (t,s) \mapsto S_{\mathbb{R}^d}(t,s)$ and $\Lambda \ni (t,s) \mapsto S_D(t,s)$. Moreover, note that $\nabla p_D(t;s) = (\text{Id} - \mathbb{P}_D)\tilde{L}_D(t)S_D(t,s)f_D$. Hence, by Poincaré's inequality we conclude that also $\{(t,s) \in \Lambda : t \neq s\} \ni (t,s) \mapsto p_D(t;s)$ is continuous. Furthermore, we also apply the properties of the operator $\mathbb{B}_{D_2}$ stated in Proposition 1.87 and the strong continuity follows. This concludes the proof. $\qquad\square$

By applying the Helmholtz projection $\mathbb{P}_\Omega$ to (4.57) we have

$$
\begin{cases}
u'(t) - L_\Omega(t)u(t) &= -\mathbb{P}_\Omega F(t,s)f \qquad t > s \geq 0, \\
u(s) &= f.
\end{cases}
\tag{4.59}
$$

It is clear that if an evolution system $\{S_\Omega(t,s)\}_{(t,s)\in\Lambda}$ exists on $L^p_\sigma(\Omega)$, then the solution $u(t)$ to the inhomogeneous equation (4.59) is given by the variation of constants formula

$$
u(t) = W(t,s)f = S_\Omega(t,s)f - \int_s^t S_\Omega(t,r)\mathbb{P}_\Omega F(r,s)f\, dr.
\tag{4.60}
$$

Based on this consideration, the idea is now to solve the integral equation

$$
S_\Omega(t,s)f = W(t,s)f + \int_s^t S_\Omega(t,r)\mathbb{P}_\Omega F(r,s)f\, dr \qquad (t,s) \in \Lambda, f \in L^p_\sigma(\Omega),
\tag{4.61}
$$

and then to show that the family of solution operators $\{S_\Omega(t,s)\}_{(t,s)\in\Lambda}$ is indeed the unique evolution system which solves (4.50). For this we need the following lemma on iterated convolutions (see also [GHH06a, Lemma 4.6] for a similar lemma in the one-parameter case).

Lemma 4.21. *Let X_1 and X_2 be two Banach spaces, $T > 0$ arbitrary but fixed and let*

$$
U : \{(t,s) \in \Lambda_T : t \neq s\} \rightarrow \mathscr{L}(X_2, X_1) \qquad and \qquad V : \{(t,s) \in \Lambda_T : t \neq s\} \rightarrow \mathscr{L}(X_2)
$$

be strongly continuous functions. Assume that

$$
\|U(t,s)\|_{\mathscr{L}(X_2,X_1)} \leq C_0(t-s)^\alpha, \qquad \|V(t,s)\|_{\mathscr{L}(X_2)} \leq C_0(t-s)^\beta, \quad 0 \leq s < t \leq T,
$$

hold for some constant $C_0 := C_0(T) > 0$ and for $\alpha, \beta > -1$. For $f \in X_2$ and $(t,s) \in \Lambda_T$ with $t \neq s$ set $S_0(t,s)f := U(t,s)f$ and

$$
S_n(t,s)f := \int_s^t S_{n-1}(t,r)V(r,s)f\, ds, \qquad n \in \mathbb{N}.
$$

Then there exists a constant $C := C(\alpha, \beta, T) > 0$ such that

$$
\sum_{n=0}^\infty \|S_n(t,s)f\|_{X_1} \leq C(t-s)^\alpha\|f\|_{X_2}, \qquad 0 \leq s < t \leq T.
$$

Moreover, if $\alpha \geq 0$ and $U : \Lambda_T \rightarrow \mathscr{L}(X_2, X_1)$ is strongly continuous, then the convergence of the series above is uniform on Λ_T.

Proof. For $f \in X_2$ and $(t,s) \in \Lambda_T$ with $t \neq s$ we have

$$
\|S_1(t,s)f\|_{X_1} \leq C_0^2 \int_s^t (t-r)^\alpha(r-s)^\beta\, dr\|f\|_{X_2} = C_0^2(t-s)^{\alpha+\beta+1}\mathrm{B}(\beta+1,\alpha+1)\|f\|_{X_2},
$$

where $\mathrm{B}(\cdot,\cdot)$ denotes the Beta function. By induction on $n \in \mathbb{N}$ we obtain

$$\|S_n(t,s)f\|_{X_1}$$
$$\leq C_0^{n+1}(t-s)^{\alpha+n(\beta+1)}\mathrm{B}(\beta+1,\alpha+1)\cdots\mathrm{B}(\beta+1,\alpha+1+(n-1)(\beta+1))\|f\|_{X_2}$$
$$= C_0^{n+1}(t-s)^{\alpha+n(\beta+1)}\Gamma(\beta+1)^n\frac{\Gamma(\alpha+1)}{\Gamma(\alpha+1+n(\beta+1))}\|f\|_{X_2}$$

for $n \in \mathbb{N}$ and $(t,s) \in \Lambda_T$ with $t \neq s$. Here $\Gamma(\cdot)$ denotes the Gamma function. By using the identity $\Gamma(x+1) = x\Gamma(x)$ for $x > -1$ we obtain

$$\frac{\Gamma(\alpha+1)}{\Gamma(\alpha+1+n(\beta+1))} \leq \frac{C}{[n(\beta+1)]!}, \qquad n \in \mathbb{N}$$

for some $C := C(\alpha) > 0$. Here $[\cdot]$ denote the Gaussian brackets. Hence,

$$\|S_n(t,s)f\|_{X_1} \leq CC_0(t-s)^{\alpha}\Gamma(\beta+1)^n C_0^n\frac{(t-s)^{n(\beta+1)}}{[n(\beta+1)]!}\|f\|_{X_2}$$
$$\leq CC_0(t-s)^{\alpha}e^{t-s}(C_0\Gamma(\beta+1))^n\frac{(t-s)^{[n(\beta+1)]}}{[n(\beta+1)]!}\|f\|_{X_2}$$

for $n \in \mathbb{N}$ and $(t,s) \in \Lambda_T$ with $t \neq s$. Since

$$\sum_{n=0}^{\infty}(C_0\Gamma(\beta+1))^n\frac{(t-s)^{[n(\beta+1)]}}{[n(\beta+1)]!} \leq Ce^{c(t-s)} \leq Ce^{cT}, \qquad 0 \leq s < t \leq T$$

for constants $C := C(\beta) > 0, c := c(\beta) > 0$, we conclude that there exists a constant $C := C(\alpha,\beta,T) > 0$ such that

$$\sum_{n=0}^{\infty}\|T_n(t,s)f\|_{X_1} \leq C(t-s)^{\alpha}\|f\|_{X_2}, \qquad 0 \leq s < t \leq T.$$

It is clear that if $\alpha \geq 0$ and $U : \Lambda_T \to \mathscr{L}(X_2, X_1)$ is strongly continuous, then the convergence of the above series is uniform on Λ_T. $\qquad\square$

Now we are in position to give the proof of Theorem 4.19.

Proof of Theorem 4.19

Let $T > 0$ and let $\gamma \in (1+\frac{1}{p}, 2)$. In the following, all constants may depend on p, γ, Ω and $T > 0$. By using Theorem 4.9, Proposition 4.15 and since $\mathrm{B}_{D_2} \in \mathscr{L}(L^p(D_2), W_0^{1,p}(D_2)^d)$, we have

$$\|W(t,s)f\|_{L^p(\Omega)} \leq C\|f\|_{L^p(\Omega)}, \qquad f \in L_\sigma^p(\Omega), \quad 0 \leq s \leq t \leq T.$$

Set $S_0(t,s)f := W(t,s)f$ and

$$S_n(t,s)f := \int_s^t S_{n-1}(t,r)\mathbb{P}_\Omega F(r,s)f\,\mathrm{d}r, \qquad n \in \mathbb{N}.$$

The first estimate in Lemma 4.20 implies that we can apply Lemma 4.21 with $U = W$,

$V = \mathbb{P}_\Omega F$, $\alpha = 0$, $\beta = -\frac{\gamma}{2}$ and $X_1 = X_2 = L^p_\sigma(\Omega)$. Thus, for any $f \in L^p_\sigma(\Omega)$ the series

$$\sum_{n=0}^\infty S_n(t, s) f$$

converges uniformly in Λ_T. Since $T > 0$ is arbitrary, we conclude that

$$S_\Omega(t, s) := \sum_{n=0}^\infty S_n(t, s) \tag{4.62}$$

is even well-defined for all $(t, s) \in \Lambda$ and $S_\Omega(t, s) \in \mathscr{L}(L^p_\sigma(\Omega))$ for $(t, s) \in \Lambda$. It is easy to see that $S_\Omega(t, s)$ satisfies the integral equation (4.61). Moreover, from the strong continuity of $\Lambda \ni (t, s) \mapsto W(t, s)$ and from Lemma 4.20 we deduce inductively that $\Lambda \ni (t, s) \mapsto S_k(t, s)$ is strongly continuous and hence, by the uniform convergence of the series we get the strong continuity of $\Lambda \ni (t, s) \mapsto S_\Omega(t, s)$. In the following we show that the two-parameter family of linear operators defined in (4.62) is indeed the unique strongly continuous evolution system which solves the non-autonomous abstract Cauchy problem (4.50). For that we proceed in four steps:

Step 1: Invariance of the regularity space. Theorem 4.9, Proposition 4.15 and Lemma 4.20 allow us to apply Lemma 4.21 with $X_1 = X_2 = \widetilde{Y}_{\Omega,p}$, $U = W$, $V = \mathbb{P}_\Omega F$, $\alpha = 0$ and $\beta = -\frac{\gamma}{2}$. Thus we obtain that $S_\Omega(t, s)f \in \widetilde{Y}_{\Omega,p}$ for all $f \in \widetilde{Y}_{\Omega,p}$ and all $(t, s) \in \Lambda$. Moreover, we apply Lemma 4.21 with $X_1 = W^{2,p}(\Omega)^d \cap W^{1,p}_0(\Omega)^d \cap L^p_\sigma(\Omega)$, $X_2 = W^{1,p}_0(\Omega)^d \cap L^p_\sigma(\Omega)$, $U = W$, $V = \mathbb{P}_\Omega F$, $\alpha = -\frac{1}{2}$ and $\beta = -\frac{\gamma}{2}$. Thus, we see that $S_\Omega(t, s)f \in W^{2,p}(\Omega)^d$ for $f \subset W^{2,p}(\Omega)^d \cap W^{1,p}_0(\Omega)^d \cap L^p_\sigma(\Omega)$ and $(t, s) \in \Lambda$. Furthermore, we conclude that

$$\sum_{n=0}^\infty \|S_n(t, s)f\|_{Y_{\Omega,p}} \le C(t - s)^{-\frac{1}{2}} \|f\|_{Y_{\Omega,p}}, \qquad f \in Y_{\Omega,p}, \quad 0 \le s < t \le T \tag{4.63}$$

for a constant $C > 0$. In particular, this shows that the two-parameter family of linear operators defined in (4.62) fulfills property (a) in Theorem 4.19.

Step 2: Differentiability with respect to t. Let us now prove that for every $f \in Y_{\Omega,p}$ and for every $s \ge 0$ fixed, the map $t \mapsto S_\Omega(t, s)f$ is differentiable on (s, ∞). First we may assume that $0 < t - s < 1$. For $f \in Y_{\Omega,p}$ we compute

$$\partial_t S_0(t, s)f = L_\Omega(t) S_0(t, s)f - \mathbb{P}_\Omega F(t, s)f$$

$$\partial_t S_1(t, s)f = L_\Omega(t) S_1(t, s)f + \mathbb{P}_\Omega F(t, s)f - \int_s^t \mathbb{P}_\Omega F(t, r) \mathbb{P}_\Omega F(r, s)f \, dr$$

$$\partial_t S_2(t, s)f = L_\Omega(t) S_2(t, s)f + \int_s^t \mathbb{P}_\Omega F(t, r) \mathbb{P}_\Omega F(r, s)f \, dr$$

$$- \int_s^t \int_{r_1}^t \mathbb{P}_\Omega F(t, r_2) \mathbb{P}_\Omega F(r_2, r_1) \mathbb{P}_\Omega F(r_1, s)f \, dr_2 \, dr_1.$$

Note that we have used here the fact that the operator $L_\Omega(t)$ is closed.

Inductively we see that

$$\partial_t \sum_{n=0}^{N} S_n(t,s)f = L_{\Omega}(t) \sum_{n=0}^{N} S_n(t,s)f - R_N(t,s)f \tag{4.64}$$

holds for $N \in \mathbb{N}$, where

$$R_N(t,s)f := \int_s^t \int_{r_1}^t \cdots \int_{r_{N-1}}^t \mathbb{P}_{\Omega}F(t,r_N)\mathbb{P}_{\Omega}F(r_N,r_{N-1})\ldots\mathbb{P}_{\Omega}F(r_1,s)f \; dr_N \ldots dr_2 \, dr_1.$$

Now we estimate the norm of the remainder terms $R_N(t,s)f$. By Lemma 4.20 we obtain

$$\|R_1(t,s)f\|_{L^p(\Omega)} \le C^2 \int_s^t (t-r)^{-\frac{\gamma}{2}}(r-s)^{-\frac{\gamma}{2}} \, dr \|f\|_{L^p(\Omega)}$$
$$= C^2 \mathrm{B}(1-\gamma/2, 1-\gamma/2)(t-s)^{1-\gamma}\|f\|_{L^p(\Omega)},$$

$$\|R_2(t,s)f\|_{L^p(\Omega)} \le C^3 \mathrm{B}(1-\gamma/2, 1-\gamma/2) \int_s^t (t-r)^{1-\gamma}(r-s)^{-\frac{\gamma}{2}} \, dr \|f\|_{L^p(\Omega)}$$
$$= C^3 \mathrm{B}(1-\gamma/2, 1-\gamma/2)\mathrm{B}(1-\gamma/2, 2-\gamma)(t-s)^{2-\frac{3\gamma}{2}}\|f\|_{L^p(\Omega)}.$$

Inductively we see that

$$\|R_N(t,s)f\|_{L^p(\Omega)} \le C^{N+1}\mathrm{B}(1-\gamma/2, 1-\gamma/2)\mathrm{B}(1-\gamma/2, 2-\gamma)\ldots$$
$$\ldots \mathrm{B}(1-\gamma/2, N-(N\gamma)/2)(t-s)^{N-\frac{(N+1)\gamma}{2}}\|f\|_{L^p(\Omega)}$$
$$\le \frac{C^{N+1}\Gamma(1-\gamma/2)^N}{[N-\frac{(N+1)\gamma}{2}]!}(t-s)^{N-\frac{(N+1)\gamma}{2}}\|f\|_{L^p(\Omega)}, \qquad 0 < t-s < 1 \tag{4.65}$$

holds for $N \in \mathbb{N}$. From estimate (4.65) it follows that $\|R_N(t,s)\|_{L^p(\Omega)}$ tends to zero as $N \to \infty$. Using the estimate (4.63) and the closedness of the operator $L_{\Omega}(t)$ implies that

$$\partial_t \sum_{n=0}^{\infty} S_n(t,s)f = L_{\Omega}(t) \sum_{n=0}^{\infty} S_n(t,s)f$$

for $t \in (s, s+1)$. The general case follows from the fact that $S_{\Omega}(t,s) = S_{\Omega}(t,r)S_{\Omega}(r,s)$ with $0 \le s < r < t$, $0 < t-r < 1$ and the fact that $S_{\Omega}(r,s)$ leaves $Y_{\Omega,p}$ invariant. This shows that the two-parameter family of linear operators defined in (4.62) fulfills property (b).

Step 3: Differentiability with respect to s. In the next step we show the differentiability of the map $s \mapsto S_{\Omega}(t,s)f$ on $[0,t)$ for $t > 0$ and $f \in Y_{\Omega,p}$. By a short calculation and by using (4.51) and (4.52) we see that for $f \in Y_{\mathbb{R}^d, p}$ we have

$$\tilde{L}_{\mathbb{R}^d}(s)f_R = (\tilde{L}_{\Omega}(s)f)_R + 2(\nabla\xi) \cdot \nabla f + (\Delta\xi + (M(s)x + c(s)) \cdot \nabla\xi)f$$
$$- \tilde{L}_{\mathbb{R}^d}(s)\mathbb{B}_{D_1}((\nabla\xi) \cdot f) + \mathbb{B}_{D_1}((\nabla\xi) \cdot \tilde{L}_{\Omega}(s)f) \tag{4.66}$$

and

$$\tilde{L}_D(s)f_D = (\tilde{L}_{\Omega}(s)f)_D + 2(\nabla\eta) \cdot \nabla f + (\Delta\eta + (M(s)x + c(s)) \cdot \nabla\eta)f$$
$$- \tilde{L}_D(s)\mathbb{B}_{D_3}((\nabla\eta) \cdot f) + \mathbb{B}_{D_3}((\nabla\eta) \cdot \tilde{L}_{\Omega}(s)f). \tag{4.67}$$

Let $\tilde{p}(s) \in \widehat{W}^{1,p}(\Omega)$ be such that $\nabla\tilde{p}(s) = (\mathrm{Id} - \mathbb{P}_\Omega)\tilde{L}_\Omega(s)f$. Similarly to Section 4.2.2, we may also assume here that $p(s) \in L_0^p(D)$. Then by using again (4.51) we conclude that

$$
\begin{aligned}
\mathbb{P}_{\mathbb{R}^d}(\tilde{L}_\Omega(s)f)_R &= \mathbb{P}_{\mathbb{R}^d}\left[\xi L_\Omega(s)f - \mathbb{B}_{D_1}((\nabla\xi)\cdot L_\Omega(s)f) + \xi\nabla\tilde{p}(s) - \mathbb{B}_{D_1}((\nabla\xi)\cdot\nabla\tilde{p}(s))\right] \\
&= (L_\Omega f)_R - \mathbb{P}_{\mathbb{R}^d}[\tilde{p}(s)\nabla\xi + \mathbb{B}_{D_1}((\nabla\xi)\cdot\nabla\tilde{p}(s))].
\end{aligned}
$$
(4.68)

We used here that $\mathbb{P}_{\mathbb{R}^d}(\xi\nabla\tilde{p}(s)) = -\mathbb{P}_{\mathbb{R}^d}(\tilde{p}(s)\nabla\xi)$ which follows easily from the relation $\xi\nabla\tilde{p}(s) = \nabla(\xi\tilde{p}(s)) - \tilde{p}(s)\nabla\xi$. Analogously, by using (4.52) we see that

$$
\mathbb{P}_D(\tilde{L}_\Omega(s)f)_D = (L_\Omega f)_D - \mathbb{P}_D[\tilde{p}(s)\nabla\eta + \mathbb{B}_{D_3}((\nabla\eta)\cdot\nabla\tilde{p}(s))].
$$
(4.69)

Thus, by using (4.66) – (4.69), Theorem 4.6 and Proposition 4.14, we obtain

$$
\begin{aligned}
\partial_s W(t,s)f &= -\varphi S_{\mathbb{R}^d}(t,s)L_{\mathbb{R}^d}(s)f_R - (1-\varphi)S_D(t,s)L_D(s)f_D \\
&\quad + \mathbb{B}_{D_2}((\nabla\varphi)\cdot(S_{\mathbb{R}^d}(t,s)L_{\mathbb{R}^d}(s)f_R - S_D(t,s)L_D(s)f_D)) \\
&= -W(t,s)L_\Omega(s)f - G(t,s)f
\end{aligned}
$$
(4.70)

for $f \in Y_{\Omega,p}$, where

$$
\begin{aligned}
G(t,s)f &:= \varphi S_{\mathbb{R}^d}(t,s)\mathbb{P}_{\mathbb{R}^d}\Big(2(\nabla\xi)\cdot\nabla f + (\Delta\xi + (M(s)x + c(s))\cdot\nabla\xi)f \\
&\quad - \tilde{L}_{\mathbb{R}^d}(s)\mathbb{B}_{D_1}((\nabla\xi)\cdot f) + \mathbb{B}_{D_1}((\nabla\xi)\cdot\tilde{L}_\Omega(s)f) - \tilde{p}(s)\nabla\xi - \mathbb{B}_{D_1}((\nabla\xi)\cdot\nabla\tilde{p}(s))\Big) \\
&\quad + (1-\varphi)S_D(t,s)\mathbb{P}_D\Big(2(\nabla\eta)\cdot\nabla f + (\Delta\eta + (M(s)x + c(s))\cdot\nabla\eta)f \\
&\quad - \tilde{L}_D(s)\mathbb{B}_{D_3}((\nabla\eta)\cdot f) + \mathbb{B}_{D_3}((\nabla\eta)\cdot L_\Omega(s)f) - \tilde{p}(s)\nabla\eta - \mathbb{B}_{D_3}((\nabla\eta)\cdot\nabla\tilde{p}(s))\Big) \\
&\quad - \mathbb{B}_{D_2}\Big((\nabla\varphi)\cdot S_{\mathbb{R}^d}(t,s)\mathbb{P}_{\mathbb{R}^d}\big(2(\nabla\xi)\cdot\nabla f + (\Delta\xi + (M(s)x + c(s))\cdot\nabla\xi)f \\
&\quad - \tilde{L}_{\mathbb{R}^d}(s)\mathbb{B}_{D_1}((\nabla\xi)\cdot f) + \mathbb{B}_{D_1}((\nabla\xi)\cdot\tilde{L}_\Omega(s)f) - \tilde{p}(s)\nabla\xi - \mathbb{B}_{D_1}((\nabla\xi)\cdot\nabla\tilde{p}(s))\big)\Big) \\
&\quad + \mathbb{B}_{D_2}\Big((\nabla\varphi)\cdot S_D(t,s)\mathbb{P}_D\big(2(\nabla\eta)\cdot\nabla f + (\Delta\eta + (M(s)x + c(s))\cdot\nabla\eta)f \\
&\quad - \tilde{L}_D(s)\mathbb{B}_{D_3}((\nabla\eta)\cdot f) + \mathbb{B}_{D_3}((\nabla\eta)\cdot\tilde{L}_\Omega(s)f) - \tilde{p}(s)\nabla\eta - \mathbb{B}_{D_3}((\nabla\eta)\cdot\nabla\tilde{p}(s))\big)\Big) \\
&=: J_1 + J_2 + J_3 + J_4.
\end{aligned}
$$

Similarly to Lemma 4.20, we need to estimate the norm of the remainder term $G(t,s)f$. At first we estimate the norm of $\tilde{p}$ in $L^p(D)^d$. The strategy is similar to the one used in the proof of Proposition 4.16. Let $\varphi \in C_c^\infty(D)$ and set $\tilde{\varphi} := \varphi - |D|^{-1}\int_D \varphi\,\mathrm{d}x$. Now we extend $\tilde{\varphi}$ to Ω by zero and we still denote the extension by $\tilde{\varphi}$. By this construction we have $\tilde{\varphi} \in L^{p'}(\Omega)$, where $\frac{1}{p} + \frac{1}{p'} = 1$, $\tilde{\varphi}(x) = 0$ for $x \in \Omega \setminus D$ and $\int_\Omega \tilde{\varphi}\,\mathrm{d}x = 0$. There exists a unique solution ψ to the classical Neumann problem

$$
\begin{cases}
\Delta\psi &= \tilde{\varphi} \quad \text{in } \Omega, \\
\nu\cdot\nabla\psi &= 0 \quad \text{on } \partial\Omega,
\end{cases}
$$
(4.71)

which satisfies the estimate

$$
\|\psi\|_{L^{p'}(D)} + \|\nabla\psi\|_{W^{1,p'}(\Omega)} \leq C\|\tilde{\varphi}\|_{L^{p'}(\Omega)} \leq 2C\|\varphi\|_{L^{p'}(D)}
$$

for some constant $C > 0$, see [SS07a, Proposition 5.6]. Then we have

$$
\begin{aligned}
\langle \tilde{\mathrm{p}}(s), \varphi \rangle_D &= \langle \tilde{\mathrm{p}}(s), \tilde{\varphi} \rangle_D = \langle \tilde{\mathrm{p}}(s), \Delta \psi \rangle_\Omega = -\langle \nabla \tilde{\mathrm{p}}(s), \nabla \psi \rangle_\Omega = -\langle (\mathrm{Id} - \mathbb{P}_\Omega) \tilde{L}_\Omega(s) f, \nabla \psi \rangle_\Omega \\
&= -\langle \Delta f, \nabla \psi \rangle_\Omega - \langle (M(s)x + c(s)) \cdot \nabla f - M(s)f, \nabla \psi \rangle_\Omega \\
&\quad + \langle b(s) \cdot \nabla f, \nabla \psi \rangle_\Omega + \langle f \cdot \nabla b(s), \nabla \psi \rangle_\Omega \\
&= -\langle \nu \cdot \nabla f, \nabla \psi \rangle_{\partial \Omega} + \langle \nabla f, \nabla^2 \psi \rangle_\Omega + \langle \mathrm{div}\,((M(s)x + c(s)) \cdot \nabla f - M(s)f), \psi \rangle_\Omega \\
&\quad - \langle \nu \cdot ((M(s)x + c(s)) \cdot \nabla f - M(s)f), \psi \rangle_{\partial \Omega} + \langle b(s) \cdot \nabla f, \nabla \psi \rangle_\Omega \\
&\quad + \langle f \cdot \nabla b(s), \nabla \psi \rangle_\Omega \\
&= -\langle \nu \cdot \nabla f, \nabla \psi \rangle_{\partial D} + \langle \nabla f, \nabla^2 \psi \rangle_\Omega - \langle \nu \cdot ((M(s)x + c(s)) \cdot \nabla f - M(s)f), \psi \rangle_{\partial D} \\
&\quad + \langle b(s) \cdot \nabla f, \nabla \psi \rangle_\Omega + \langle f \cdot \nabla b(s), \nabla \psi \rangle_\Omega.
\end{aligned}
$$

Here we used that $\partial \Omega \subset \partial D$ and that $\mathrm{div}\,((M(s)x + c(s)) \cdot \nabla f - M(s)f) = 0$, see the calculations in the proof of Lemma 1.86. By using the embedding $W_p^{1/p+\varepsilon}(D) \hookrightarrow L^p(\partial D)$ for $0 < \varepsilon \le 1 - \frac{1}{p}$, see Proposition 1.10, we obtain

$$
\begin{aligned}
|\langle \tilde{\mathrm{p}}(s), \varphi \rangle_D| &\le C \left(\|\nu \cdot \nabla f\|_{L^p(\partial D)} + \|\nabla f\|_{L^p(\Omega)} + \|f\|_{L^p(\Omega)} \right) \|\nabla \psi\|_{W^{1,p'}(\Omega)} \\
&\quad + C \left(\|\nu \cdot \nabla f\|_{L^p(\partial D)} + \|f\|_{L^p(\partial D)} \right) \|\psi\|_{W^{1,p'}(D)} \\
&\le C \|f\|_{W_p^{1+1/p+\varepsilon}(\Omega)} \|\varphi\|_{L^{p'}(\Omega)}.
\end{aligned}
$$

Now we take $\varepsilon < 1 - \frac{1}{p}$ and set $\gamma = 1 + \frac{1}{p} + \varepsilon$. Then we conclude that

$$
\|\tilde{\mathrm{p}}(s)\|_{L^p(D)} \le C \|f\|_{W_p^\gamma(\Omega)} \tag{4.72}
$$

holds for some $\gamma \in (1 + \frac{1}{p}, 2)$.

Fix $\gamma \in (1 + \frac{1}{p}, 2)$ from estimate (4.72). Moreover, let $T > 0$ and $(t, s) \in \Lambda_T$ with $t \ne s$. We shall now use arguments very similar to the ones used in the proof of Lemma 4.20. Note that Proposition 1.87 yields $\tilde{L}_{\mathbb{R}^d}(s) \mathbb{B}_{D_1} \in \mathscr{L}(W_0^{1,p}(D_1), L^p(D_1)^d)$ and $\mathbb{B}_{D_1} \in \mathscr{L}(W_0^{-1,p}(D_1), L^p(D_1)^d)$. Moreover, recall that $W_p^\gamma(\Omega)^d$ is the real interpolation space between $L^p(\Omega)^d$ and $W^{2,p}(\Omega)^d$ and therefore in the following we shall always apply the interpolation estimate from Remark 1.8. Thus, by applying Theorem 4.9 and estimate (4.72) we conclude that

$$
\|J_1\|_{W_p^\gamma(\Omega)} \le C(t-s)^{-\frac{\gamma}{2}} \|f\|_{W_p^\gamma(\Omega)}.
$$

Similarly, Proposition 4.15 yields

$$
\|J_2\|_{W_p^\gamma(\Omega)} \le C(t-s)^{-\frac{\gamma}{2}} \|f\|_{W_p^\gamma(\Omega)}.
$$

Moreover, by using that $\mathbb{B}_{D_2} \in \mathscr{L}(W_0^{1,p}(D_2), W_0^{2,p}(D_2)^d)$ and by applying Theorem 4.9 and Proposition 4.15 we obtain

$$
\|J_3\|_{W_p^\gamma(\Omega)} + \|J_4\|_{W_p^\gamma(\Omega)} \le C(t-s)^{-\frac{\gamma}{2}} \|f\|_{W_p^\gamma(\Omega)}.
$$

Summing up, we have

$$
\|G(t,s)f\|_{W_p^\gamma(\Omega)} \le C(t-s)^{-\frac{\gamma}{2}} \|f\|_{W_p^\gamma(\Omega)} \tag{4.73}
$$

for some constant $C > 0$ and $f \in Y_{\Omega,p}$. Density arguments imply that (4.73) even holds for all $f \in W_p^\gamma(\Omega)^d \cap L_\sigma^p(\Omega)$ and that $G(t,s) \in \mathcal{L}(W_p^\gamma(\Omega)^d \cap L_\sigma^p(\Omega))$. It is easy to see that $\{(t,s) \in \Lambda : t \neq s\} \ni (t,s) \mapsto G(t,s)$ is strongly continuous. For $f \in W_p^\gamma(\Omega)^d \cap L_\sigma^p(\Omega)$ and $(t,s) \in \Lambda_T$ with $t \neq s$, set $Z_0(t,s)f = G(t,s)f$ and

$$Z_{n+1}(t,s)f = -\int_s^t Z_n(t,r)G(r,s)f, \qquad n \in \mathbb{N}.$$

Since $T > 0$ is arbitrary, applying Lemma 4.21 with $X_1 = X_2 = W_p^\gamma(\Omega)^d \cap L_\sigma^p(\Omega)$, $U = V = G$ and $\alpha = \beta = -\frac{\gamma}{2}$ yields that the series

$$Z(t,s)f := \sum_{n=0}^\infty Z_n(t,s)f, \qquad (t,s) \in \Lambda, t \neq s,$$

is well-defined.

Moreover for fixed $T > 0$ and $f \in W_p^\gamma(\Omega)^d \cap L_\sigma^p(\Omega)$ we have

$$\|Z(t,s)f\|_{W_p^\gamma(\Omega)} \leq C(t-s)^{-\frac{\gamma}{2}} \|f\|_{W_p^\gamma(\Omega)}, \quad (t,s) \in \Lambda_T, t \neq s. \tag{4.74}$$

It is easy to see that $Z(\cdot,\cdot)$ satisfies the integral equation

$$Z(t,s)f = G(t,s)f - \int_s^t Z(t,r)G(r,s)f \, \mathrm{d}r, \quad (t,s) \in \Lambda, t \neq s, \ f \in W_p^\gamma(\Omega)^d \cap L_\sigma^p(\Omega). \tag{4.75}$$

In particular $r \mapsto Z(t,r)f$ is continuous on $[0,t)$ with respect to the L^p-norm for any function $f \in W_p^\gamma(\Omega)^d \cap L_\sigma^p(\Omega)$ and any $t > 0$. Now for $f \in L_\sigma^p(\Omega)$ and $(t,s) \in \Lambda$, set

$$\tilde{S}(t,s)f := W(t,s)f - \int_s^t Z(t,r)W(r,s)f \, \mathrm{d}r.$$

Computing the derivative with respect to $s \in [0,t)$ yields

$$\partial_s \tilde{S}(t,s)f = -W(t,s)L_\Omega(s)f - G(t,s)f + Z(t,s)f + \int_s^t Z(t,r)W(r,s)L_\Omega(s)f \, \mathrm{d}r$$

$$+ \int_s^t Z(t,r)G(r,s)f \, \mathrm{d}r \tag{4.76}$$

$$= -\tilde{S}(t,s)L_\Omega(s)f$$

for any $f \in Y_{\Omega,p}$, due to relations (4.70) and (4.75).

For $(t,s) \in \Lambda$ with $t \neq s$ and for $f \in Y_{\Omega,p}$, we set $u(r) := \tilde{S}(t,r)S_\Omega(r,s)f$, $s \leq r \leq t$. Then by using (4.76) and the results from Step 1 and Step 2 we conclude that

$$\partial_r u(r) = -\tilde{S}(t,r)L_\Omega(r)S(r,s)f + \tilde{S}(t,r)L_\Omega(r)S(r,s)f = 0$$

for $r \in (s,t)$. Hence, the map $r \mapsto u(r)$ is constant on (s,t), and by continuity we may conclude that $u(t) = \tilde{S}(t,s)f = S_\Omega(t,s)f = u(s)$. Thus, by the density of $Y_{\Omega,p}$ in $L_\sigma^p(\Omega)$, it follows that $\tilde{S}(t,s)f = S_\Omega(t,s)f$ holds for all $f \in L_\sigma^p(\Omega)$, $(t,s) \in \Lambda$. This shows that the two-parameter family of linear operators defined in (4.62) fulfills property (c) in Theorem 4.19.

Step 4: Uniqueness. Assume there exists another two-parameter family $\{V(t,s)\}_{(t,s)\in\Lambda}$ such that (a) – (c) holds. Then for $(t,s)\in\Lambda$ with $t\neq s$ and for $f\in Y_{\Omega,p}$, we set $u(r):=V(t,r)S_\Omega(r,s)f$, $s\leq r\leq t$. Then, using properties (a) – (c), we see that

$$\partial_r u(r) = -V(t,r)L_\Omega(r)S(r,s)f + V(t,r)L_\Omega(r)S(r,s)f = 0$$

for $r\in(s,t)$. Hence, the map $r\mapsto u(r)$ is constant on (s,t), and by continuity we may conclude that $u(s) = V(t,s)f = S_\Omega(t,s)f = u(t)$ holds. This shows the uniqueness of $\{S(t,s)\}_{(t,s)\in\Lambda}$. The uniqueness directly implies that

$$S(t,s)f = S(t,r)S(r,s)f, \qquad s\leq r\leq t$$

for all $f\in Y_{\Omega,p}$ and by density even for all $f\in L^p_\sigma(\Omega)$. Hence, $\{S(t,s)\}_{(t,s)\in\Lambda}$ is indeed an evolution system.

The well-posedness of (4.50) follows from Proposition 1.53. This concludes the proof. □

4.3.2 L^p-L^q-estimates

In this section we derive certain smoothing properties and norm estimates for the evolution system $\{S_\Omega(t,s)\}_{(t,s)\in\Lambda}$. In particular, we prove L^p-L^q-estimates and gradient estimates. As in the whole space situation, see Section 4.1.2, such estimates do not follow from the general theory. However, since $S_\Omega(t,s)$ is defined via the series in (4.62) we may apply Lemma 4.21 to obtain the estimates from the corresponding estimates for $\{W(t,s)\}_{(t,s)\in\Lambda}$. For that we can then use the results obtained in Theorem 4.9 and Proposition 4.15.

Theorem 4.22. *Let $T > 0$. Moreover, let $\{S_{\Omega,p}(t,s)\}_{(t,s)\in\Lambda}$ be the evolution system from Theorem 4.19. Then there exists a constant $C := C(p,q,\Omega,T) > 0$ such that for every $f\in L^p_\sigma(\Omega)$*

$$\|S_{\Omega,p}(t,s)f\|_{L^q(\Omega)} \leq C(t-s)^{-\frac{d}{2}\left(\frac{1}{p}-\frac{1}{q}\right)}\|f\|_{L^p(\Omega)}, \qquad 0\leq s < t \leq T, \qquad (4.77)$$

$$\|\nabla S_{\Omega,p}(t,s)f\|_{L^q(\Omega)} \leq C(t-s)^{-\frac{d}{2}\left(\frac{1}{p}-\frac{1}{q}\right)-\frac{1}{2}}\|f\|_{L^p(\Omega)}, \qquad 0\leq s < t \leq T, \qquad (4.78)$$

holds. Moreover there exists a constant $C := C(p,\Omega,T) > 0$ such that

$$\|S_{\Omega,p}(t,s)f\|_{W^{1,p}(\Omega)} \leq C\|f\|_{W^{1,p}(\Omega)}, \qquad 0\leq s \leq t \leq T, \qquad (4.79)$$

for all $f\in W^{1,p}_0(\Omega)^d\cap L^p_\sigma(\Omega)$ and

$$\|S_{\Omega,p}(t,s)f\|_{W^{2,p}(\Omega)} \leq C(t-s)^{-\frac{1}{2}}\|f\|_{W^{1,p}(\Omega)}, \qquad 0\leq s < t \leq T, \qquad (4.80)$$

for all $f\in W^{1,p}_0(\Omega)^d\cap L^p_\sigma(\Omega)$.

Proof. Let $T > 0$ be arbitrary but fixed and let $\gamma(1+\frac{1}{p},2)$. In the following all constants may depend on p,q,γ,Ω and $T > 0$. The family of operators $\{W(t,s)\}_{(t,s)\in\Lambda}$ is defined in (4.53) and for $(t,s)\in\Lambda$ with $t\neq s$ and $f\in L^p_\sigma(\Omega)$ the function $F(t,s)f$ is defined as in

(4.58). By Theorem 4.9, Proposition 4.15 and since $B_{D_2} \in \mathscr{L}(L^p(D_2), W_0^{1,p}(D_2)^d)$, we know that

$$\|W(t,s)f\|_{L^q(\Omega)} \leq C(t-s)^{-\frac{d}{2}\left(\frac{1}{p}-\frac{1}{q}\right)}\|f\|_{L^p(\Omega)}, \qquad 0 \leq s < t \leq T,$$

$$\|\nabla W(t,s)f\|_{L^q(\Omega)} \leq C(t-s)^{-\frac{d}{2}\left(\frac{1}{p}-\frac{1}{q}\right)-\frac{1}{2}}\|f\|_{L^p(\Omega)}, \qquad 0 \leq s < t \leq T,$$

for all $f \in L_\sigma^p(\Omega)$. Moreover, Lemma 4.20 states that

$$\|F(t,s)f\|_{L^p(\Omega)} \leq C(t-s)^{-\frac{\gamma}{2}}\|f\|_{L^p(\Omega)}, \quad 0 \leq s < t \leq T,$$

for all $f \in L_\sigma^p(\Omega)$ and

$$\|F(t,s)f\|_{W^{1,p}(\Omega)} \leq C(t-s)^{-\frac{\gamma}{2}}\|f\|_{W^{1,p}(\Omega)}, \quad 0 \leq s < t \leq T,$$

for all $f \in W_0^{1,p}(\Omega)^d \cap L_\sigma^p(\Omega)$. First, we show the estimate (4.77) for $0 \leq 1/p - 1/q < 2/d$. The estimates for $W(t,s)f$ and $F(t,s)f$ allow us to apply Lemma 4.21 with $X_1 = X_2 = L_\sigma^p(\Omega)$, $U = W$, $V = \mathbb{P}_\Omega F$, $\alpha = -d/2\,(1/p - 1/q)$, $\beta = -\gamma/2$, and thus we obtain

$$\|S_{\Omega,p}(t,s)f\|_{L^q(\Omega)} \leq C(t-s)^{-\frac{d}{2}\left(\frac{1}{p}-\frac{1}{q}\right)}\|f\|_{L^p(\Omega)}, \qquad 0 \leq s < t \leq T.$$

To show the general case $1 < p \leq q < \infty$ we use an iteration argument as done in the proof of Proposition 4.15 . Assume that $1 < p < q < \infty$ such that $2/d \leq 1/p - 1/q < 4/d$. Let $\varepsilon > 0$ be sufficiently small and set $1/r = 1/q + 2/d - \varepsilon$ such that $1 < p < r < q < \infty$, $1/r - 1/q < 2/d$ and $0 < 1/p - 1/r \leq 2/d$. So, by using the algebraic property of evolution systems and the first step we obtain

$$
\begin{aligned}
\|S_\Omega(t,s)f\|_{L^q(\Omega)} &= \left\|S_\Omega\left(t, s+\tfrac{t-s}{2}\right)S_\Omega\left(s+\tfrac{t-s}{2}, s\right)f\right\|_{L^q(\Omega)} \\
&\leq C\left(\frac{t-s}{2}\right)^{-\frac{d}{2}\left(\frac{1}{r}-\frac{1}{q}\right)}\left\|S_\Omega\left(s+\tfrac{t-s}{2}, s\right)f\right\|_{L^r(\Omega)} \\
&\leq C\left(\frac{t-s}{2}\right)^{-\frac{d}{2}\left(\frac{1}{r}-\frac{1}{q}\right)}\left(\frac{t-s}{2}\right)^{-\frac{d}{2}\left(\frac{1}{p}-\frac{1}{r}\right)}\|f\|_{L^p(\Omega)} \\
&\leq C(t-s)^{-\frac{d}{2}\left(\frac{1}{p}-\frac{1}{q}\right)}\|f\|_{L^p(\Omega)}, \qquad 0 \leq s < t \leq T.
\end{aligned}
$$

By iterating this argument we obtain (4.77) for all $1 < p \leq q < \infty$. Estimate (4.78) for $p = q$ and $f \in W_0^{1,p}(\Omega)^d \cap L_\sigma^p(\Omega)$ follows by applying Lemma 4.21 with $X_1 = W^{1,p}(\Omega)^d \cap L_\sigma^p(\Omega)$, $X_2 = L_\sigma^p(\Omega)$, $U = W$, $V = \mathbb{P}_\Omega F$, $\alpha = -1/2$, $\beta = -\gamma/2$. By the denseness of $W_0^{1,p}(\Omega)^d \cap L_\sigma^p(\Omega)$ in $L_\sigma^p(\Omega)$ it directly follows that (4.78) even holds for all $f \in L_\sigma^p(\Omega)$. By using the algebraic property of evolution and (4.77) we then have

$$
\begin{aligned}
\|\nabla S_\Omega(t,s)f\|_{L^q(\Omega)} &= \left\|\nabla S_\Omega\left(t, s+\tfrac{t-s}{2}\right)S_\Omega\left(s+\tfrac{t-s}{2}, s\right)f\right\|_{L^q(\Omega)} \\
&\leq C\left(\frac{t-s}{2}\right)^{-\frac{1}{2}}\left\|S_\Omega\left(s+\tfrac{t-s}{2}, s\right)f\right\|_{L^q(\Omega)} \\
&\leq C\left(\frac{t-s}{2}\right)^{-\frac{1}{2}}\left(\frac{t-s}{2}\right)^{-\frac{d}{2}\left(\frac{1}{p}-\frac{1}{p}\right)}\|f\|_{L^p(\Omega)} \\
&\leq C(t-s)^{-\frac{d}{2}\left(\frac{1}{p}-\frac{1}{q}\right)-\frac{1}{2}}\|f\|_{L^p(\Omega)}, \qquad 0 \leq s < t \leq T.
\end{aligned}
$$

So (4.78) is true for arbitrary $1 < p \leq q < \infty$. It remains to prove (4.79) and (4.80). By Theorem 4.9, Proposition 4.15 and sine $\mathbb{B}_{D_2} \in \mathscr{L}(L^p(D_2), W_0^{1,p}(D_2)^d)$, we know that

$$\|W(t,s)f\|_{W^{1,p}(\Omega)} \leq C\|f\|_{W^{1,p}(\Omega)}, \qquad 0 \leq s \leq t \leq T$$

for all $f \in W_0^{1,p}(\Omega)^d \cap L_\sigma^p(\Omega)$, and

$$\|W(t,s)f\|_{W^{2,p}(\Omega)} \leq C(t-s)^{-\frac{1}{2}}\|f\|_{W^{1,p}(\Omega)}, \quad 0 \leq s < t \leq T$$

for all $f \in W_0^{1,p}(\Omega)^d \cap L_\sigma^p(\Omega)$. To show (4.79) we apply Lemma 4.21 with $X_1 = X_2 = W_0^{1,p}(\Omega)^d \cap L_\sigma^p(\Omega)$, $U = W$, $V = \mathbb{P}_\Omega F$, $\alpha = 0$, $\beta = -\gamma/2$. Similarly, (4.80) follows by applying Lemma 4.21 with $X_1 = W^{2,p}(\Omega)^d \cap W^{1,p}(\Omega)_0^d \cap L_\sigma^p(\Omega)$, $X_2 = W_0^{1,p}(\Omega)^d \cap L_\sigma^p(\Omega)$, $U = W$, $V = \mathbb{P}_\Omega F$, $\alpha = -1/2$, $\beta = -\gamma/2$. $\qquad \square$

Proposition 4.23. *Let $1 < p < q < \infty$ and $f \in L_\sigma^p(\Omega)$. Then the following holds:*

$$\begin{aligned}
(t-s)^{\frac{d}{2}\left(\frac{1}{p}-\frac{1}{q}\right)}\|S_{\Omega,p}(t,s)f\|_{L^q(\Omega)} \to 0 \quad &as \quad t \to s, \\
(t-s)^{\frac{1}{2}}\|\nabla S_{\Omega,p}(t,s)f\|_{L^p(\Omega)} \to 0 \quad &as \quad t \to s.
\end{aligned} \tag{4.81}$$

Proof. The proof is analogous to the one of Proposition 4.10. Let $0 < t - s \leq 1$ and take a sequence $\{f_n\} \subset C_{c,\sigma}^\infty(\Omega) \subset L_\sigma^p(\Omega) \cap L_\sigma^q(\Omega)$ such that $f_n \to f$ in $L_\sigma^p(\Omega)$ as $n \to \infty$. The L^p-L^q-estimates (4.77) imply that there exist constants $C_1, C_2 > 0$ such that

$$\begin{aligned}
&(t-s)^{\frac{d}{2}\left(\frac{1}{p}-\frac{1}{q}\right)}\|S_\Omega(t,s)f\|_{L^q(\Omega)} \\
&\leq (t-s)^{\frac{d}{2}\left(\frac{1}{p}-\frac{1}{q}\right)}\|S_\Omega(t,s)f - S_\Omega(t,s)f_n\|_{L^q(\Omega)} + (t-s)^{\frac{d}{2}\left(\frac{1}{p}-\frac{1}{q}\right)}\|S_\Omega(t,s)f_n\|_{L^q(\Omega)} \\
&\leq C_1\|f - f_n\|_{L^p(\Omega)} + C_2(t-s)^{\frac{d}{2}\left(\frac{1}{p}-\frac{1}{q}\right)}\|f_n\|_{L^q(\Omega)}.
\end{aligned}$$

Now by arguing as in the proof of Proposition 4.10 the first assertion. Similarly, by using (4.78) and (4.79), we obtain

$$\begin{aligned}
&(t-s)^{\frac{1}{2}}\|\nabla S_\Omega(t,s)f\|_{L^p(\Omega)} \\
&\leq (t-s)^{\frac{1}{2}}\|\nabla S_\Omega(t,s)f - \nabla S_\Omega(t,s)f_n\|_{L^p(\Omega)} + (t-s)^{\frac{1}{2}}\|\nabla S_\Omega(t,s)f_n\|_{L^p(\Omega)} \\
&\leq C_1\|f - f_n\|_{L^p(\Omega)} + C_2(t-s)^{\frac{1}{2}}\|f_n\|_{W^{1,p}(\Omega)}.
\end{aligned}$$

Now the second assertion follows again by arguing as in the proof of Proposition 4.10. $\quad \square$

4.4 The non-linear problem: Existence of mild solutions

In this section we come back to the starting point of this chapter, i.e. to the non-linear problem (4.6), or equivalently to the non-linear problem (4.9). To make the linear results from Section 4.3 applicable, we rewrite (4.9) in abstract form. Therefore we formally apply the Helmholtz projection $\mathbb{P}_\Omega$ to (4.9) and we obtain the following abstract Cauchy problem:

$$\begin{cases}
u'(t) - L_\Omega(t)u(t) + \mathbb{P}_\Omega(u \cdot \nabla u)(t) = \mathbb{P}_\Omega \widetilde{F}(t), & t > 0, \\
u(0) = a.
\end{cases} \tag{4.82}$$

Here $a := u_0 - b(0)$, u_0 is the given initial value for (4.6), $b(0)$ is given by the solenoidal boundary extension b defined in (4.8) and $\widetilde{F}$ is given in (4.10). Recall that the family of linear operators $\{L_\Omega(t)\}_{t \geq 0}$ on $L^p_\sigma(\Omega)$, $1 < p < \infty$, is defined as follows:

$$\begin{aligned}
\mathcal{D}(L_\Omega(t)) &:= \{u \in W^{2,p}(\Omega)^d \cap W^{1,p}_0(\Omega)^d \cap L^p_\sigma(\Omega) : M(t)x \cdot \nabla u \in L^p(\Omega)^d\}, \\
L_\Omega(t)u &:= \mathbb{P}_\Omega \left(\Delta u + M(t)x \cdot \nabla u + c(t) \cdot \nabla u - M(t)u - b \cdot \nabla u - u \cdot \nabla b \right).
\end{aligned}$$

For the formulation of (4.6) and (4.9) we imposed the following conditions:

Assumption 4.24. (i) $\Omega \subset \mathbb{R}^d$, $d \geq 2$, is an exterior domain with boundary of class C^2.

(ii) $M \in C^1([0,\infty); \mathbb{R}^{d \times d})$ with $M(t)^\top = -M(t)$ for all $t \in [0,\infty)$.

(iii) $c \in C^1([0,\infty); \mathbb{R}^{d \times d})$.

(iv) $u_0 \in L^p(\Omega)^d$, $1 < p < \infty$, with $\operatorname{div} u_0 = 0$ and $u_0 \cdot \nu = (M(0)x + c(0)) \cdot \nu$ on $\partial \Omega$.

Note that Assumption 4.24 (iv) ensures that $a := u_0 - b(0) \in L^p_\sigma(\Omega)$. Therefore (4.82) is indeed an abstract Cauchy problem on $L^p_\sigma(\Omega)$.

In the following, let $\{S_\Omega(t,s)\}_{(t,s) \in \Lambda}$ denote the strongly continuous evolution system on $L^p_\sigma(\Omega)$, $1 < p < \infty$, obtained in Theorem 4.19. Formally, by using the variation of constants formula, problem (4.82) can be reduced to the following integral equation on $L^p_\sigma(\Omega)$:

$$u(t) = S_\Omega(t,0)a - \int_0^t S_\Omega(t,s)\mathbb{P}_\Omega(u \cdot \nabla u)(s)\,\mathrm{d}s + \int_0^t S_\Omega(t,s)\mathbb{P}_\Omega \widetilde{F}(s)\,\mathrm{d}s, \qquad t \geq 0. \quad (4.83)$$

For given $0 < T_0 \leq \infty$ we call a function $u \in C([0,T_0); L^p_\sigma(\Omega))$ a *mild solution* to (4.82) if the right-hand side in (4.83) is well-defined and if u satisfies the integral equation (4.83).

The following main result states the existence of a unique local mild solution to (4.82) in $L^p_\sigma(\Omega)$ if $p \geq d$ and if Assumption 4.24 holds.

Theorem 4.25. *Let $2 \leq d \leq p \leq q < \infty$ and suppose that Assumption 4.24 holds. Then there exists a $T > 0$ and a unique mild solution $u \in C([0,T]; L^p_\sigma(\Omega))$ to (4.82) which has the properties*

$$\begin{aligned}
t &\mapsto t^{\frac{d}{2}\left(\frac{1}{p} - \frac{1}{q}\right)} u(t) \in C([0,T]; L^q_\sigma(\Omega)), \\
t &\mapsto t^{\frac{d}{2}\left(\frac{1}{p} - \frac{1}{q}\right) + \frac{1}{2}} \nabla u(t) \in C([0,T]; L^q(\Omega)^{d \times d}).
\end{aligned} \quad (4.84)$$

Remarks 4.26. 1. Theorem 4.25 includes the special case $M(t) \equiv M \in \mathbb{R}^{d \times d}$ and $c \equiv 0$ which was considered by Geissert, Heck and Hieber in [GHH06a, Theorem 5.1].

2. Let $p < q$. As a consequence of (4.84) we see that

$$t^{\frac{d}{2}\left(\frac{1}{p} - \frac{1}{q}\right)} \|u(t)\|_{L^q(\Omega)} + t^{\frac{1}{2}} \|\nabla u(t)\|_{L^p(\Omega)} \to 0 \qquad \text{as } t \to 0^+. \quad (4.85)$$

The proof of Theorem 4.25 is based on a variant of Kato's iteration scheme, see [Kat84, Gig86]. See also the survey article [Wie99]. The basic ingredients for the proof are the results for the evolution system $\{S_\Omega(t,s)\}_{(t,s)\in\Lambda}$ stated in Theorem 4.22 and Proposition 4.23. Moreover, in the proof we shall frequently make use of the following estimate.

Remark 4.27. Let $0 < \alpha, \beta < 1$. Then the estimate

$$\int_0^t (t-s)^{-\alpha}s^{-\beta}\,\mathrm{d}s = \int_{t/2}^t (t-s)^{-\alpha}s^{-\beta}\,\mathrm{d}s + \int_0^{t/2} (t-s)^{-\alpha}s^{-\beta}\,\mathrm{d}s$$

$$\leq \left(\frac{t}{2}\right)^{-\beta}\int_{t/2}^t (t-s)^{-\alpha}\,\mathrm{d}s + \left(\frac{t}{2}\right)^{-\alpha}\int_0^{t/2} s^{-\beta}\,\mathrm{d}s$$

$$\leq \left(\frac{t}{2}\right)^{1-\beta-\alpha}\left(\frac{1}{1-\alpha} + \frac{1}{1-\beta}\right)$$

holds.

Proof of Theorem 4.25. **Step 1: Existence.** For $t \geq 0$ we define $u_1(t) := S_\Omega(t,0)a$ and

$$u_{n+1}(t) = u_1(t) - \int_0^t S_\Omega(t,s)\mathbb{P}_\Omega(u_n \cdot \nabla u_n)(s)\,\mathrm{d}s + \int_0^t S_\Omega(t,s)\mathbb{P}_\Omega\widetilde{F}(s)\,\mathrm{d}s, \qquad n \in \mathbb{N},$$

$$=: u_1(t) - Gu_n(t).$$

Fix $0 < T_0 < \infty$. In the following, all constants may depend on T_0. Let $0 < T \leq T_0$, $n \in \mathbb{N}$ and set $\beta := \frac{d}{2}\left(\frac{1}{p} - \frac{1}{q}\right)$ so that $\beta < \frac{1}{2}$. Now we define

$$K_n(q,T) := \sup_{t\in(0,T]} t^\beta \|u_n(t)\|_{L^q(\Omega)}, \qquad K_n'(p,T) := \sup_{t\in(0,T]} t^{\frac{1}{2}}\|\nabla u_n\|_{L^p(\Omega)}$$

and for simplicity we abbreviate $K_n := K_n(q,T)$ and $K_n' := K_n'(p,T)$.

Let $d \leq p < q$ and let $1 < r < \infty$ be such that $\frac{1}{r} = \frac{1}{p} + \frac{1}{q}$. From the L^r-L^q-estimates (4.77) and the boundedness of Helmholtz projection $\mathbb{P}_\Omega$ we obtain

$$\|u_{n+1}(t)\|_{L^q(\Omega)} \leq \|u_1(t)\|_{L^q(\Omega)} + \int_0^t \|S_\Omega(t,s)\mathbb{P}_\Omega(u_n \cdot \nabla u_n)(s)\|_{L^q(\Omega)}\,\mathrm{d}s$$

$$+ \int_0^t \|S_\Omega(t,s)\mathbb{P}_\Omega\widetilde{F}(s)\|_{L^q(\Omega)}\,\mathrm{d}s$$

$$\leq t^{-\beta}K_1 + C\int_0^t (t-s)^{-\frac{d}{2}\left(\frac{1}{r}-\frac{1}{q}\right)}\|u_n(s) \cdot \nabla u_n(s)\|_{L^r(\Omega)}\,\mathrm{d}s$$

$$+ C\int_0^t \|\widetilde{F}(s)\|_{L^q(\Omega)}\,\mathrm{d}s$$

for $0 < t \leq T \leq T_0$ and some constant $C := C(T_0) > 0$ independent of $n \in \mathbb{N}$ and $0 < T \leq T_0$.

Similarly, by using the gradient L^r-L^p-estimates (4.78) we have

$$\|\nabla u_{n+1}(t)\|_{L^p(\Omega)} \leq t^{-\frac{1}{2}} K_1' + C \int_0^t (t-s)^{-\frac{d}{2}(\frac{1}{r}-\frac{1}{p})-\frac{1}{2}} \|u_n(s) \cdot \nabla u_n(s)\|_{L^r(\Omega)} \, ds$$

$$+ C \int_0^t (t-s)^{-\frac{1}{2}} \|\widetilde{F}(s)\|_{L^p(\Omega)} \, ds$$

for $0 < t \leq T \leq T_0$ and some constant $C := C(T_0) > 0$ independent of $n \in \mathbb{N}$ and $0 < T \leq T_0$. By Hölder's inequality we conclude that

$$\|u_n(s) \cdot \nabla u_n(s)\|_{L^r(\Omega)} \leq \|u_n(s)\|_{L^q(\Omega)} \|\nabla u_n(s)\|_{L^p(\Omega)} \leq K_n K_n' s^{-\beta-\frac{1}{2}}.$$

Hence, we obtain

$$\|u_{n+1}(t)\|_{L^q(\Omega)} \leq t^{-\beta} K_1 + C K_n K_n' \int_0^t (t-s)^{-\frac{d}{2p}} s^{-\beta-\frac{1}{2}} \, ds + C \int_0^t \|\widetilde{F}(s)\|_{L^q(\Omega)} \, ds \quad (4.86)$$

and

$$\|\nabla u_{n+1}(t)\|_{L^p(\Omega)} \leq t^{-\frac{1}{2}} K_1' + C K_n K_n' \int_0^t (t-s)^{-\frac{d}{2q}-\frac{1}{2}} s^{-\beta-\frac{1}{2}} \, ds$$

$$+ C \int_0^t (t-s)^{-\frac{1}{2}} \|\widetilde{F}(s)\|_{L^p(\Omega)} \, ds \quad (4.87)$$

for $0 < t \leq T \leq T_0$ and some constant $C := C(T_0) > 0$ independent of $n \in \mathbb{N}$ and $0 < T \leq T_0$. Since $\beta < \frac{1}{2}$ and $q > d$ we conclude by Remark 4.27 that

$$\int_0^t (t-s)^{-\frac{d}{2p}} s^{-\beta-\frac{1}{2}} \, ds \leq C t^{\frac{1}{2}+\frac{d}{2q}-\frac{d}{p}} \quad \text{and} \quad \int_0^t (t-s)^{-\frac{d}{2q}-\frac{1}{2}} s^{-\beta-\frac{1}{2}} \, ds \leq C t^{-\frac{d}{2p}}$$

for a constant $C > 0$. Hence, multiplying (4.86) and (4.87) with t^β and $t^{\frac{1}{2}}$, respectively, and then taking the supremum over $t \in (0, T]$ yield

$$\begin{aligned} K_{n+1} &\leq K_1 + C_1 K_n K_n' + C_2 T, \\ K_{n+1}' &\leq K_1' + C_3 K_n K_n' + C_4 T, \end{aligned} \quad (4.88)$$

where $C_1 := C_1(T_0)$, $C_2 := C_2(T_0)$, $C_3 := C_3(T_0)$, $C_4 := C_4(T_0) > 0$ are constants independent of $n \in \mathbb{N}$ and $0 < T \leq T_0$. Next, we set $R_n := R_n(T) := \max\{K_n, K_n'\}$. It follows from (4.88) that

$$R_{n+1} \leq R_1 + c_1 R_n^2 + c_2 T \quad (4.89)$$

for constants $c_1 := c_1(T_0)$, $c_2 := c_2(T_0) \geq 1$ independent of $n \in \mathbb{N}$ and $0 < T \leq T_0$. By Proposition 4.23 we conclude that for every $\varepsilon > 0$ there is a $\widetilde{T} := \widetilde{T}(\varepsilon) > 0$ such that $R_1(T) \leq \varepsilon$ if $T \leq \widetilde{T}$. So for $\varepsilon \leq \frac{1}{8c_1}$ it follows by induction that

$$R_n(T) \leq 4\varepsilon, \quad n \in \mathbb{N}, \quad \text{if } T \leq \min\left\{\widetilde{T}(\varepsilon), \frac{\varepsilon}{c_2}, T_0\right\} =: T^*. \quad (4.90)$$

Thus, the sequences

$$\{t \mapsto t^\beta u_n(t)\}_{n \in \mathbb{N}} \quad \text{and} \quad \{t \mapsto t^{\frac{1}{2}} \nabla u_n(t)\}_{n \in \mathbb{N}}$$

are uniformly bounded on $[0, T]$ with values in $L_\sigma^q(\Omega)$ and $L^p(\Omega)^{d \times d}$, respectively.

Next, we consider the functions $w_1(t) := u_1(t)$, $w_n(t) := u_{n+1}(t) - u_n(t)$, $n \in \mathbb{N}$. We set

$$M_n(q, T) := \sup_{t \in (0,T]} t^\beta \|w_n(t)\|_{L^q(\Omega)}, \qquad M_n'(p, T) := \sup_{t \in (0,T]} t^{\frac{1}{2}} \|\nabla w_n(t)\|_{L^p(\Omega)}$$

and for simplicity we write $M_n := M_n(q, T)$ and $M_n' := M_n'(p, T)$. It is clear that

$$\|w_1(t)\|_{L^q(\Omega)} = \|u_1(t)\|_{L^q(\Omega)}$$

and easy to see that

$$\|w_2(t)\|_{L^q(\Omega)} \le C \int_0^t (t-s)^{-\frac{d}{2p}} \|u_1(t)\|_{L^q(\Omega)} \|\nabla u_1(t)\|_{L^p(\Omega)} + C \int_0^t \|\widetilde{F}(s)\|_{L^q(\Omega)} \, ds.$$

Similarly, we obtain estimates for $\nabla w_1(t)$ and $\nabla w_2(t)$. Now let $n \ge 2$. Note that

$$u_n \cdot \nabla u_n - u_{n-1} \cdot \nabla u_{n-1} = u_n \cdot \nabla w_n + w_n \cdot \nabla u_{n-1}$$

holds. By similar computations as above, we obtain

$$\|w_{n+1}(t)\|_{L^q(\Omega)} \le \int_0^t \|S_\Omega(t, s)\mathbb{P}_\Omega \left((u_n \cdot \nabla u_n)(s) - (u_{n-1} \cdot \nabla u_{n-1})(s)\right)\|_{L^q(\Omega)} \, ds$$

$$\le C \int_0^t (t-s)^{-\frac{d}{2p}} \Big(\|u_n(s)\|_{L^q(\Omega)} \|\nabla w_n(s)\|_{L^p(\Omega)}$$

$$+ \|w_n(s)\|_{L^q(\Omega)} \|\nabla u_{n-1}(s)\|_{L^p(\Omega)} \Big) \, ds,$$

$$\|\nabla w_{n+1}(t)\|_{L^p(\Omega)} \le \int_0^t \|\nabla S_\Omega(t, s)\mathbb{P}_\Omega \left((u_n \cdot \nabla u_n)(s) - (u_{n-1} \cdot \nabla u_{n-1})(s)\right)\|_{L^p(\Omega)} \, ds$$

$$\le C \int_0^t (t-s)^{-\frac{d}{2q}-\frac{1}{2}} \Big(\|u_n(s)\|_{L^q(\Omega)} \|\nabla w_n(s)\|_{L^p(\Omega)}$$

$$+ \|w_n(s)\|_{L^q(\Omega)} \|\nabla u_{n-1}(s)\|_{L^p(\Omega)} \Big) \, ds$$

for $0 < t \le T \le T_0$ and for a constant $C := C(T_0) > 0$ independent of $n \in \mathbb{N}$ and $0 < T \le T_0$. Therefore, similarly to (4.88) we conclude that for $n \ge 2$ we have

$$M_{n+1} \le C_5(M_n' K_n + M_n K_{n-1}'),$$
$$M_{n+1}' \le C_6(M_n' K_n + M_n K_{n-1}'),$$

where $C_5 := C_5(T_0)$, $C_6 := C_6(T_0) > 0$ are constants independent of $n \in \mathbb{N}$ and $0 < T \le T_0$. We set $\tilde{R}_n := \tilde{R}_n(T) := \max\{M_n, M_n'\}$. Hence, by using (4.90), we have

$$\tilde{R}_{n+1} \le 8\varepsilon(C_5 + C_6)\tilde{R}_n, \qquad n \ge 2$$

if $T \le T^*$, where T^* is chosen as in (4.90). From that we conclude that

$$\tilde{R}_{n+1} \le \frac{1}{2}\tilde{R}_n, \qquad n \ge 2 \tag{4.91}$$

for $\varepsilon \le \min\left\{\frac{1}{16(C_5+C_6)}, \frac{1}{8c_1}\right\}$ and if $T \le T^*$. Hence the series $\sum_{n=1}^\infty \tilde{R}_n$ converges. This yields in particular that the sequence

$$\left\{t \mapsto t^{\frac{1}{2}}\nabla u_n(t)\right\}_{n\in\mathbb{N}}$$

is a Cauchy sequence in $C([0,T]; L^p(\Omega)^{d\times d})$, provided we show that ∇u_n remains in this space (this is done below). Therefore, it tends to a unique limit functions

$$\left(t \mapsto t^{\frac{1}{2}}v(t)\right) \in C([0,T]; L^p(\Omega)^{d\times d}).$$

Next, by similar computations as above, we obtain

$$\|w_{n+1}(t)\|_{L^p(\Omega)} \le C\int_0^t (t-s)^{-\frac{d}{2q}}\Big(\|u_n(s)\|_{L^q(\Omega)}\|\nabla w_n(s)\|_{L^p(\Omega)}$$

$$+ \|w_n(s)\|_{L^q(\Omega)}\|\nabla u_{n-1}(s)\|_{L^p(\Omega)}\Big)\,\mathrm{d}s$$

$$\le C\left(\tilde{R}_n R_n + \tilde{R}_n R_{n-1}\right) \le C\left(\frac{1}{2}\right)^n$$

for some constant $C := C(T_0) > 0$ that may change from line to line and for $0 < t \le T \le T^*$. Since $u_n(t) = \sum_{k=1}^n w_k(t)$, we see that $\{t \mapsto u_n(t)\}_{n\in\mathbb{N}}$ is uniformly bounded on $[0,T]$ with values in $L^p_\sigma(\Omega)$ and that $\{t \mapsto u_n(t)\}_{n\in\mathbb{N}}$ is even a Cauchy sequence in $C([0,T]; L^p_\sigma(\Omega))$, provided u_n remains in this space (this is shown below). Therefore, it tends to a unique limit function $(t \mapsto u(t)) \in C([0,T]; L^p_\sigma(\Omega))$. By construction $v(t) = \nabla u(t)$ and $u \in C([0,T]; L^p_\sigma(\Omega))$ is a mild solution to (4.82).

To conclude the existence part it remains to show the continuity of the iterated functions $t \mapsto u_n(t)$ and $t \mapsto t^{\frac{1}{2}}\nabla u_n(t)$ mapping $[0,T]$ to $L^p(\Omega)$ and $L^p(\Omega)^{d\times d}$, respectively. We start with the continuity at $t = 0$. Let $0 < \delta, \lambda < 1$ and define

$$E_n := \sup_{0\le t\le\delta} \|u_n(t) - S_\Omega(\lambda,0)a\|_{L^p(\Omega)}, \qquad n \in \mathbb{N}.$$

Fix some $q > d$. Then, by using similar calculations as above we estimate

$$E_{n+1} \le \sup_{0\le t\le\delta}\|S_\Omega(t,0)a - S_\Omega(\lambda,0)a\|_{L^p(\Omega)} + \sup_{0\le t\le\delta}\left\|\int_0^t S_\Omega(t,s)\mathbb{P}_\Omega(u_n\cdot\nabla u_n)(s)\,\mathrm{d}s\right\|_{L^p(\Omega)}$$

$$+ \sup_{0\le t\le\delta}\left\|\int_0^t S_\Omega(t,s)\mathbb{P}_\Omega\widetilde{F}(s)\,\mathrm{d}s\right\|_{L^p(\Omega)}$$

$$\le E_1 + \sup_{0\le t\le\delta}\left\|\int_0^t S_\Omega(t,s)\mathbb{P}_\Omega([u_n - S_\Omega(\lambda,0)a]\cdot\nabla u_n)(s)\,\mathrm{d}s\right\|_{L^p(\Omega)}$$

$$+ \sup_{0\le t\le\delta}\left\|\int_0^t S_\Omega(t,s)\mathbb{P}_\Omega(S_\Omega(\lambda,0)a\cdot\nabla u_n)(s)\,\mathrm{d}s\right\|_{L^p(\Omega)} + C\delta$$

$$\le E_1 + CE_n K_n'(p,\delta)\int_0^t (t-s)^{-\frac{d}{2p}}s^{-\frac{1}{2}}\,\mathrm{d}s$$

$$+ CK_n'(p,\delta)\int_0^t (t-s)^{-\frac{d}{2q}}s^{\frac{1}{2}}\,\mathrm{d}s\,\|S_\Omega(\lambda,0)a\|_{L^q(\Omega)} + C\delta$$

$$\le E_1 + CE_n K_n'(p,\delta) + CK_n'(p,\delta)\left(\frac{\delta}{\lambda}\right)^{\frac{1}{2}-\frac{d}{2q}}\|a\|_{L^p(\Omega)} + C\delta$$

113

for some constant $C > 0$ that may change from line to line. Note that it follows from (4.90) that $K_n'(\delta)$ can be made arbitrary small if we choose $\delta > 0$ small. Therefore, we can assume that δ is chosen sufficiently small such that

$$E_{n+1} \leq E_1 + \frac{1}{3}E_n + \left(\frac{\delta}{\lambda}\right)^{\frac{1}{2}-\frac{d}{2q}} \|a\|_{L^p(\Omega)} + C\delta.$$

By induction we see that

$$E_n \leq 2E_1 + 2\left(\frac{\delta}{\lambda}\right)^{\frac{1}{2}-\frac{d}{2q}} \|a\|_{L^p(\Omega)} + 2C\delta. \tag{4.92}$$

Now let $\varepsilon > 0$. Since

$$E_1 \leq \sup_{0 \leq t \leq \delta} \|S_\Omega(t,0)a - a\|_{L^p(\Omega)} + \|a - S_\Omega(\lambda,0)a\|_{L^p(\Omega)}$$

and

$$\sup_{0 \leq t \leq \delta} \|u_n(t) - a\|_{L^p(\Omega)} \leq E_n + \|S_\Omega(\lambda,0)a - a\|_{L^p(\Omega)},$$

we conclude from (4.92) that

$$\sup_{0 \leq t \leq \delta} \|u_n(t) - a\|_{L^p(\Omega)} \leq \varepsilon$$

if we first choose $\lambda > 0$ sufficiently small and then $\delta > 0$ sufficiently small. This implies the continuity of $t \mapsto u_n(t)$ in $L_\sigma^p(\Omega)$ at $t = 0$.

Let us now show the continuity at $t > 0$. This is in fact easier than the continuity at $t = 0$. Let $\delta < \tau < t \leq T$. Then by using the algebraic property of evolution systems we obtain

$$\|Gu_n(t) - Gu_n(\tau)\|_{L^p(\Omega)}$$

$$\leq \left\| (S_\Omega(t,\tau) - \mathrm{Id}) \left(\int_0^\tau S_\Omega(\tau,s)\mathbb{P}_\Omega(u_n \cdot \nabla u_n)(s)\,\mathrm{d}s - \int_0^\tau S_\Omega(\tau,s)\mathbb{P}_\Omega\widetilde{F}(s)\,\mathrm{d}s \right) \right\|_{L^p(\Omega)}$$

$$+ \left\| \int_\tau^t S_\Omega(t,s)\mathbb{P}_\Omega(u_n \cdot \nabla u_n)(s)\,\mathrm{d}s \right\|_{L^p(\Omega)} + \left\| \int_\tau^t S_\Omega(t,s)\mathbb{P}_\Omega\widetilde{F}(s)\,\mathrm{d}s \right\|_{L^p(\Omega)}$$

$$= \left\| (S_\Omega(t,\tau) - \mathrm{Id})\,(u_n(\tau) - u_1(\tau)) \right\|_{L^p(\Omega)} + \left\| \int_\tau^t S_\Omega(t,s)\mathbb{P}_\Omega(u_n \cdot \nabla u_n)(s)\,\mathrm{d}s \right\|_{L^p(\Omega)}$$

$$+ \left\| \int_\tau^t S_\Omega(t,s)\mathbb{P}_\Omega\widetilde{F}(s)\,\mathrm{d}s \right\|_{L^p(\Omega)}.$$

As a consequence of the strong continuity of $\{S_\Omega(t,s)\}_{(t,s)\in\Lambda}$ and since we showed above that the L^p-norm of $u_n(t)$ is uniformly bounded in $[0,T]$, it is clear that the first term on the right-hand side above gets small if $t - \tau$ is small. For the second term we use similar calculations as above. Let $q > p \geq d$. Then we obtain

$$\left\| \int_\tau^t S_\Omega(\tau,s)\mathbb{P}_\Omega(u_n \cdot \nabla u_n)(s)\,\mathrm{d}s \right\|_{L^p(\Omega)} \leq CK_n K_n' \delta^{-\beta-\frac{1}{2}}(t-\tau)^{1-\frac{d}{2q}}.$$

Hence, this term can be made small if $t - \tau$ is small. The third term above is treated analogously. Hence by taking everything together we obtain the continuity of $t \mapsto u_n(t)$ on $[0, T]$ with values in $L^p_\sigma(\Omega)$, i.e. $\{t \mapsto u_n(t)\}_{n \in \mathbb{N}} \subset C([0, T]; L^p_\sigma(\Omega))$. Analogously, we can also show that $\{t \mapsto t^{\frac{1}{2}} \nabla u_n(t)\}_{n \in \mathbb{N}} \subset C([0, T]; L^p(\Omega)^{d \times d})$.

This concludes the existence part. The properties stated in (4.84) are clear by construction and the calculations above.

Step 2: Uniqueness: Let u, v be two mild solutions of (4.82) satisfying (4.84). Moreover, let $q > p$, $0 < \tilde{T} \leq T$ and define

$$K := K(\tilde{T}) := \max\left\{ \sup_{0 < t \leq \tilde{T}} t^\beta \|u(t)\|_{L^q(\Omega)}, \ \sup_{0 < t \leq \tilde{T}} t^{\frac{1}{2}} \|\nabla v(t)\|_{L^p(\Omega)} \right\}.$$

Since u and v both solve the integral equation (4.83), we obtain similarly as above

$$\|u(t) - v(t)\|_{L^q(\Omega)} \leq KC\left(\int_0^t (t - s)^{-\frac{d}{2p} s^{-\beta - \frac{1}{2}}} ds \right)$$
$$\sup_{0 < \tau \leq \tilde{T}} \left(\tau^\beta \|u(\tau) - v(\tau)\|_{L^q(\Omega)} + \tau^{\frac{1}{2}} \|\nabla(u(\tau) - v(\tau))\|_{L^p(\Omega)} \right)$$

and

$$\|\nabla u(t) - \nabla v(t)\|_{L^p(\Omega)} \leq KC\left(\int_0^t (t - s)^{-\frac{d}{2q} - \frac{1}{2} s^{-\beta - \frac{1}{2}}} ds \right).$$
$$\sup_{0 < \tau \leq \tilde{T}} \left(\tau^\beta \|u(\tau) - v(\tau)\|_{L^q(\Omega)} + \tau^{\frac{1}{2}} \|\nabla(u(\tau) - v(\tau))\|_{L^p(\Omega)} \right)$$

for $0 < t \leq \tilde{T}$. Thus for $0 < t \leq \tilde{T}$ we have

$$t^\beta \|u(t) - v(t)\|_{L^q(\Omega)} + t^{\frac{1}{2}} \|\nabla(u(t) - v(t))\|_{L^p(\Omega)} \tag{4.93}$$
$$\leq 2KC\tilde{T}^{1 - \frac{d}{2p} - \frac{1}{2}} \sup_{0 < \tau \leq \tilde{T}} \left(\tau^\beta \|u(\tau) - v(\tau)\|_{L^q(\Omega)} + \tau^{\frac{1}{2}} \|\nabla(u(\tau) - v(\tau))\|_{L^p(\Omega)} \right).$$

In the case $p > d$ we can choose $\tilde{T}$ small so that $2KC\tilde{T}^{1 - \frac{d}{2p} - \frac{1}{2}} < 1$. This directly implies $u \equiv v$ on $[0, \tilde{T}]$. Since $u, v \in C([\varepsilon, T]; L^q_\sigma(\mathbb{R}^d))$ for every $\varepsilon > 0$, the above argument with initial data $u(\varepsilon) = v(\varepsilon)$ yields that the set $\{t \in (0, T] : u(t) = v(t)\}$ is open. The continuity of u, v and the connectedness of $(0, T]$ imply that $u \equiv v$ on $[0, T]$.

It remains to prove the uniqueness in the case $p = d$. In this case (4.93) reads as follows:

$$t^\beta \|u(t) - v(t)\|_{L^q(\Omega)} + t^{\frac{1}{2}} \|\nabla(u(t) - v(t))\|_{L^p(\Omega)}$$
$$\leq 2KC \sup_{0 < \tau \leq \tilde{T}} \left(\tau^\beta \|u(\tau) - v(\tau)\|_{L^q(\Omega)} + \tau^{\frac{1}{2}} \|\nabla(u(\tau) - v(\tau))\|_{L^p(\Omega)} \right).$$

By (4.85) the constant $K := K(\tilde{T})$ tends to zero as $\tilde{T} \to 0$. Thus, we can choose $\tilde{T}$ small so that $2KC < 1$. This shows $u \equiv v$ on $[0, \tilde{T}]$. Since $u, v \in C([\tilde{T}/2, T]; L^q_\sigma(\mathbb{R}^d))$ for $q > d$

with $u(\tilde{T}/2) = v(\tilde{T}/2)$, the uniqueness in the case $p > d$ implies $u \equiv v$ on $[\tilde{T}/2, T]$. This concludes the proof. $\qquad\square$

In Theorem 4.25 we showed that for arbitrarily given data we can find a (possibly small) time $T > 0$ and a mild solution u on $[0, T]$. Note that we cannot expect to obtain global mild solutions for small data in the present situation. This is due to two reasons: First of all the L^p-L^q-estimates in Theorem 4.22 only hold on finite time intervals and not on $(0, \infty)$. However, in order to obtain solutions on $(0, \infty)$ by this method, it is essential to have L^p-L^q-estimates on $(0, \infty)$. It is in fact an interesting *open question* if the estimates in Theorem 4.22 hold on $(0, \infty)$. Second, the function $\tilde{F}$, caused by the boundary extension, cannot be treated in the iteration procedure on $(0, \infty)$ without any further decay assumptions on $M(\cdot), c(\cdot)$.

By a simple modification of the proof of Theorem 4.25 we can still show long-time existence in the sense that for an arbitrarily fixed $0 < T_0 < \infty$ we obtain a mild solution on $[0, T_0]$ if the given data $u_0, M(\cdot)$ and $c(\cdot)$ are sufficiently small (here the smallness depends on the actual value of $T_0 > 0$). To see this let us fix a $T_0 > 0$ and set

$$N_0 := N_0(T_0) := \max_{t \in [0, T_0]} \Big(|M(t)| + |M'(t)| + |c(t)| + |c'(t)| \Big).$$

Then by using Proposition 1.87 for the Bogovskiĭ operator we can estimate the L^p-norm of the function $\tilde{F}$ defined in (4.10) as follows:

$$\|\tilde{F}(t)\|_{L^p(\Omega)} \leq C N_0, \qquad t \in [0, T_0].$$

Here the constant $C > 0$ is independent of N_0 and $T_0 > 0$. Then similarly to (4.89) we can show that

$$R_{n+1} \leq R_1 + c_1 R_n^2 + c_2 N_0 T_0$$

for some constants $c_1, c_2 \geq 1$ independent of $n \in \mathbb{N}$, but depending on $T_0 > 0$. By Theorem 4.22 we know that $R_1 \leq C_0(\|u_0\|_{L^d(\Omega)} + N_0)$ for some constant $C_0 := C_0(T_0) > 0$. Hence, we conclude that for a given $\varepsilon > 0$ we have $R_1 \leq \varepsilon$ if $\|u_0\|_{L^d(\Omega)} + N_0 \leq \tilde{\lambda}$, where $\tilde{\lambda} := \tilde{\lambda}(\varepsilon) := \varepsilon/C_0$. So for $\varepsilon \leq \frac{1}{8c_1}$ it follows by induction that

$$R_n \leq 4\varepsilon, \quad n \in \mathbb{N}, \qquad \text{if } \|u_0\|_{L^d(\Omega)} + N_0 \leq \lambda, \text{ where } \lambda := \lambda(\varepsilon) := \min\left\{ \tilde{\lambda}, \frac{\varepsilon}{c_2 T_0} \right\}.$$

This shows that by imposing a smallness assumption on N_0 and u_0 and by closely following the proof of Theorem 4.25 we obtain the following *long-time existence result*.

Theorem 4.28. *Let $2 \leq d \leq p \leq q < \infty$ and suppose that Assumption 4.24 holds. For every $0 < T_0 < \infty$ there exists a $\lambda := \lambda(T_0)$ such that if $\|u_0\|_{L^p(\Omega)} + N_0 \leq \lambda$, then there exists a unique mild solution $u \in C([0, T_0]; L^p_\sigma(\Omega))$ to (4.82) which has the properties*

$$t \mapsto t^{\frac{d}{2}\left(\frac{1}{p}-\frac{1}{q}\right)} u(t) \in C([0, T_0]; L^q_\sigma(\Omega)),$$
$$t \mapsto t^{\frac{d}{2}\left(\frac{1}{p}-\frac{1}{q}\right)+\frac{1}{2}} \nabla u(t) \in C([0, T_0]; L^q(\Omega)^{d \times d}).$$

For the proof of Theorem 4.28 it is necessary to have a smallness assumption on N_0. It is quite remarkable that such a smallness assumption is not needed when treating a related problem in the whole space $\mathbb{R}^d$. In this case we have global L^p-L^q-estimates at hand, see Theorem 4.9, and therefore we can even expect to obtain global mild solutions. To conclude this section we want to make this more precise.

Let $1 < p < \infty$ and let $\{L_{\mathbb{R}^d}(t)\}_{t \geq 0}$ denote the family of operators on $L_\sigma^p(\mathbb{R}^d)$ introduced in Section 4.1, i.e.

$$\mathcal{D}(L_{\mathbb{R}^d}(t)) := \{u \in W^{2,p}(\mathbb{R}^d)^d \cap L_\sigma^p(\mathbb{R}^d) : M(t)x \cdot \nabla u \in L^p(\mathbb{R}^d)^d\},$$

$$L_{\mathbb{R}^d}(t)u := \Delta u + M(t)x \cdot \nabla u + c(t) \cdot \nabla u - M(t)u.$$

As in Section 4.1 we shall suppose that Assumption 4.1 holds, i.e. $M \in C([0, \infty); \mathbb{R}^{d \times d})$ and $c \in C([0, \infty); \mathbb{R}^d)$. By $\{U(t, s)\}_{t,s \geq 0}$ we denote the evolution system on $\mathbb{R}^d$ generated by the family of matrices $\{-M(t)\}_{t \geq 0}$, i.c.

$$\begin{cases} \partial_t U(t, s) = -M(t)U(t, s), & t, s \geq 0 \\ U(s, s) = \text{Id}. \end{cases}$$

For an initial value $u_0 \in L_\sigma^p(\mathbb{R}^d)$, $1 < p < \infty$, we consider now the following abstract Cauchy problem:

$$\begin{cases} u'(t) - L_{\mathbb{R}^d}(t)u(t) = -\mathbb{P}_{\mathbb{R}^d}(u \cdot \nabla u)(t), & t > 0, \\ u(0) = u_0. \end{cases} \tag{4.94}$$

The strongly continuous evolution system on $L_\sigma^p(\mathbb{R}^d)$ obtained in Theorem 4.6 is denoted by $\{S_{\mathbb{R}^d}(t, s)\}_{(t,s) \in \Lambda}$.

Formally, by using the variation of constants formula, problem (4.94) can be reduced to the following integral equation on $L_\sigma^p(\mathbb{R}^d)$:

$$u(t) = S_{\mathbb{R}^d}(t, 0)u_0 - \int_0^t S_{\mathbb{R}^d}(t, s)\mathbb{P}_{\mathbb{R}^d}(u \cdot \nabla u)(s)\, ds, \qquad t \geq 0. \tag{4.95}$$

For given $0 < T_0 \leq \infty$ we call a function $u \in C([0, T_0); L_\sigma^p(\mathbb{R}^d))$ a *mild solution* to (4.94) if the right-hand side of (4.95) is well-defined and if u satisfies the integral equation (4.95).

The following *main result* states the existence of a unique global mild solution if the initial value $u_0 \in L_\sigma^d(\mathbb{R}^d)$ is sufficiently small and if $\{U(t, s)\}_{t,s \geq 0}$ is uniformly bounded, i.e. there exists a constant $M_\infty > 0$ such that $|U(t, s)| \leq M_\infty$ for all $t, s \geq 0$.

Theorem 4.29. *Let $2 \leq d \leq q < \infty$ and $u_0 \in L_\sigma^d(\mathbb{R}^d)$. Moreover we suppose that $\{U(t, s)\}_{t,s \geq 0}$ is uniformly bounded. There exists a $\lambda > 0$ such that if $\|u_0\|_{L^d(\mathbb{R}^d)} \leq \lambda$, then there exists a unique mild solution $u \in C([0, \infty); L_\sigma^d(\mathbb{R}^d))$ to (4.82) which has the properties*

$$t \mapsto t^{\frac{d}{2}\left(\frac{1}{d} - \frac{1}{q}\right)}u(t) \in C([0, \infty); L_\sigma^q(\mathbb{R}^d)),$$

$$t \mapsto t^{\frac{d}{2}\left(\frac{1}{d} - \frac{1}{q}\right) + \frac{1}{2}}\nabla u(t) \in C([0, \infty); L^q(\mathbb{R}^d)^{d \times d}).$$

117

Remark 4.30. The uniform boundedness of $\{U(t,s)\}_{t,s\geq 0}$ is no smallness assumption on $M(\cdot)$. For example in the case $d = 3$ and for $\omega \in C([0,\infty);\mathbb{R}^3)$, let $M(t) := \omega(t) \times \cdot$ represent the linear mapping $x \mapsto \omega(t) \times x$. In this case $M(t)$ is skew-symmetric for all $t \in [0,\infty)$, i.e. $M(t)^\top = -M(t)$. Then it is easy to see that $\{U(t,s)\}_{t,s\geq 0}$ is a family of orthogonal matrices, see Remark 4.7. In this case we have $|U(t,s)| = 1$ for all $t,s \geq 0$. Therefore the assumptions in Theorem 4.29 are satisfied and we obtain a global mild solution if the initial value $u_0 \in L^d_\sigma(\mathbb{R}^d)$ is sufficiently small. This result holds even if $\omega(t) \to \infty$ as $t \to \infty$.

Proof. The proof is very similar to the proof of Theorem 4.25. It follows again the iteration scheme introduced in [Kat84, Gig86]. Therefore we restrict ourselves to the important steps in order to obtain a global solution. In the following let $M_\infty > 0$ be such that $|U(t,s)| \leq M_\infty$ for all $t,s \geq 0$

For $t \geq 0$ define $u_1(t) := S_{\mathbb{R}^d}(t,0)u_0$ and

$$u_{n+1}(t) = u_1(t) - \int_0^t S_\Omega(t,s)\mathbb{P}_{\mathbb{R}^d}(u_n \cdot \nabla u_n)(s)\, ds, \qquad n \in \mathbb{N}.$$

Set $w_1(t) = u_1(t)$ and $w_{n+1}(t) = u_{n+1}(t) - u_n(t)$, $n \in \mathbb{N}$. Let $q > d$ and set $\beta := \frac{d}{2}\left(\frac{1}{d} - \frac{1}{q}\right)$. For $n \in \mathbb{N}$ define

$$K_n := \sup_{t\in(0,\infty)} t^\beta \|u_n(t)\|_{L^q(\mathbb{R}^d)}, \qquad K_n' := \sup_{t\in(0,\infty)} t^{\frac{1}{2}} \|\nabla u_n\|_{L^d(\mathbb{R}^d)}$$

and

$$M_n := \sup_{t\in(0,\infty)} t^\beta \|w_n\|_{L^q(\mathbb{R}^d)}, \qquad M_n' := \sup_{t\in(0,\infty)} t^{\frac{1}{2}} \|\nabla w_n(t)\|_{L^d(\mathbb{R}^d)}.$$

As in the proof of Theorem 4.25 we can conclude

$$\|u_{n+1}(t)\|_{L^q(\Omega)} \leq t^{-\beta} K_1 + C K_n K_n' \int_0^t (t-s)^{-\frac{1}{2}} s^{-\beta-\frac{1}{2}}\, ds \tag{4.96}$$

and

$$\|\nabla u_{n+1}(t)\|_{L^p(\Omega)} \leq t^{-\frac{1}{2}} K_1' + C K_n K_n' \int_0^t (t-s)^{-\frac{d}{2q}-\frac{1}{2}} s^{-\beta-\frac{1}{2}}\, ds \tag{4.97}$$

for all $t \in (0,\infty)$, where $C := C(M_\infty) > 0$, see Theorem 4.9. Since $\beta < \frac{1}{2}$ and $q > d$ we conclude by Remark 4.27 that

$$\int_0^t (t-s)^{-\frac{1}{2}} s^{-\beta-\frac{1}{2}}\, ds \leq C t^{-\frac{1}{2}+\frac{d}{2q}} \qquad \text{and} \qquad \int_0^t (t-s)^{-\frac{d}{2q}-\frac{1}{2}} s^{-\beta-\frac{1}{2}}\, ds \leq C t^{-\frac{1}{2}}$$

for some constant $C > 0$. Hence multiplying (4.96) and (4.97) with t^β and $t^{\frac{1}{2}}$, respectively, and then taking the supremum over $t \in (0,\infty)$ yields

$$\begin{aligned} K_{n+1} &\leq K_1 + C_1 K_n K_n', \\ K_{n+1}' &\leq K_1' + C_2 K_n K_n', \end{aligned} \tag{4.98}$$

where $C_1 := C_1(M_\infty)$, $C_2 := C_2(M_\infty) > 0$ are constants independent of $n \in \mathbb{N}$. Next, set $R_n := \max\{K_n, K'_n\}$. It follows from (4.88) that

$$R_{n+1} \leq R_1 + c_1 R_n^2$$

for some constant $c_1 := c_1(M_\infty) \geq 1$ independent of $n \in \mathbb{N}$. By Theorem 4.9 we know that

$$R_1 \leq C_0 \|u_0\|_{L^d(\mathbb{R}^d)}$$

for some constant $C_0 := C_0(M_\infty) > 0$. Hence we can conclude that for any $\varepsilon > 0$ we have $R_1 \leq \varepsilon$ if $\|u_0\|_{L^d(\mathbb{R}^d)} \leq \lambda$, where $\lambda := \lambda(\varepsilon) := \varepsilon/C_0$. So for $\varepsilon \leq \frac{1}{8c_1}$ it follows by induction that

$$R_n \leq 4\varepsilon, \quad n \in \mathbb{N}, \qquad \text{if } \|u_0\|_{L^d(\mathbb{R}^d)} \leq \lambda, \text{ where } \lambda := \lambda(\varepsilon) := \varepsilon/C_0. \tag{4.99}$$

Moreover, as in Theorem 4.25 we can show that

$$\|w_{n+1}(t)\|_{L^q(\mathbb{R}^d)} \leq C \int_0^t (t-s)^{-\frac{1}{2}} \Big(\|u_n(s)\|_{L^q(\mathbb{R}^d)} \|\nabla w_n(s)\|_{L^d(\mathbb{R}^d)} $$
$$+ \|w_n(s)\|_{L^q(\mathbb{R}^d)} \|\nabla u_{n-1}(s)\|_{L^d(\mathbb{R}^d)} \Big) \, \mathrm{d}s$$

and

$$\|\nabla w_{n+1}(t)\|_{L^d(\mathbb{R}^d)} \leq C \int_0^t (t-s)^{-\frac{d}{2q}-\frac{1}{2}} \Big(\|u_n(s)\|_{L^q(\mathbb{R}^d)} \|\nabla w_n(s)\|_{L^d(\mathbb{R}^d)} $$
$$+ \|w_n(s)\|_{L^q(\mathbb{R}^d)} \|\nabla u_{n-1}(s)\|_{L^d(\mathbb{R}^d)} \Big) \, \mathrm{d}s.$$

Therefore similarly to (4.98) we conclude that

$$M_{n+1} \leq C_3(M'_{n-1} K_n + M_{n-1} K'_{n-1}),$$
$$M'_{n+1} \leq C_4(M'_{n-1} K_n + M_{n-1} K'_{n-1}),$$

where $C_3 := C_3(M_\infty)$, $C_4 := C_4(M_\infty) > 0$ are constants independent of $n \in \mathbb{N}$. Hence, if we set $\tilde{R} := \max\{M_n, M'_n\}$ and use (4.99), then we obtain

$$\tilde{R}_{n+1} \leq 8\varepsilon(C_3 + C_4)\tilde{R}_n \leq \frac{1}{2}\tilde{R}_n$$

for $\varepsilon \leq \min\left\{\frac{1}{8(C_3+C_4)}, \frac{1}{8c_1}\right\}$ and $\|u_0\|_{L^d(\mathbb{R}^d)} \leq \lambda$, where λ is chosen as in (4.99), cf. (4.91).

Now, by arguing analogously to the proof of Theorem 4.25 we obtain that the sequences

$$\{t \mapsto u_n(t)\}_{n \in \mathbb{N}} \quad \text{and} \quad \{t \mapsto t^{\frac{1}{2}} \nabla u_n(t)\}_{n \in \mathbb{N}}$$

are Cauchy sequences in $C([0, T]; L^d_\sigma(\Omega))$ and $C([0, T]; L^d(\Omega)^{d \times d})$, respectively. Therefore, they tend to unique limit functions

$$(t \mapsto u(t)) \in C([0, T]; L^d_\sigma(\Omega)) \quad \text{and} \quad (t \mapsto t^{\frac{1}{2}} v(t)) \in C([0, T]; L^d(\Omega)^{d \times d}),$$

respectively. By construction $v(t) = \nabla u(t)$ and $u \in C([0, T]; L^p_\sigma(\Omega))$ is a mild solution to (4.82). The stated properties and the uniqueness of the mild solution follow easily as in the proof of Theorem 4.25. $\qquad \square$

5 Density-dependent Navier-Stokes flows around a rotating obstacle

In this chapter we study density-dependent Navier-Stokes flows in the exterior of a rotating obstacle, and present a local existence and uniqueness result.

To describe the situation and the problem of this chapter more precisely, let $\mathcal{O} \subset \mathbb{R}^3$ be a compact set, also referred to as the *obstacle*, with boundary $\Gamma := \partial\mathcal{O}$. Let $\Omega := \mathbb{R}^3 \setminus \mathcal{O}$ be an exterior domain of class C^3. The obstacle $\mathcal{O}$ is assumed to be rotating with a prescribed constant angular velocity $\omega \in \mathbb{R}^3$. For notational purpose, let $M \in \mathbb{R}^{3\times3}$ denote the matrix that describes the linear map $y \mapsto \omega \times y$. Note that the matrix M is *skew-symmetric*, i.e. $M^\top = -M$. Since the rotation is prescribed, the obstacle at time $t \geq 0$ and its boundary at time $t \geq 0$ are represented by

$$\mathcal{O}(t) := \{y(t) = \mathrm{e}^{tM}x : x \in \mathcal{O}\}, \qquad \text{and} \qquad \Gamma(t) := \{y(t) = \mathrm{e}^{tM}x : x \in \Gamma\},$$

respectively. The exterior of the rotated obstacle at time $t \geq 0$ is represented by

$$\Omega(t) := \mathbb{R}^3 \setminus \mathcal{O}(t) = \{y(t) = \mathrm{e}^{tM}x : x \in \Omega\}.$$

Note that the matrices e^{tM} are *orthogonal* due to the fact that M is skew-symmetric.

The exterior of the rotating obstacle is filled with a viscous, incompressible fluid with variable density. The motion of such a fluid is decribed by the density-dependent Navier-Stokes equations on the time-dependent domain $\Omega(t)$ with the usual no-slip boundary condition:

$$\begin{cases}
\tilde{\rho}_t + v \cdot \nabla\tilde{\rho} &= 0 & \text{for } t \in \mathbb{R}_+,\, y \in \Omega(t), \\
\tilde{\rho}v_t + \tilde{\rho}v \cdot \nabla v + \nabla\mathrm{q} - \Delta v &= \tilde{\rho}\tilde{F} & \text{for } t \in \mathbb{R}_+,\, y \in \Omega(t), \\
\operatorname{div} v &= 0 & \text{for } t \in \mathbb{R}_+,\, y \in \Omega(t), \\
v(t,y) &= My & \text{for } t \in \mathbb{R}_+,\, y \in \Gamma(t), \\
\lim_{|y|\to\infty} \tilde{\rho}(t,y) &= \rho_\infty & \text{for } t \in \mathbb{R}_+, \\
(\tilde{\rho},v)|_{t=0} &= (\rho_0,u_0) & \text{for } y \in \Omega.
\end{cases} \tag{5.1}$$

Here $\rho_\infty \in \mathbb{R}_+$ is a fixed constant, $\tilde{\rho}$ stands for the unknown density of the fluid and v and q are the unknown velocity field and pressure of the fluid, respectively. Moreover, $\rho_0 : \Omega \to \mathbb{R}$ is a given initial density, $u_0 : \Omega \to \mathbb{R}^3$ a given initial velocity field and $\tilde{F}$ is a given external force per unit volume acting on the fluid.

To overcome the difficulty that $\Omega(t)$ varies with time, we rewrite (5.1) as a new problem on the fixed exterior domain Ω by using a linear coordinate transformation. Set

$$x = \mathrm{e}^{-tM}y, \qquad w(t,x) = \mathrm{e}^{-tM}v(t,y), \quad \mathrm{p}(t,x) = \mathrm{q}(t,y),$$
$$\rho(t,x) = \tilde{\rho}(t,y), \quad F(t,x) = \mathrm{e}^{-tM}\tilde{F}(t,y),$$

(5.2)

see also [His99a, GHH06a]. Here for simplicity, we shall assume that $F = F(x)$ does not depend on time, that is $\tilde{F}(t,y) = \mathrm{e}^{tM}F(\mathrm{e}^{-tM}y)$ is a periodic force. A simple computation yields

$$\partial_t \tilde{\rho}(t,y) = \partial_t \rho(t,x) - Mx \cdot \nabla \rho(t,x),$$
$$v \cdot \nabla_y \tilde{\rho} = w \cdot \nabla_x \rho,$$
$$\partial_t v(t,y) = \mathrm{e}^{tM}\left[\partial_t w(t,x) - Mx \cdot \nabla_x w(t,x) + Mw(t,x)\right],$$
$$\Delta_y v(t,y) = \mathrm{e}^{tM}\Delta_x w(t,x),$$
$$\nabla_y \mathrm{q}(t,y) = \mathrm{e}^{tM}\nabla_x \mathrm{p}(t,x),$$
$$v \cdot \nabla_y v = \mathrm{e}^{tM}\left(w \cdot \nabla_x w\right),$$
$$\operatorname{div} v(t,y) = \operatorname{div} w(t,x).$$

It is now easily seen that $(v, \mathrm{q}, \tilde{\rho})$ solves the original problem (5.1) if (w, p, ρ) solves the following new system of equations on the fixed domain Ω:

$$\begin{cases} \rho_t + w \cdot \nabla \rho - Mx \cdot \nabla \rho &= 0 & \text{in } \mathbb{R}_+ \times \Omega, \\ \rho w_t + \rho w \cdot \nabla w - \rho Mx \cdot \nabla w + \rho Mw + \nabla \mathrm{p} - \Delta w &= \rho F & \text{in } \mathbb{R}_+ \times \Omega, \\ \operatorname{div} w &= 0 & \text{in } \mathbb{R}_+ \times \Omega, \\ w &= Mx & \text{on } \mathbb{R}_+ \times \Gamma, \\ \lim_{|x| \to \infty} \rho(t,x) &= \rho_\infty & \text{in } \mathbb{R}_+, \\ (\rho, w)|_{t=0} &= (\rho_0, u_0) & \text{in } \Omega. \end{cases}$$

(5.3)

This new system of equations contains terms of the form $Mx \cdot \nabla \rho$ and $\rho Mx \cdot \nabla w$. The coefficients of these terms grow linearly in the space variable x and are therefore unbounded. This is the main particularity of the system (5.3) and here lies the main difficulty.

Density-dependent Navier-Stokes equations in bounded domains were studied e.g. by Danchin in [Dan06], where existence of local strong solutions and even global strong solutions for small data in the L^p-framework are proved. The main tools in Danchin's approach are maximal L^q-regularity properties of the classical Stokes operator. Danchin indicates that his techniques also work for exterior domains. However, this approach cannot work in our situation due to the unbounded term $\rho Mx \cdot \nabla w$. As explained in Section 1.5, the Stokes operator with rotating effect $L_{\Omega,p}$, defined by $L_{\Omega,p}w := \mathbb{P}_{\Omega,p}\left(\Delta w + Mx \cdot \nabla w - Mw\right)$, generates a C_0-semigroup which is *not analytic*, see Proposition 1.85. Therefore, this operator cannot have the property of maximal L^q-regularity, see Remark 1.46.

Recently, a variant of (5.3) was studied in the whole space case $\Omega = \mathbb{R}^d$ by Fang, Hieber and Zhang [FHZ10]. Similarly to [Dan04], they use techniques from Fourier analysis to obtain local existence of strong solutions. These techniques are applicable in the whole space $\mathbb{R}^d$ but do not carry over to exterior domains Ω.

Another widely used approach to density-dependent Navier-Stokes equations is based on a Faedo-Galerkin approximation in the L^2-setting, see e.g. [Kim87, BRMFC03, CK03, CK04]. Such a method was also used by Galdi and Silvestre, see [GS02, GS05], to study homogeneous Navier-Stokes flows around a rotating obstacle. This indicates that this technique seems suitable to study system (5.3).

Due to technical reasons it is more convenient to work with zero boundary data. Therefore we construct a boundary extension $b \in C_{c,\sigma}^{\infty}(\mathbb{R}^3)$ given by

$$b(x) := \zeta(x)Mx - \mathbb{B}_{\Omega_b}((\nabla\zeta) \cdot Mx), \tag{5.4}$$

where ξ is a smooth cut-off function which is supported only close to the boundary Γ and Ω_b is a bounded domain, see Section 1.6.2 for details. By setting $w := u + b$ problem (5.3) is equivalent to

$$
\begin{cases}
\rho_t + u \cdot \nabla\rho + b \cdot \nabla\rho - Mx \cdot \nabla\rho &= 0 & \text{in } \mathbb{R}_+ \times \Omega, \\
\left.\begin{array}{r}\rho u_t + \rho u \cdot \nabla u + \rho b \cdot \nabla u + \rho u \cdot \nabla b \\ - \rho Mx \cdot \nabla u + \rho Mu + \nabla\mathrm{p} - \Delta u\end{array}\right\} &= \rho f + \Delta b & \text{in } \mathbb{R}_+ \times \Omega, \\
\operatorname{div} u &= 0 & \text{in } \mathbb{R}_+ \times \Omega, \\
u &= 0 & \text{on } \mathbb{R}_+ \times \Gamma, \\
\lim_{|x|\to\infty} \rho(t,x) &= \rho_\infty & \text{in } \mathbb{R}_+, \\
(\rho, u)|_{t=0} &= (\rho_0, a) & \text{in } \Omega,
\end{cases}
\tag{5.5}
$$

where

$$a := u_0 - b, \qquad \text{and} \quad f := F + Mx \cdot \nabla b - Mb - b \cdot \nabla b. \tag{5.6}$$

5.1 Main result

In this section we state the main result of this chapter which concerns the existence and uniqueness of a local strong solution to (5.5). This directly yields a strong solution to (5.3). For notational simplicity we make the following assumption.

Assumption 5.1. For simplicity we shall suppose that $\rho_\infty = 1$. However, the result in this chapter also holds for arbitrary $\rho_\infty \in \mathbb{R}_+$.

For the approach in this chapter, based on a Faedo-Galerkin method and suitable a priori estimates, we need the following assumptions on the data (ρ_0, u_0).

Assumption 5.2. Throughout the rest of this chapter we suppose that the following holds:

$$F \in H^1(\Omega)^3,$$

$$\rho_0 \in C(\overline{\Omega}), \quad \underline{\rho} \le \rho_0(x) \le \bar{\rho} \quad \text{for all } x \in \overline{\Omega}, \quad 0 < \underline{\rho} \le 1,$$

$$\rho_0 - 1 \in H^2(\Omega), \quad (\rho_0 - 1)Mx \in H^1(\Omega)^3,$$

$$u_0 \in H^2(\Omega)^3, \quad \operatorname{div} u_0 = 0, \quad u_0|_\Gamma = Mx, \quad Mx \cdot \nabla u_0 \in L^2(\Omega)^3.$$

Remarks 5.3. 1. Clearly, when dealing with arbitrary $\rho_\infty \in \mathbb{R}_+$ we need to suppose $\rho_0 - \rho_\infty \in H^2(\Omega)$ and $(\rho_0 - \rho_\infty)Mx \in H^1(\Omega)^3$.

2. The condition $\rho_0 \geq \underline{\rho} > 0$ avoids the presence of regions of vacuum[1] in the fluid. This seems to be a physically reasonable assumption. We refer to [CK03, CK04] where existence results for density-dependent Navier-Stokes equations are proved for initial densities that may vanish on certain parts of the fluid domain.

3. Let $a := u_0 - b$ be the initial velocity field for the system (5.5). Assumption 5.2 ensures that $a \in H^2(\Omega)^3 \cap H^1_{0,\sigma}(\Omega)$ and $Mx \cdot \nabla a \in L^2(\Omega)^3$.

In order to fit the notion of a strong solution, introduced in Definition 1.44, to the setting of this chapter let us recall that in Section 1.5 we defined the Stokes operator with rotating effect $L_{\Omega,p}$ by

$$\mathcal{D}(L_{\Omega,p}) := \{u \in W^{2,p}(\Omega)^3 \cap W^{1,p}_0(\Omega)^3 \cap L^p_\sigma(\Omega) : Mx \cdot \nabla u \in L^p(\Omega)^3\},$$
$$L_{\Omega,p}u := \mathbb{P}_{\Omega,p}(\Delta u + Mx \cdot \nabla u - Mu),$$

where $1 < p < \infty$.

Let $T > 0$. By a *strong solution* to problem (5.5) on some interval $(0, T)$ we mean a triple of functions (u, p, ρ) with the properties

$$u \in H^1((0, T); L^2_\sigma(\Omega)) \cap L^2((0, T); \mathcal{D}(L_{\Omega,2})),$$
$$\nabla \mathrm{p} \in L^2((0, T); L^2(\Omega)),$$
$$\rho - 1 \in H^1((0, T); L^2(\Omega)) \cap L^2((0, T); H^1(\Omega)),$$

and such that (u, p, ρ) solves (5.5) (almost everywhere in $(0, T) \times \Omega$). Note that by the Sobolev embedding $H^1(0, T); L^2_\sigma(\Omega)) \hookrightarrow C([0, T]; L^2_\sigma(\Omega))$ it makes sense to write $(\rho(0), u(0))$.

Now we are in position to state the main result of this chapter.

Theorem 5.4. *Assume that the conditions on (ρ_0, u_0, F) stated in Assumption 5.2 hold. Then there exists a $T_0 > 0$ and a unique strong solution (u, p, ρ) to (5.5) defined on $(0, T_0)$ satisfying*

$$u \in C([0, T_0]; H^1_{0,\sigma}(\Omega)) \cap L^\infty((0, T_0); \mathcal{D}(L_{\Omega,2})) \cap L^2((0, T_0); \mathcal{D}(L_{\Omega,6})),$$

$$u_t \in L^\infty((0, T_0); L^2_\sigma(\Omega)) \cap L^2((0, T_0); H^1_{0,\sigma}(\Omega)),$$

$$\nabla \mathrm{p} \in L^\infty((0, T_0); L^2(\Omega)),$$

$$\rho \in L^\infty((0, T_0) \times \Omega),$$

$$\rho - 1 \in C([0, T_0]; L^2(\Omega)) \cap L^\infty((0, T_0); H^2(\Omega)),$$

$$\rho_t \in L^\infty((0, T_0); L^2(\Omega)),$$

$$\underline{\rho} \leq \rho(t, x) \leq \bar{\rho} \qquad \text{for almost all } (t, x) \in (0, T_0) \times \Omega.$$

---

[1] By regions of vacuum we mean open subsets of Ω where the density ρ vanishes.

Remarks 5.5.   1. It is clear that the result in Theorem 5.4 directly yields a strong solution (w, p, ρ) to (5.3). Moreover, by the transformation (5.2), we also obtain a solution $(v, \mathrm{q}, \bar{\rho})$ to the original problem (5.1).

2. Theorem 5.4 includes in particular the special case $\rho_0 \equiv 1$ and therefore the local existence result [GS05, Theorem 1] of Galdi and Silvestre is a special case of our result.

5.2 Outline of the proof of the main result

As the proof of the main result, Theorem 5.4, involves a fair amount of technicalities, we shall give an outline of the proof in this section. We present the main steps and ideas and indicate some of the difficulties. All technical details and the rigorous analysis are provided in later sections.

The main technique used in the proof of Theorem 5.4 is the Faedo-Galerkin method in the L^2-framework. In principle, this is a standard and widely used method. We refer e.g. to Heywood [Hey80], Kim [Kim87], Boldrini, Rojas-Medar and Fernández-Cara [BRMFC03], Choe and Kim [CK03], Cho and Kim [CK04] and Galdi and Silvestre [GS02, GS05] for similar approaches in various situations. The crucial point of such a Faedo-Galerkin method is to derive suitable (uniform) *a priori estimates* for the constructed sequence of approximating solutions. Here actually lies the main difficulty of our proof.

We shall first construct suitable weak solutions on some bounded domains, and then we show that these solutions have even higher regularity. Then, by deriving a priori estimates which do not depend on the actual size of the bounded domains, we even obtain a strong solution on the unbounded domain Ω. More precisely, we shall proceed as follows:

1. Construct approximating solutions on bounded domains.

2. Show existence of weak solutions on bounded domains.

3. Show that these weak solutions are even strong solutions.

4. Derive uniform estimates for the strong solutions on bounded domains, which do not depend on the actual size of the domain. Then, based on these uniform estimates, strong solution on the unbounded domain Ω are obtained.

In the course of this chapter it is particularly important to express on which parameters all appearing constants depend. To express the dependence of constants on the given data $a := u_0 - b$, ρ_0 and on the matrix M we use a parameter N_0 defined by

$$
\begin{aligned}
N_0 := \|F\|_{H^1(\Omega)} + \|a\|_{H^2(\Omega)} &+ \|Mx \cdot \nabla a\|_{L^2(\Omega)} + \|\rho_0 - 1\|_{H^2(\Omega)} \\
&+ \|(\rho_0 - 1)Mx\|_{H^1(\Omega)} + \underline{\rho} + \bar{\rho} + |M|.
\end{aligned} \tag{5.7}
$$

All the norms in (5.7) are finite as a consequence of Assumption 5.2, see Remarks 5.3.

Remark 5.6. Let $1 < p \leq \infty$. By using the construction (5.4) and Proposition 1.87 we conclude that there exists a constant $C := C(p) > 0$ such that

$$\|b\|_{W^{2,p}(\mathbb{R}^3)} \leq C|M|.$$

Moreover, using this estimate for b and since $F \in H^1(\Omega)^3$ we conclude by Sobolev's embedding, see Proposition 1.14, that for all $2 \leq q \leq 6$ there exists a constant $C := C(N_0, q) > 0$ such that

$$\|f\|_{L^q(\Omega)} \leq C \qquad \text{and} \qquad \|f\|_{H^1(\Omega)} \leq C.$$

In the following these estimates are frequently used without any further reference.

Step 1: Construction of approximating solutions on bounded domains

We start with constructing a suitable sequence of approximating solutions. As in [Hey80] or [GS05] we shall not work directly on the unbounded domain Ω, but we shall construct approximating solutions on a sequence of bounded domains. For that purpose we fix some radius $R_0 > 4$ sufficiently large such that

$$\overline{\operatorname{supp} b} \subset B_{R_0-4}.$$

This is possible since b is supported only close to the boundary Γ, see the construction in Section 1.6.2 for details. The radius R_0 will be fixed throughout the entire chapter. We shall first work on a bounded domain $\Omega_R := \Omega \cap B_R$ for some ball B_R with radius $R > R_0$. Then, later in the proof, we shall let $R \to \infty$ to obtain a solution on the exterior domain Ω. The

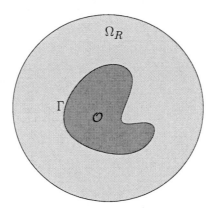

Figure 5.1: The bounded domain Ω_R.

advantage of working on a bounded domain Ω_R is that by Proposition 1.70 there exists an orthonormal basis $\{\psi_k\}_{k\in\mathbb{N}}$ of $L^2_\sigma(\Omega_R)$ consisting of eigenvectors of the Stokes operator A_{Ω_R} in $L^2_\sigma(\Omega_R)$, i.e.

$$A_{\Omega_R}\psi_k = -\lambda_k \psi_k, \qquad k \in \mathbb{N},$$

where $\{-\lambda_k\}_{k \in \mathbb{N}}$ are the corresponding eigenvalues of the Stokes operator. The existence of such an orthonormal bases will help us later to obtain norm estimates of higher order which are necessary in order to obtain a strong solution. In particular, we use the fact that if $u \in \text{span}\{\psi_1, \ldots, \psi_n\}$, then $A_{\Omega_R} u \in \text{span}\{\psi_1, \ldots, \psi_n\}$.

Remark 5.7. Since by assumption the domain Ω_R is of class C^3, we conclude by Proposition 2.3 that the eigenfunctions ψ_k even belong to $H^3(\Omega_R)^3$.

In the following denote by $\mathbb{P}^n_{\Omega_R}$ the *orthogonal projection* onto $\text{span}\{\psi_1, \ldots, \psi_n\}$, i.e.

$$\mathbb{P}^n_{\Omega_R} u := \sum_{k=1}^n \langle u, \psi_k \rangle \psi_k, \qquad u \in L^2(\Omega_R)^3.$$

Remark 5.8. It is clear that $\|\mathbb{P}^n_{\Omega_R} u\|_{L^2(\Omega_R)} \leq \|u\|_{L^2(\Omega_R)}$ holds since $\mathbb{P}^n_{\Omega_R}$ is an orthogonal projection. Moreover, $\mathbb{P}^n_{\Omega_R} u \to \mathbb{P}_{\Omega_R} u$ as $n \to \infty$ for all $u \in L^2(\Omega_R)^3$.

The relevant equations on the bounded domain Ω_R are given by

$$\begin{cases} \rho_t^R + u^R \cdot \nabla \rho^R + b \cdot \nabla \rho^R - Mx \cdot \nabla \rho^R = 0 & \text{in } \mathbb{R}_+ \times \Omega_R, \\ \left.\begin{array}{r} \rho^R u_t^R + \rho^R u^R \cdot \nabla u^R + \rho^R b \cdot \nabla u^R \\ + \rho^R u^R \cdot \nabla b - \rho^R Mx \cdot \nabla u^R \\ + \rho^R M u^R + \nabla \mathrm{p}^R - \Delta u^R \end{array}\right\} = \rho^R f^R + \Delta b & \text{in } \mathbb{R}_+ \times \Omega_R, \\ \text{div } u^R = 0 & \text{in } \mathbb{R}_+ \times \Omega_R, \\ u^R = 0 & \text{on } \mathbb{R}_+ \times \partial\Omega_R, \\ (\rho^R, u^R)|_{t=0} = (\rho_0^R, a^R) & \text{in } \Omega_R, \end{cases} \tag{5.8}$$

where (ρ_0^R, a^R) are suitably chosen initial values on Ω_R that correspond to the initial values (ρ_0, a) on Ω. See Lemma 5.18 and Remark 5.21 for details. In our case ρ_0^R can be just taken as the restriction of ρ_0 to Ω_R and a^R is chosen such that $a^R \to a$ in H^2-norm as $R \to \infty$ and such that the estimates

$$\|a^R\|_{L^2(\Omega_R)} \leq C \|a\|_{L^2(\Omega)}, \quad \|\nabla a^R\|_{L^2(\Omega_R)} \leq C \|\nabla a\|_{L^2(\Omega)}, \quad \|a^R\|_{H^2(\Omega_R)} \leq C \|a\|_{H^2(\Omega)},$$

and

$$\|Mx \cdot \nabla a^R\|_{L^2(\Omega_R)} \leq C \left(\|Mx \cdot \nabla a\|_{L^2(\Omega)} + \|a\|_{L^2(\Omega)} \right),$$

hold for some constant $C > 0$ independent of R. Furthermore, the function f^R is the restriction of f to Ω_R. The function b is supported in B_{R_0-4} only and thus can be considered as a function on Ω_R. By $u^R : \mathbb{R}_+ \times \Omega_R \to \mathbb{R}^3$ and $\mathrm{p}^R : \mathbb{R}_+ \times \Omega_R \to \mathbb{R}$ we denote the unknown velocity field and pressure of the fluid, respectively, and $\rho^R : \mathbb{R}_+ \times \Omega_R \to \mathbb{R}$ is the unknown density.

To simplify notation we drop the superscript R most of the time when it is clear from the context that we work on Ω_R. In particular, we will just write f instead of f^R. However, we shall keep track of all appearing constants and we shall always state if the constants depend on R or not. This will be very important later.

127

For fixed $n \in \mathbb{N}$ we consider now the following approximating problems on Ω_R:

$$\begin{cases} \rho_t^n + (u^n + b - Mx) \cdot \nabla \rho^n = 0 & \text{in } \mathbb{R}_+ \times \Omega_R, \\ \rho^n|_{t=0} = \rho_0^{n,R} & \text{in } \Omega_R, \end{cases} \tag{5.9}$$

and

$$\begin{cases} \mathbb{P}_{\Omega_R}^n \left(\rho^n u_t^n + \rho^n u^n \cdot \nabla u^n \right. \\ \left. \quad + \rho^n u^n \cdot \nabla b - \rho^n Mx \cdot \nabla u^n \right\} = \mathbb{P}_{\Omega_R}^n (\rho^n f + \Delta b) & \text{in } \mathbb{R}_+ \times \Omega_R, \\ \left. \quad + \rho^n b \cdot \nabla u^n + \rho^n M u^n - \Delta u^n \right) \\ \qquad\qquad\qquad\qquad \operatorname{div} u^n = 0 & \text{in } \mathbb{R}_+ \times \Omega_R, \\ \qquad\qquad\qquad\qquad u^n = 0 & \text{on } \mathbb{R}_+ \times \partial\Omega_R, \\ \qquad\qquad\qquad\qquad u^n|_{t=0} = a^{n,R} & \text{in } \Omega_R. \end{cases} \tag{5.10}$$

Here $\rho_0^{n,R} \in C^1(\overline{\Omega}_R)$ is a regularization of ρ_0^R and $a^{n,R} := \mathbb{P}_{\Omega_R}^n a^R$, see Lemma 5.19, Lemma 5.20 and Remark 5.21. There is no pressure p^n in (5.10) because of the projection $\mathbb{P}_{\Omega_R}^n$.

Since (5.10) is a problem in span $\{\psi_1, \ldots, \psi_n\}$, we make the ansatz

$$u^n(t) = \sum_{k=1}^{n} \alpha_{nk}(t) \psi_k, \qquad n \in \mathbb{N},$$

where the coefficients $\alpha_{nk}(\cdot)$ have to be chosen such that (5.10) holds. With this ansatz the system of equations (5.10) can be equivalently written as a system of ordinary differential equations for the coefficients $(\alpha_{nk}(\cdot))_{k=1}^{n}$:

$$\sum_{k=1}^{n} A_{ik}^n(t) \frac{\mathrm{d}}{\mathrm{d}t} \alpha_{nk}(t) + \sum_{k,l=1}^{n} B_{ikl}^n(t) \alpha_{nk}(t) \alpha_{nl}(t) + \sum_{k=1}^{n} C_{ik}^n(t) \alpha_{nk}(t) + \lambda_i \alpha_{ni}(t)$$
$$= d_i^n(t), \quad t \in \mathbb{R}_+, \quad i = 1, \ldots, n, \tag{5.11}$$

with initial conditions

$$\alpha_{nk}(0) = \int_{\Omega_R} a^{n,R}(x) \cdot \psi_k \, \mathrm{d}x, \qquad k = 1, \ldots, n,$$

and with

$$A_{ik}^n(t) := \int_{\Omega_R} \rho^n(t,x) \psi_k(x) \cdot \psi_i(x) \, \mathrm{d}x,$$

$$B_{ikl}^n(t) := \int_{\Omega_R} (\rho^n(t,x) \psi_k(x) \cdot \nabla \psi_l(x)) \cdot \psi_i(x) \, \mathrm{d}x,$$

$$C_{ik}^n(t) := \int_{\Omega_R} \rho^n(t,x) \left(b \cdot \nabla \psi_k(x) + \psi_k(x) \cdot \nabla b - Mx \cdot \nabla \psi_k(x) \right. \tag{5.12}$$
$$\left. + M\psi_k(x) \right) \cdot \psi_i(x) \, \mathrm{d}x,$$

$$d_i^n(t) := \int_{\Omega_R} (\rho^n(t,x) f(x) + \Delta b(x)) \cdot \psi_i(x) \, \mathrm{d}x.$$

The coupled problem (5.9) – (5.10) can be solved by combining classical techniques for ordinary differential equations, the method of characteristics for linear transport equations and a fixed point argument. The proof of the following result is presented in Section 5.4.

Proposition 5.9. *For every $R > R_0$, there exists a time $T_0 := T_0(N_0, R) > 0$, independent of n, and a pair of functions $(\rho^n, u^n) := (\rho^{n,R}, u^{n,R})$ satisfying*

$$\rho^n \in C^1([0, T_0] \times \overline{\Omega}_R), \qquad u^n \in C^2([0, T_0]; \mathbb{P}^n_{\Omega_R} L^2(\Omega_R)^3)$$

and such that (ρ^n, u^n) solves the coupled problem (5.9) – (5.10) on $[0, T_0]$. Moreover, there exists a constant $C := C(N_0, R, T_0) > 0$ also independent of n such that

$$\|\nabla u^n(t)\|^2_{L^2(\Omega_R)} + \int_0^t \|u_t^n(s)\|^2_{L^2(\Omega_R)} \, \mathrm{d}s + \int_0^t \|A_{\Omega_R} u^n(s)\|^2_{L^2(\Omega_R)} \, \mathrm{d}s \leq C$$

for all $t \in [0, T_0]$.

Step 2: Existence of weak solutions on bounded domains

In this step we shall prove the existence of a local weak solution $(\rho, u) := (\rho^R, u^R)$ to the system of equations (5.8) on the bounded domain Ω_R. The definition of a weak solution is given in Definition 5.10 below, see also [Kim87, Definition 1.1]. The notion of a weak solution is motivated by testing (5.8) in space and time with suitable test functions and by doing integration by parts with respect to x and t.

Definition 5.10. Let $T > 0$. A *weak solution* to problem (5.8) on the interval $(0, T)$ is a pair of functions (u, ρ) such that

(i) $u \in L^2((0, T); H^1_{0,\sigma}(\Omega_R))$, $\quad \rho \in L^\infty((0, T) \times \Omega_R)$, $\quad$ and

(ii) for all $\Phi \in C^1([0, T]; H^1_{0,\sigma}(\Omega_R))$ and $\Psi \in C^1([0, T]; H^1(\Omega_R))$ satisfying $\Phi(T, x) = \Psi(T, x) = 0$ almost everywhere in Ω_R, it holds that

$$-\int_0^T \int_{\Omega_R} \rho u \cdot \partial_t \Phi \, \mathrm{d}x \, \mathrm{d}t - \sum_{i=1}^3 \int_0^T \int_{\Omega_R} \rho u_i u \cdot \partial_i \Phi \, \mathrm{d}x \, \mathrm{d}t + \int_0^T \int_{\Omega_R} \nabla u : \nabla \Phi \, \mathrm{d}x \, \mathrm{d}t$$

$$+ \int_0^T \int_{\Omega_R} [\rho b \cdot \nabla u + \rho u \cdot \nabla b] \cdot \Phi \, \mathrm{d}x \, \mathrm{d}t - \int_0^T \int_{\Omega_R} \rho(Mx \cdot \nabla u - Mu) \cdot \Phi \, \mathrm{d}x \, \mathrm{d}t$$

$$= \int_0^T \int_{\Omega_R} (\rho f + \Delta b) \cdot \Phi \, \mathrm{d}x \, \mathrm{d}t + \int_{\Omega_R} \rho_0^R(x) a^R(x) \cdot \Phi(0, x) \, \mathrm{d}x,$$

and

$$-\int_0^T \int_{\Omega_R} \rho \cdot \partial_t \Psi \, \mathrm{d}x \, \mathrm{d}t - \int_0^T \int_{\Omega_R} \rho \cdot [(u + b - Mx) \cdot \nabla \Psi] \, \mathrm{d}x \, \mathrm{d}t$$

$$= \int_{\Omega_R} \rho_0^R(x) \cdot \Psi(0, x) \, \mathrm{d}x.$$

Proposition 5.11. *For every $R > R_0$, there exist a time $T_0 := T_0(N_0, R) > 0$ and a weak solution $(\rho, u) := (\rho^R, u^R)$ to problem (5.8) on $(0, T_0)$ such that*

$$u \in L^\infty((0, T_0); H^2(\Omega_R)^3 \cap H^1_{0,\sigma}(\Omega_R)),$$

$$u \in C([0, T_0]; H^1_{0,\sigma}(\Omega_R)),$$

$$u_t \in L^\infty((0, T_0); L^2_\sigma(\Omega_R)) \cap L^2((0, T_0); H^1_{0,\sigma}(\Omega_R)),$$

$$\rho \in C([0, T_0]; L^2(\Omega_R)).$$

Moreover, there exists a constant $C := C(N_0, R, T_0) > 0$ such that

$$\|u(t)\|^2_{L^2(\Omega_R)} + \|\nabla u(t)\|^2_{L^2(\Omega_R)} + \int_0^t \|\nabla^2 u(s)\|^2_{L^2(\Omega_R)} \, \mathrm{d}s \leq C,$$

$$\|u_t(t)\|^2_{L^2(\Omega_R)} + \|\nabla^2 u(t)\|^2_{L^2(\Omega_R)} + \int_0^t \|\nabla u_t(s)\|^2_{L^2(\Omega_R)} \, \mathrm{d}s \leq C,$$

hold for almost all $t \in (0, T_0)$.

Remark 5.12. The regularity of u in Proposition 5.11 is higher than required in Definition 5.10. However, for the density we just have $\rho \in L^\infty((0, T_0) \times \Omega_R)$. Thus, in order to obtain a strong solution later on, we need to improve the regularity of ρ.

The proof of Proposition 5.11 is presented in Section 5.6. It heavily relies on the following uniform estimates for the approximating solutions.

Proposition 5.13. *Let $(\rho^n, u^n) := (\rho^{n,R}, u^{n,R})$ be the approximating solution from Proposition 5.9. Then, for every $R > R_0$, there exists a $T_0 := T_0(N_0, R) > 0$, independent of n, and a constant $C := C(N_0, R, T_0) > 0$, independent of n, such that*

$$\underline{\rho} \leq \rho^n(t, x) \leq \bar{\rho} \quad \text{for all} \quad x \in \overline{\Omega}_R,$$

$$\|u^n(t)\|^2_{L^2(\Omega_R)} + \|\nabla u^n(t)\|^2_{L^2(\Omega_R)} + \int_0^t \|\nabla^2 u^n(s)\|^2_{L^2(\Omega_R)} \, \mathrm{d}s \leq C,$$

$$\|u_t^n(t)\|^2_{L^2(\Omega_R)} + \|\nabla^2 u^n(t)\|^2_{L^2(\Omega_R)} + \int_0^t \|\nabla u_t^n(s)\|^2_{L^2(\Omega_R)} \, \mathrm{d}s \leq C,$$

hold for all $t \in [0, T_0]$.

Step 3: Existence of strong solutions on bounded domains

Next, we show that the weak solution $(\rho, u) := (\rho^R, u^R)$ from Proposition 5.11 has even higher regularity. For that purpose, as mentioned in Remark 5.12, we need to prove higher regularity for the density ρ.

In order to sketch the idea how to get higher regularity for ρ, let us make first some formal considerations: We assume that ρ and u are as smooth as we want and that ρ solves

$$\begin{cases} \rho_t + (u + b - Mx) \cdot \nabla\rho &= 0 \quad \text{in } \mathbb{R}_+ \times \Omega_R, \\ \rho|_{t=0} &= \rho_0 \quad \text{in } \Omega_R \end{cases}$$

in the classical sense. Applying the derivative ∂_i to the transport equation, we obtain

$$\partial_t(\partial_i\rho) + (u + b - Mx) \cdot \nabla(\partial_i\rho) + (\partial_i u + \partial_i b - M_{\cdot i}) \cdot \nabla\rho = 0 \quad \text{in } \mathbb{R}_+ \times \Omega_R$$

for $i = 1, 2, 3$. Now, we test this equation with $\partial_i\rho$ and integrate by parts to obtain

$$\frac{d}{dt} \int_{\Omega_R} \frac{|\partial_i\rho|^2}{2} \, dx = - \int_{\Omega_R} [(\partial_i u + \partial_i b - M_{\cdot i}) \cdot \nabla\rho] \cdot \partial_i\rho \, dx.$$

Thus,

$$\frac{d}{dt} \int_{\Omega_R} \frac{|\partial_i\rho|^2}{2} \, dx \leq \left(\|\nabla u\|_{L^\infty(\Omega_R)} + \|\nabla b\|_{L^\infty(\Omega_R)} + |M| \right) \|\nabla\rho\|^2_{L^2(\Omega_R)}.$$

By Gronwall's inequality we can now control the L^2-norm of $\nabla\rho$ if we can control the L^∞-norm of ∇u. These calculations are done rigorously in Lemma 5.26.

The consideration above and the Sobolev embedding $L^\infty(\Omega_R) \hookrightarrow W^{1,q}(\Omega_R)$ for $q > 3$ indicate that in order to get higher regularity of the density ρ we need to control the velocity field u in $W^{2,q}(\Omega_R)^3$ for some $q > 3$. In the following we shall do this for $q = 6$.

Proposition 5.14. *Let $R > R_0$ and let $(\rho, u) := (\rho^R, u^R)$ be the weak solution from Proposition 5.11. Then there exist a time $T_0 := T_0(N_0, R) > 0$ and a constant $C := C(N_0, R, T_0) > 0$ such that*

$$\int_0^t \|u(s)\|^2_{W^{2,6}(\Omega_R)} \, ds \leq C$$

for all $t \in [0, T_0]$.

The basic idea in order to prove Proposition 5.14 is to use elliptic theory for the Stokes system. One can, at least formally, consider the momentum equation in (5.8) as a stationary Stokes equation, i.e. for fixed t we consider the equation

$$\begin{cases} -\Delta u(t) + \nabla \mathrm{p}(t) &= g(t) \quad \text{in } \Omega_R, \\ \operatorname{div} u(t) &= 0 \quad \text{in } \Omega_R, \\ u(t) &= 0 \quad \text{on } \partial\Omega_R, \end{cases} \tag{5.13}$$

where

$$g(t) := -\rho u_t - \rho u \cdot \nabla u + \ldots + \rho(Mx \cdot \nabla u - Mu) + \rho f + \Delta b.$$

By classical elliptic theory for the Stokes system, see Chapter 2, it is clear that $\nabla^2 u$ in the L^6-norm can be estimated by the right hand side g in the L^6-norm. The estimates in Proposition 5.11 and the Sobolev embedding $H^1(\Omega_R) \hookrightarrow L^6(\Omega_R)$ allow to estimate g in the L^6-norm.

By using the estimate stated in Proposition 5.14 and by Lemma 5.26 on linear transport equations it is now easy to show that (ρ, u) is even a strong solution to (5.8).

Proposition 5.15. *For every $R > R_0$, there exist a time $T_0 := T_0(N_0, R) > 0$ and a unique strong solution $(\rho, u) := (\rho^R, u^R)$ to problem (5.8) on $(0, T_0)$ such that*

$$\underline{\rho} \leq \rho(t, x) \leq \bar{\rho} \quad \text{for all } (t, x) \in [0, T_0] \times \Omega,$$

$$\rho \in L^\infty((0, T_0) \times \Omega_R),$$

$$\rho \in C([0, T_0]; H^2(\Omega_R)),$$

$$\rho_t \in C([0, T_0]; L^2(\Omega_R)),$$

$$u \in C([0, T_0]; H^2(\Omega_R)^3 \cap H^1_{0,\sigma}(\Omega_R)) \cap L^2((0, T_0); W^{2,6}(\Omega_R)),$$

$$u_t \in C([0, T_0]; L^2(\Omega_R)^3) \cap L^2((0, T_0); H^1_{0,\sigma}(\Omega_R)).$$

Moreover, the estimates stated in Proposition 5.11 and Proposition 5.14 also hold for u.

Step 4: Uniform estimates in R and strong solutions on Ω

This is the crucial part of the proof with the major difficulties. All the a priori estimates for the velocity field $u := u^R$ stated in Proposition 5.11 and Proposition 5.14 depend on the actual value of $R > R_0$. This is mainly because in the proofs of Proposition 5.11 and Proposition 5.14 we estimate Mx simply in the L^∞-norm. This certainly contributes to a constant depending on $R > R_0$. However, in order to pass to the limit $R \to \infty$ we actually need estimates uniformly in $R > R_0$. Therefore, we need to show that the constants can indeed be chosen independently of the actual value of $R > R_0$. Note that it is due to technical reasons that we first prove Proposition 5.11 and Proposition 5.14 with constants depending on $R > R_0$, and then separately the following estimates uniformly in $R > R_0$. This is pointed out again in Section 5.5.

Proposition 5.16. *Let $R > R_0$ and let $(\rho, u) := (\rho^R, u^R)$ be the strong solution to problem (5.8) on the bounded domain Ω_R obtained in Proposition 5.15. Then there exist a time $T_0 := T_0(N_0) > 0$, independent of R, and a constant $C := C(N_0, T_0) > 0$, also independent of R, such that*

$$\|u(t)\|^2_{L^2(\Omega_R)} + \int_0^t \|\nabla u(s)\|^2_{L^2(\Omega_R)} \, ds \leq C,$$

$$\|\nabla u(t)\|^2_{L^2(\Omega_R)} + \int_0^t \|\nabla^2 u(s)\|^2_{L^2(\Omega_R)} \, ds \leq C,$$

$$\|u_t(t)\|^2_{L^2(\Omega_R)} + \int_0^t \|\nabla u_t(s)\|^2_{L^2(\Omega_R)} \, ds \leq C,$$

$$\|\nabla^2 u(t)\|^2_{L^2(\Omega_R)} + \|Mx \cdot \nabla u(t) - Mu(t)\|^2_{L^2(\Omega_R)} \leq C,$$

$$\int_0^t \|\nabla^2 u(s)\|^2_{L^6(\Omega_R)} \, ds + \int_0^t \|Mx \cdot \nabla u(s) - Mu(s)\|^2_{L^6(\Omega_R)} \, ds \leq C,$$

$$\|\rho(t) - 1\|^2_{H^2(\Omega_R)} + \|\rho_t(t)\|^2_{L^2(\Omega_R)} + \|(\rho(t) - 1)Mx\|^2_{H^1(\Omega_R)} \leq C,$$

hold for all $t \in [0, T_0]$.

Since the system of equations (5.8) contains terms of the form $Mx \cdot \nabla \rho$ and $Mx \cdot \nabla u$, where the size of the coefficients actually grows with R, it is by far not trivial to obtain estimates uniformly in $R > R_0$. For that one needs to use properties of functions of the form $Mx \cdot \nabla u$, see Lemma 5.22 and Lemma 5.23, which were already used by Galdi and Silvestre in [GS05] to prove Theorem 5.4 in the special case $\rho_0 \equiv 1$. One of the main differences between constant and non-constant densities is that in order to get higher regularity for the density ρ we need to control u in $W^{2,6}(\Omega_R)^3$. Such a higher estimate in the L^p-framework was not needed in the special situation of Galdi and Silvestre. As explained above, to get an estimate for $\nabla^2 u$ in $L^6(\Omega_R)^3$ one uses elliptic theory for the Stokes system. However, to get estimates which do not depend on $R > R_0$, it is not suitable to put the term $Mx \cdot \nabla u$ on the right hand side, as done in (5.13), and then use elliptic estimates for the classical Stokes system. One rather needs to consider the "bad" term $Mx \cdot \nabla u$ as part of the operator. More precisely, we need the following elliptic estimate for the modified Stokes system with rotating effect.

Proposition 5.17. *Let $R > R_0$, $g \in L^6(\Omega_R)^3$ and let $(u, \mathrm{p}) \in W^{2,6}(\Omega_R)^3 \times W^{1,6}(\Omega_R)$ be a solution to the problem*

$$
\begin{cases}
-\Delta u - Mx \cdot \nabla u + Mu + \nabla \mathrm{p} &= g \quad \text{in } \Omega_R, \\
\operatorname{div} u &= 0 \quad \text{in } \Omega_R, \\
u &= 0 \quad \text{on } \partial\Omega_R.
\end{cases}
$$

Then there exists a constant $C := C(|M|) > 0$, independent of R, such that

$$
\|\nabla^2 u\|_{L^6(\Omega_R)} + \|Mx \cdot \nabla u - Mu\|_{L^6(\Omega_R)} + \|\nabla \mathrm{p}\|_{L^6(\Omega_R)} \leq C\Big(\|g\|_{L^6(\Omega_R)} + \|\nabla u\|_{L^2(\Omega_R)} \Big).
$$

The proof of Proposition 5.17 is presented in Section 5.3.3. Taking the special form of domain $\Omega_R := \Omega \cap B_R$ into account, we can use a localization (cut-off) argument in order to reduce the problem on Ω_R to a problem close to the boundary Γ and to a problem in the ball B_R. For the problem in the ball B_R we can apply the estimate stated in Theorem 2.12.

Based on the uniform estimates stated in Proposition 5.16 we let $R \to \infty$ and extract a strong solution (u, ρ) to the exterior domain problem (5.5). This procedure is again more or less standard and is also used e.g. in [Hey80] and [GS05]. The uniqueness of the strong solution (u, ρ) follows also by standard arguments. As it is usual for incompressible fluids, the pressure term $\nabla \mathrm{p}$ can be reconstructed from the solution (u, ρ).

5.3 Preparatory results

This section is of preparatory character. We collect technical lemmas, we prove results for linear transport equations and we give the full proof of the elliptic L^6-estimate stated in Proposition 5.17.

5.3.1 Technical lemmas

As it is outlined in Section 5.2, we often work on bounded domains $\Omega_R := \Omega \cap B_R$ for $R > R_0$. Thus, it is necessary to adapt the initial values (ρ, a) "living" on Ω to this given situation. For a proof of the following two lemmas we refer to [GS05, Lemma 4].

Lemma 5.18. *Let $a \in H^2(\Omega)^3 \cap H^1_{0,\sigma}(\Omega)$ with $Mx \cdot \nabla a \in L^2(\Omega)^3$ be the initial velocity field for the system of equations (5.5) and let $R > R_0$. There exits a function a^R and a constant $C := C(\Omega) > 0$, independent of R, such that*

(i) $a^R \in H^2(\Omega_R)^3 \cap H^1_{0,\sigma}(\Omega_R)$ and $a^R(x) = 0$ for all $x \in \Omega \setminus \Omega_{R-2}$,

(ii) $\|a^R\|_{L^2(\Omega_R)} \leq C\|a\|_{L^2(\Omega)}, \quad \|\nabla a^R\|_{L^2(\Omega_R)} \leq C\|\nabla a\|_{L^2(\Omega)}, \quad \|a^R\|_{H^2(\Omega_R)} \leq C\|a\|_{H^2(\Omega)},$

(iii) $\|Mx \cdot \nabla a^R\|_{L^2(\Omega_R)} \leq C\left(\|Mx \cdot \nabla a\|_{L^2(\Omega)} + \|a\|_{L^2(\Omega)}\right).$

Moreover, $a^R \to a$ in $H^2(\Omega)^3 \cap H^1_{0,\sigma}(\Omega)$ as $R \to \infty$.

Lemma 5.19. *Let $n \in \mathbb{N}$ and let a^R be the function from Lemma 5.18. Set $a^{n,R} := \mathbb{P}^n_{\Omega_R} a^R$. There exists a constant $C := C(\Omega) > 0$, independent of R and n, and a number $N := N(R) \in \mathbb{N}$ such that*

(i) $\|a^{n,R}\|_{H^1(\Omega_R)} \leq C\|a\|_{H^1(\Omega)}, \quad \|A_{\Omega_R} a^{n,R}\|_{L^2(\Omega_R)} \leq C\|a\|_{H^2(\Omega)} \quad$ for all $n \in \mathbb{N}$,

(ii) $\|Mx \cdot \nabla a^{n,R}\|_{L^2(\Omega_R)} \leq C\left(\|Mx \cdot \nabla a\|_{L^2(\Omega)} + \|a\|_{L^2(\Omega)}\right)$ for all $n > N$.

Moreover, $a^{n,R} \to a^R$ in $H^2(\Omega_R)^3 \cap H^1_{0,\sigma}(\Omega)$ as $n \to \infty$.

Lemma 5.20. *Let $\rho_0 \in C(\overline{\Omega})$ be the initial density for the system of equations (5.5) with the properties stated in Assumption 5.2. There exits a sequence of functions $\{\rho_0^n\}_{n \in \mathbb{N}}$ such that*

$$\rho_0^n \in C^1(\overline{\Omega}), \qquad \tfrac{1}{2}\underline{\rho} \leq \rho_0^n(x) \leq 2\bar{\rho} \quad \text{for all } x \in \overline{\Omega},$$

and $\rho_0^n \to \rho_0$ in $C(\overline{\Omega})$ as $n \to \infty$.

Proof. In the following let $\psi \in C_c^\infty(\mathbb{R}^3)$ with $\operatorname{supp}\psi \in B_1$, $\psi \geq 0$, $\int_{\mathbb{R}^3} \psi \, \mathrm{d}x = 1$ and consider the mollifier $\{\psi^n\}_{n \in \mathbb{N}}$ given by

$$\psi^n(x) := \frac{1}{n^3}\psi(nx), \qquad x \in \mathbb{R}^3, \qquad n \in \mathbb{N}.$$

Using the extension operator $E : H^2(\Omega) \to H^2(\mathbb{R}^3)$, we extend $\rho_0 - 1$ to a function $\tilde{\rho}_0 - 1 \in H^2(\mathbb{R}^3)$. By the Sobolev embedding $H^2(\mathbb{R}^3) \hookrightarrow C^{0,\frac{1}{2}}(\mathbb{R}^3)$ we conclude that $\tilde{\rho}_0$ is uniformly continuous. Now set $\tilde{\rho}_0^n = \rho_0 * \psi^n$ and denote by ρ_0^n the restriction of $\tilde{\rho}_0^n$ to $\overline{\Omega}_R$. It is clear that $\rho_0^n \in C^1(\overline{\Omega})$ and that $\rho_0^n \to \rho_0$ in $C(\overline{\Omega})$ as $n \to \infty$. We conclude that there exists a $N \in \mathbb{N}$ such that $\frac{1}{2}\underline{\rho} \leq \rho_0^n(x) \leq 2\bar{\rho}$ for all $x \in \overline{\Omega}$ provided $n > N$. After relabeling the sequence the assertion follows. $\qquad\square$

Remark 5.21. Let $R > R_0$. If we define ρ_0^R by taking the restriction of ρ_0 to the bounded domain $\overline{\Omega}_R$, then it is clear that $\rho_0^R \in C(\overline{\Omega}_R)$, $\underline{\rho} \leq \rho_0^R(x) \leq \bar{\rho}$ for all $x \in \overline{\Omega}_R$ and that

$$\|\rho_0^R - 1\|_{H^2(\Omega_R)} \leq \|\rho_0 - 1\|_{H^2(\Omega)}, \qquad \|(\rho_0^R - 1)Mx\|_{H^1(\Omega_R)} \leq \|(\rho_0 - 1)Mx\|._{H^1(\Omega)}$$

Analogously, for fixed $n \in \mathbb{N}$, we define $\rho_0^{n,R}$ as the restriction of ρ_0^n, defined in Lemma 5.20, to the bounded domain Ω_R. Then we have a sequence of regularized initial densities $\{\rho_0^{n,R}\}_{n \in \mathbb{N}} \subset C^1(\overline{\Omega}_R)$ and by Lemma 5.20 it is clear that $\rho_0^{n,R} \to \rho_0^R$ in $C(\overline{\Omega}_R)$ as $n \to \infty$.

The next two lemmas provide properties of functions of the form $Mx \cdot \nabla u - Mu$ defined on $\Omega_R := \Omega \cap B_R$. These lemmas were proved by Galdi and Silvestre, see [GS02, GS05] and [Sil04], and they are crucial for deriving the uniform a priori estimates in Section 5.5.2.

Lemma 5.22. *Let $R > R_0$ and let $u \in \mathcal{D}(A_{\Omega_R})$. Then it holds that*

$$Mx \cdot \nu|_{\partial B_R} = 0, \qquad \text{and} \qquad Mx \cdot \nabla u|_{\partial B_R} = 0.$$

Proof. The outer normal vector at $x \in \partial B_R$ is given by $\nu(x) = \frac{x}{R}$. Hence, for $x \in \partial B_R$ we have

$$Mx \cdot \nu(x) = \frac{1}{R} \sum_{i,j=1}^{3} M_{ij} x_j x_i.$$

Since M is skew symmetric, i.e. $M_{ij} = -M_{ji}$, we see that $Mx \cdot \nu(x) = 0$ holds for $x \in \partial B_R$. For a proof of the second assertion we refer to [GS05, Lemma 3]. □

Lemma 5.23. *For $u \in \mathcal{D}(A_{\Omega_R})$ it holds that $Mx \cdot \nabla u - Mu \in L_\sigma^2(\Omega_R)$ and*

$$\int_{\Omega_R} (-Mx \cdot \nabla u + Mu) \cdot A_{\Omega_R} u \, dx = \int_{\Gamma} \left(-(\nu \cdot \nabla u) \cdot (Mx \cdot \nabla u) + \frac{1}{2} |\nabla u|^2 Mx \cdot \nu \right) d\sigma$$

$$- \int_{\Omega_R} \nabla Mu : \nabla u \, dx.$$

Proof. Here we follow [Sil04, Lemma 2.1, Lemma 2.2]. Take a sequence $\{v_k\}_{k \in \mathbb{N}} \subset C_{c,\sigma}^\infty(\Omega_R)$ such that $v_k \to u$ in $H_{0,\sigma}^1(\Omega_R)$ as $k \to \infty$. Then an easy computation shows that

$$\text{div}\,(Mx \cdot \nabla v_k - Mv_k) = Mx \cdot \nabla(\text{div}\,v_k) + \text{div}\,Mv_k - \text{div}\,Mv_k = 0.$$

Thus, $\Phi_k := Mx \cdot \nabla v_k - Mv_k \in C_{c,\sigma}^\infty(\Omega_R)$. Since $v_k \to u$ in $H_0^1(\Omega_R)^3$ as $k \to \infty$, we conclude that $\Phi_k \to Mx \cdot \nabla u - Mu$ in $L^2(\Omega_R)^3$ as $k \to \infty$. The first assertion follows. Since $Mx \cdot \nabla u - Mu \in L_\sigma^2(\Omega_R)$, $Mx \cdot \nabla u|_{\partial B_R} = 0$ and $Mu|_{\partial \Omega_R} = 0$, we obtain

$$\int_{\Omega_R} (-Mx \cdot \nabla u + Mu) \cdot A_{\Omega_R} u \, dx = \int_{\Omega_R} (-Mx \cdot \nabla u + Mu) \cdot \Delta u \, dx$$

$$= \int_{\Omega_R} \nabla(Mx \cdot \nabla u - Mu) : \nabla u \, dx - \int_{\Gamma} (\nu \cdot \nabla u) \cdot (Mx \cdot \nabla u) \, d\sigma.$$

Integrating by parts, and using that $\operatorname{div} Mx = 0$, $Mx \cdot \nu|_{\partial B_R} = 0$ and $M^\top = -M$, we obtain

$$\int_{\Omega_R} \nabla(Mx \cdot \nabla u) : \nabla u \, dx$$

$$= \sum_{k=1}^{3} \sum_{i=1}^{3} \int_{\Omega_R} \nabla M_{i.}x D_i u_k \cdot \nabla u_k \, dx$$

$$= \sum_{k=1}^{3} \int_{\Omega_R} \nabla u_k M \cdot \nabla u_k \, dx + \sum_{k=1}^{3} \sum_{i=1}^{3} \int_{\Omega_R} M_{i.}x D_i \nabla u_k \cdot \nabla u_k \, dx$$

$$= \sum_{k=1}^{3} \int_{\Omega_R} Mx \cdot \nabla \frac{|\nabla u_k|^2}{2} \, dx = \int_\Gamma Mx \cdot \nu \frac{|\nabla u|^2}{2} \, dx.$$

Thus, also the second assertion is proved. $\qquad\square$

5.3.2 The linear transport equation

Here we collect existence and regularity results for linear transport equations of the form

$$\begin{cases} \rho_t + (u + b - Mx) \cdot \nabla\rho &= 0, \quad \text{in } [0, T] \times \Omega_R, \\ \rho|_{t=0} &= \rho_0, \quad \text{in } \Omega_R, \end{cases} \tag{5.14}$$

where $T > 0$ is arbitrary but fixed, ρ_0 is a given initial density, u is a given solenoidal vector field, $\Omega_R := \Omega \cap B_R$ for $R > R_0$, and $b \in C^\infty_{c,\sigma}(\mathbb{R}^3)$ is the boundary extension in (5.4), i.e.

$$b|_\Gamma = Mx, \qquad \overline{\operatorname{supp} b} \subset B_{R_0-4}, \qquad \text{and} \qquad \operatorname{div} b = 0.$$

The first result concerns the classical solvability of (5.14) for smooth data, cf. Lemma 5.46.

Lemma 5.24. *Let $u \in C^1([0,T]; H^3(\Omega_R)^3 \cap H^1_{0,\sigma}(\Omega_R))$ and $\rho_0 \in C^1(\overline{\Omega}_R)$. Then problem (5.14) admits a unique, classical solution $\rho \in C^1([0,T] \times \overline{\Omega}_R)$ given by*

$$\rho(t, x) = \rho_0(U(0, t, x)),$$

where $U \in C^1([0,T] \times [0,T] \times \overline{\Omega}_R)^3$ is the unique solution to the initial value problem

$$\begin{cases} \partial_t U(t, s, x) &= u(t, U(t, s, x)) + b(U(t, s, x)) - MU(t, s, x), \quad 0 \le t \le T, \\ U(s, s, x) &= x \in \overline{\Omega}_R, \qquad\qquad\qquad\qquad\qquad\qquad\qquad\qquad\quad 0 \le s \le T. \end{cases} \tag{5.15}$$

Moreover, if $\underline{\rho} \le \rho_0(x) \le \bar{\rho}$ for all $x \in \overline{\Omega}_R$, then $\underline{\rho} \le \rho(t, x) \le \bar{\rho}$ for all $(t, x) \in [0,T] \times \overline{\Omega}_R$.

Proof. The proof is based on the method of characteristics which is described in detail in the supplement to this chapter. The only main difference to the setting of Lemma 5.46 is that Mx does not vanish at ∂B_R. Therefore, we only discuss the parts of the proof that need extra considerations. As in Lemma 5.46, we conclude that $u \in C^1([0,T] \times \overline{\Omega}_R)^3$ and we extend u

to a function $\widetilde{u} \in C^1([0,T] \times \mathbb{R}^3)^3$. The classical Picard-Lindelöf theorem yields that there exists times T_0, T_1 with $0 \le T_0 < T_1 \le T$ and a unique solution $U(\cdot, s, x) \in C^1([T_0, T_1])^3$ of

$$\begin{cases} \partial_t U(t,s,x) & = \ \widetilde{u}(t, U(t,s,x)) + b(U(t,s,x)) - MU(t,s,x), \quad T_0 \le t \le T_1, \\ U(s,s,x) & = \ x \in \overline{\Omega}_R, \qquad\qquad\qquad\qquad\qquad\qquad\qquad\quad 0 \le s \le T. \end{cases}$$

If $x \in \Gamma$, then the unique solution is given by $U(\cdot, s, x) = x$, since $\widetilde{u}(t,x) + b(x) - Mx = 0$ for all $(t,x) \in [T_0, T_1] \times \Gamma$. Now let us consider the case $x \in \partial B_R$. By using the facts that b is zero close to ∂B_R and that M is a skew-symmetric matrix there exist times T_0', T_1' with $T_0 \le T_0' < T_1' \le T_1$ such that

$$\frac{1}{2}\frac{\mathrm{d}}{\mathrm{dt}}\|U(t,s,x)\|^2 = \widetilde{u}(t, U(t,s,x)) \cdot U(t,s,x) - MU(t,s,x) \cdot U(t,s,x)$$
$$= \widetilde{u}(t, U(t,s,x)) \cdot U(t,s,x)$$

for all $t \in [T_0', T_1']$. If we consider $U(\cdot, s, x)$ in polar coordinates, i.e. $U(\cdot, s, x) := r(\cdot)v(\cdot)$ where $r : [T_0, T_1] \to [0, \infty)$ and $v(\cdot)$ is a function on the unit sphere ∂B_1, then we obtain the following ordinary differential equation for $r(\cdot)$:

$$\frac{1}{2}\frac{\mathrm{d}}{\mathrm{dt}}r(t)^2 = \widetilde{u}(t, r(t)v(t)) \cdot r(t)v(t), \qquad r(s) = R, \qquad t \in [T_0', T_1']. \tag{5.16}$$

It is clear that (5.16) has a unique solution. Since $\widetilde{u}|_{\partial B_R} = 0$, it is easy to see that $r(t) \equiv R$ solves (5.16). The uniqueness of the solution implies that $\|U(t,s,x)\| = R$ for all $t \in [T_0', T_1']$ if $x \in \partial B_R$. In particular, by repeating this argument we conclude that $U(\cdot, s, x)$ stays in $\overline{\Omega}_R$ if $x \in \overline{\Omega}_R$, and that it solves (5.15) on $[0,T]$. Since $u \in C^1([0,T] \times \overline{\Omega}_R)$, classical arguments for ordinary differential equations yield that $U \in C^1([0,T] \times [0,T] \times \overline{\Omega}_R)$. The unique classical solution is now given by $\rho(t,x) := \rho_0(U(0,t,x))$, see Lemma 5.46 for details. $\qquad\square$

Next let us briefly discuss weaker forms of solutions to transport equations: Assume that $u \in L^2((0,T); H^1_{0,\sigma}(\Omega_R))$ and $\rho_0 \in L^\infty(\Omega_R)$. We say a function $\rho \in L^\infty((0,T) \times \Omega_R)$ is a *weak solution* to (5.14) if

$$\int_0^T \int_{\Omega_R} \rho \cdot \partial_t \Psi \, \mathrm{d}x \, \mathrm{dt} - \int_0^T \int_{\Omega_R} \rho \cdot [(u + b - Mx) \cdot \nabla\Psi] \, \mathrm{d}x \, \mathrm{dt}$$
$$= \int_{\Omega_R} \rho_0(x) \cdot \Psi(0,x) \, \mathrm{d}x \tag{5.17}$$

holds for all $\Psi \in C^1([0,T]; H^1(\Omega_R))$ with $\Psi(T,x) = 0$ for almost all $x \in \Omega_R$, cf. Definition 5.10. This notion of a solution is motivated by testing (5.14) in space and time with Ψ and integrating by parts with respect to x and t.

Lemma 5.25. *Let $\rho_0 \in L^\infty(\Omega_R)$ and let $\rho \in L^\infty((0,T) \times \Omega_R)$ be a weak solution to (5.14). Then $\|\rho(t)\|_{L^2(\Omega_R)} = \|\rho_0\|_{L^2(\Omega_R)}$ for all $t \in [0,T]$.*

Proof. We extend $u \in L^2((0,T); H^1_{0,\sigma}(\Omega_R))$ by zero to a function $\widetilde{u} \in L^2((0,T); H^1_{0,\sigma}(\mathbb{R}^3))$. Moreover, we extend $\rho_0 \in L^\infty(\Omega_R)$ by zero to $\widetilde{\rho}_0 \in L^p(\mathbb{R}^3)$ for all $1 \le p \le \infty$ and $\rho \in$

$L^\infty((0,T) \times \Omega_R)$ by zero to $\tilde{\rho} \in L^\infty((0,T); L^p(\mathbb{R}^3))$ for all $1 \le p \le \infty$. Since ρ is a weak solution to (5.14) it is clear that

$$- \int_0^T \int_{\mathbb{R}^3} \tilde{\rho} \cdot \partial_t \Psi \, dx \, dt - \int_0^T \int_{\mathbb{R}^3} \tilde{\rho} \cdot [(\tilde{u} + b - Mx) \cdot \nabla \Psi] \, dx \, dt = 0$$

holds for all $\Psi \in C_c^\infty((0,T) \times \mathbb{R}^3)$. Then by [DL89, Corollary II.2] we see that

$$- \int_0^T \int_{\mathbb{R}^3} |\tilde{\rho}|^2 \cdot \partial_t \Psi \, dx \, dt - \int_0^T \int_{\mathbb{R}^3} |\tilde{\rho}|^2 \cdot [(\tilde{u} + b - Mx) \cdot \nabla \Psi] \, dx \, dt = 0$$

holds for all $\Psi \in C_c^\infty((0,T) \times \mathbb{R}^3)$.

Let $\psi \in C_c^\infty((0,T))$ and $\phi \in C_c^\infty(\mathbb{R}^3)$ with $\phi \ge 0$, $\operatorname{supp} \phi \subset B_2$ and $\phi \equiv 1$ on B_1. Set $\phi_\varepsilon(x) := \phi(\varepsilon x)$ for $\varepsilon > 0$ and $x \in \mathbb{R}^3$. Then,

$$- \int_0^T \int_{\mathbb{R}^3} |\tilde{\rho}|^2 \cdot \psi_t \, \phi_\varepsilon \, dx \, dt = \int_0^T \int_{\mathbb{R}^3} |\tilde{\rho}|^2 \cdot [(\tilde{u} + b - Mx) \cdot \nabla \phi_\varepsilon] \psi \, dx \, dt.$$

The term on the left-hand side converges to

$$- \int_0^T \left(\int_{\mathbb{R}^3} |\tilde{\rho}|^2 \, dx \right) \psi_t \, dt$$

as $\varepsilon \to 0$. For the term on the right-hand side we note that

$$\left| \int_{\mathbb{R}^3} |\tilde{\rho}|^2 \cdot [(\tilde{u} + b - Mx) \cdot \nabla \phi_\varepsilon] \, dx \right| \le C\varepsilon \int_{\frac{1}{\varepsilon} \le |x| \le \frac{2}{\varepsilon}} |\tilde{u} + b| \, dx + C|M| \int_{\frac{1}{\varepsilon} \le |x| \le \frac{2}{\varepsilon}} |\tilde{\rho}|^2 \, dx.$$

Since $\tilde{u}$, b and $\tilde{\rho}$ are zero outside of Ω_R, the right-hand side is zero if ε is sufficiently small. Hence, we conclude that

$$\frac{d}{dt} \int_{\mathbb{R}^3} |\tilde{\rho}|^2 \, dx = 0$$

on $(0,T)$ in the sense of scalar distributions. It follows now from classical arguments that

$$\|\tilde{\rho}(t)\|_{L^p(\mathbb{R}^d)} = \|\tilde{\rho}_0\|_{L^p(\mathbb{R}^d)}.$$

for all $t \in [0,T]$. This yields the assertion. □

Lemma 5.25 yields that weak solutions to (5.14) are unique; see also [DL89, Theorem II.2].

We say a function $\rho \in H^1((0,T); L^2(\Omega_R)) \cap L^2((0,T); H^1(\Omega_R))$ is a *strong solution* to (5.14) if ρ solves (5.14) (almost everywhere in $(0,T) \times \Omega_R$).

The next lemma concerns the existence of strong solutions. Moreover, it collects some a priori estimates for ρ. See e.g. [CK03, Lemma 7] or [CK06, Lemma 6] for similar results.

Lemma 5.26. *Let $\rho_0 \in H^2(\Omega_R) \cap C(\overline{\Omega}_R)$ with $\underline{\rho} \le \rho_0(x) \le \bar{\rho}$ for all $x \in \overline{\Omega}_R$, where $\underline{\rho} > 0$. Moreover, let*

$$u \in C([0,T]; H^1_{0,\sigma}(\Omega_R)) \cap L^\infty((0,T); H^2(\Omega_R)^3) \cap L^2((0,T); W^{2,6}(\Omega_R)^3)$$

Then problem (5.14) has a unique, strong solution ρ with the following properties:

$$\rho \in C([0,T]; H^2(\Omega_R)), \quad \rho \in C([0,T] \times \Omega_R), \quad \rho_t \in C([0,T]; L^2(\Omega_R)) \cap L^2((0,T); H^1(\Omega_R)).$$

Moreover, $\underline{\rho} \le \rho(t,x) \le \bar{\rho}$ holds for all $(t,x) \in [0,T] \times \overline{\Omega}_R$ and the solution ρ satisfies

$$\|\rho(t)\|_{L^2(\Omega_R)} = \|\rho_0\|_{L^2(\Omega_R)}, \tag{5.18}$$

$$\|\rho(t)\|^2_{H^1(\Omega_R)} \le \|\rho_0\|^2_{H^1(\Omega_R)} \exp\left(C \int_0^t \left(\|u(s)\|_{W^{2,6}(\Omega_R)} + \|b\|_{W^{2,6}(\Omega_R)} + |M|\right) \mathrm{d}s\right), \tag{5.19}$$

$$\|\rho(t)\|^2_{H^2(\Omega_R)} \le \|\rho_0\|^2_{H^2(\Omega_R)} \exp\left(C \int_0^t \left(\|u(s)\|_{W^{2,6}(\Omega_R)} + \|b\|_{W^{2,6}(\Omega_R)} + |M|\right) \mathrm{d}s\right), \tag{5.20}$$

for all $t \in [0,T]$ and for some constant $C > 0$ independent of $R > R_0$.

Proof. The proof is divided into 4 steps. The principal approach is first to regularize the data such that we can apply the classical method of characteristics to construct a sequence of approximating solutions which are sufficiently smooth. Then, for this sequence of solutions, we derive suitable a priori estimates which allow us to extract a strong solution to (5.14) by classical compactness arguments.

Step 1: Regularizing the data. We construct suitable sequences $\{\rho_0^n\}_{n \in \mathbb{N}}$ and $\{u^n\}_{n \in \mathbb{N}}$ of smooth scalar functions and smooth vector fields, respectively, such that

$$\begin{aligned}
&\rho_0^n \in C^2(\overline{\Omega}_R), \quad u^n \in C^2([0,T]; H^4(\Omega_R)^3 \cap H^1_{0,\sigma}(\Omega_R)), \\
&\rho_0^n \to \rho_0 \quad \text{in } C(\overline{\Omega}_R) \cap H^2(\Omega_R), \quad u^n \to u \quad \text{in } L^2((0,T); W^{2,6}(\Omega_R)^3 \cap H^1_{0,\sigma}(\Omega_R)),
\end{aligned} \tag{5.21}$$

as $n \to \infty$. The construction used here is similar to the one used in the proof of Lemma 5.20.

In the following let $\psi \in C_c^\infty(\mathbb{R}^3)$ with $\operatorname{supp} \psi \in B_1$, $\psi \ge 0$, $\int_{\mathbb{R}^3} \psi \, \mathrm{d}x = 1$ and consider the mollifier $\{\psi^n\}_{n \in \mathbb{N}}$ given by

$$\psi^n(x) := \frac{1}{n^3} \psi(nx), \quad x \in \mathbb{R}^3, \quad n \in \mathbb{N}.$$

Analogously, we take some function $\phi \in C_c^\infty(\mathbb{R}^4)$ with $\operatorname{supp} \phi \in B_1$, $\phi \ge 0$, $\int_{\mathbb{R}^4} \phi \, \mathrm{d}x = 1$ and consider the mollifier $\{\phi^n\}_{n \in \mathbb{N}}$ given by

$$\phi^n(t,x) := \frac{1}{n^4} \phi(nt, nx), \quad (t,x) \in \mathbb{R} \times \mathbb{R}^3, \quad n \in \mathbb{N}.$$

Using the extension operator $E : H^2(\Omega_R) \to H^2(\mathbb{R}^3)$, we may extend ρ_0 to a function $\tilde{\rho}_0 \in H^2(\mathbb{R}^3)$. By the Sobolev embedding $H^2(\mathbb{R}^3) \hookrightarrow C^{0,\frac{1}{2}}(\mathbb{R}^3)$ we conclude that $\tilde{\rho}_0$ is in uniformly continuous. Now set $\tilde{\rho}_0^n = \rho_0 * \psi^n$ and denote by ρ_0^n the restriction of $\tilde{\rho}_0^n$ to $\overline{\Omega}_R$. It is clear that $\rho_0^n \in C^2(\overline{\Omega}_R)$, $\rho_0^n \to \rho_0$ in $H^2(\Omega_R)$ and $\rho_0^n \to \rho_0$ in $C(\overline{\Omega}_R)$ as $n \to \infty$.

Similarly, we extend u in time by 0 and in space by the extension operator $E : W^{2,6}(\Omega_R)^3 \to W^{2,6}(\mathbb{R}^3)^3$ to obtain some function $\tilde{w} \in L^2(\mathbb{R}; W^{2,6}(\mathbb{R}^3)^3)$. For $i = 1, 2, 3$, set $\tilde{w}_i^n = \tilde{w}_i * \phi^n$. By w^n we denote the restriction of $\tilde{w}^n$ to $[0, T] \times \overline{\Omega}_R$. Hence, we conclude that $w^n \in C^\infty([0, T] \times (\Omega_R))^3$. Moreover, we have $w^n \to u$ in $L^2((0, T); W^{2,6}(\Omega_R)^3)$ as $n \to \infty$. For fixed $t \in [0, T]$, let $v^n(t) \in H^2(\Omega_R)^3 \cap H_{0,\sigma}^1(\Omega_R)$, together with a suitable pressure $\mathrm{p}(t)$, be the unique solution to the stationary Stokes problem

$$\begin{cases} -\Delta v^n(t) + \nabla \mathrm{p}(t) &= -\Delta w^n(t) & \text{in } \Omega_R, \\ \operatorname{div} v^n(t) &= 0 & \text{in } \Omega_R, \\ v^n(t) &= 0 & \text{on } \partial\Omega_R. \end{cases}$$

Since $w^n(t) \in H^m(\Omega_R)^3$ for all $m \in \mathbb{N}$ and all $t \in [0, T]$, we can conclude by classical elliptic theory for the Stokes system, see Proposition 2.3, that we have $v^n(t) \in H^4(\Omega_R)^3 \cap H_{0,\sigma}^1(\Omega_R)$ for all $t \in [0, T]$ and $v^n \in L^2((0, T); H^4(\Omega_R)^3 \cap H_{0,\sigma}^1(\Omega_R)) \cap L^2((0, T); W^{2,6}(\Omega_R)^3 \cap H_{0,\sigma}^1(\Omega_R))$. Furthermore, by using again an elliptic estimate we see that

$$\int_0^T \|v^n(t) - u(t)\|_{W^{2,6}(\Omega_R)}^2 \, dt \leq C \int_0^T \|\Delta w^n(t) - \Delta u(t)\|_{L^6(\Omega_R)}^2 \, dt$$

for some suitable constant $C > 0$ depending on Ω_R. It directly follows that $v^n \to u$ in $L^2((0, T); W^{2,6}(\Omega_R)^3 \cap H_{0,\sigma}^1(\Omega_R)^3)$ as $n \to \infty$. Since the space $C^\infty([0, T]; H^4(\Omega_R)^3 \cap H_{0,\sigma}^1(\Omega_R))$ is dense in $L^2((0, T); H^4(\Omega_R)^3 \cap H_{0,\sigma}^1(\Omega_R))$, we can now conclude that there exists a sequence $\{u^n\}_{n \in \mathbb{N}} \subset C^\infty([0, T]; H^4(\Omega_R)^3 \cap H_{0,\sigma}^1(\Omega_R))$ such that $u^n \to u$ in $L^2((0, T); W^{2,6}(\Omega_R)^3 \cap H_{0,\sigma}^1(\Omega_R))$ as $n \to \infty$. Consequently, (5.21) is proved.

Step 2: Solving the regularized problem. Now for $n \in \mathbb{N}$ we consider the following regularized problem

$$\begin{cases} \rho_t + (u^n + b - Mx) \cdot \nabla\rho &= 0 & \text{in } [0, T] \times \Omega_R, \\ \rho|_{t=0} &= \rho_0^n & \text{in } \Omega_R. \end{cases} \tag{5.22}$$

By the Sobolev embbeding $H^4(\Omega_R) \hookrightarrow C^2(\overline{\Omega}_R)$ we conclude that $u^n \in C^2([0, T] \times \overline{\Omega}_R)$. Then, by arguing analogously to the proof of Lemma 5.24, see also Lemma 5.46, and by using we obtain a solution $\rho^n \in C^2([0, T] \times \overline{\Omega}_R)$ to (5.22) represented by

$$\rho^n(t, x) = \rho_0^n(U^n(0, t, x)), \tag{5.23}$$

where $U^n \in C^2([0, T] \times [0, T] \times \overline{\Omega}_R)$ solves the ordinary differential equation

$$\begin{cases} \partial_t U^n(t, s, x) &= u^n(t, U^n(t, s, x)) + b(U^n(t, s, x)) - MU^n(t, s, x), & 0 \leq t \leq T, \\ U(s, s, x) &= x \in \overline{\Omega}_R, & 0 \leq s \leq T. \end{cases}$$

Next we show that the sequence of solutions $\{\rho^n\}_{n \in \mathbb{N}}$ converges to a solution – at least in the weak sense – to the original problem (5.14). For that purpose note that for $n, m \in \mathbb{N}$, there exists a constant $C := C(|M|) > 0$ such that

$$|U^n(t,s,x) - U^m(t,s,x)|$$

$$\leq \int_s^t |u^n(\tau, U^n(\tau,s,x)) - u^m(\tau, U^m(\tau,s,x))|\, d\tau + \int_s^t |b(U^n(\tau,s,x)) - b(U^m(\tau,s,x))|\, d\tau$$

$$+ \int_s^t |MU^n(\tau,s,x) - MU^m(\tau,s,x)|\, d\tau$$

$$\leq \int_s^t \|u^n(\tau) - u^m(\tau)\|_{L^\infty(\Omega_R)}\, d\tau + C \int_s^t \|\nabla u^m\|_{L^\infty(\Omega_R)} |U^n(\tau,s,x) - U^m(\tau,s,x)|\, d\tau$$

$$+ C \int_s^t \|\nabla b\|_{L^\infty(\Omega_R)} |U^n(\tau,s,x) - U^m(\tau,s,x)|\, d\tau + C \int_s^t |U^n(\tau,s,x) - U^m(\tau,s,x)|\, d\tau.$$

By Gronwall's inequality and the Sobolev embedding $W^{1,6}(\Omega_R) \hookrightarrow L^\infty(\Omega_R)$ we obtain

$$|U^n(t,s,x) - U^m(t,s,x)|$$

$$\leq \int_0^T \|u^n(\tau) - u^m(\tau)\|_{W^{1,6}(\Omega_R)}\, d\tau \, \exp\left(\int_0^T \|\nabla u^m(\tau)\|_{W^{1,6}(\Omega_R)}\, d\tau + CT \right).$$

Thus, since $u^n \to u$ in $L^2((0,T); W^{2,6}(\Omega_R)^3 \cap H^1_{0,\sigma}(\Omega_R))$ as $n \to \infty$, we conclude that $\{U^n(t,s,x)\}_{n\in\mathbb{N}}$ is a Cauchy sequence in $C([0,T] \times [0,T] \times \overline{\Omega}_R)$. From formula (5.23) and the fact that $\rho_0^n \to \rho_0$ in $C(\overline{\Omega}_R)$, it now follows that there exists some $\rho \in C([0,T] \times \overline{\Omega}_R)$ such that $\rho^n \to \rho$ in $C([0,T] \times \overline{\Omega}_R)$ as $n \to \infty$. Moreover, since $\underline{\rho} \leq \rho_0(x) \leq \bar{\rho}$ for all $x \in \overline{\Omega}_R$, it follows that also $\underline{\rho} \leq \rho(t,x) \leq \bar{\rho}$ for all $(t,x) \in [0,T] \times \overline{\Omega}_R$.

Next, let us show that ρ is a weak solution to (5.14). Let $\Psi \in C^1([0,T]; H^1(\Omega_R))$ with $\Psi(T,x) = 0$ for almost all $x \in \Omega_R$. It is clear that

$$\int_0^T \int_{\Omega_R} \rho^n \cdot \partial_t \Psi\, dx\, dt \to \int_0^T \int_{\Omega_R} \rho \cdot \partial_t \Psi\, dx\, dt,$$

$$\int_0^T \int_{\Omega_R} \rho^n \cdot [(b - Mx) \cdot \nabla \Psi]\, dx\, dt \to \int_0^T \int_{\Omega_R} \rho \cdot [(b - Mx) \cdot \nabla \Psi]\, dx\, dt,$$

and

$$\int_{\Omega_R} \rho_0^n \cdot \Psi(0,x)\, dx \to \int_{\Omega_R} \rho_0 \cdot \Psi(0,x)\, dx,$$

as $n \to \infty$. Thus, it just remains to show that

$$\int_0^T \int_{\Omega_R} \rho^n \cdot (u^n \cdot \nabla \Psi)\, dx\, dt \to \int_0^T \int_{\Omega_R} \rho \cdot (u \cdot \nabla \Psi)\, dx\, dt \tag{5.24}$$

as $n \to \infty$. We note first that

$$\int_0^T \int_{\Omega_R} [\rho^n \cdot (u^n \cdot \nabla \Psi) - \rho \cdot (u \cdot \nabla \Psi)]\, dx\, dt$$

$$= \int_0^T \int_{\Omega_R} (\rho^n - \rho) \cdot (u^n \cdot \nabla \Psi)\, dx\, dt + \int_0^T \int_{\Omega_R} \rho \cdot ([u^n - u] \cdot \nabla \Psi)\, dx\, dt$$

$$=: I_1 + I_2.$$

141

Then, by using the convergence of ρ^n in $C([0, T] \times \overline{\Omega}_R)$ and of u^n in $L^2((0, T); W^{2,6}(\Omega_R)^3 \cap H^1_{0,\sigma}(\Omega_R))$, we see that

$$|I_1| \leq C \int_0^T \|\rho^n - \rho\|_{L^2(\Omega_R)} \|u^n\|_{L^2(\Omega_R)} \, dt \to 0,$$

and

$$|I_2| \leq C \int_0^T \|\rho\|_{L^2(\Omega_R)} \|u^n - u\|_{L^2(\Omega_R)} \, dt \to 0,$$

as $n \to \infty$. Hence (5.24) follows and we can conclude that ρ solves (5.14) in the weak sense.

Step 3: Uniform estimates for ρ^n. In order to show that ρ has even higher regularity we need to derive uniform estimates for the approximating sequence $\{\rho^n\}_{n \in \mathbb{N}}$ in higher order norms (up to the H^2-norm).

To start with, we test equation (5.22) with ρ^n and integrate by parts to obtain

$$\frac{d}{dt} \int_{\Omega_R} \frac{|\rho^n|^2}{2} \, dx = -\int_{\Omega_R} [(u^n + b - Mx) \cdot \nabla \rho^n] \cdot \rho^n \, dx = 0. \tag{5.25}$$

Here we have used that

$$\int_{\Omega_R} [(u^n + b - Mx) \cdot \nabla \rho^n] \cdot \rho^n \, dx = \int_{\Omega_R} (u^n + b - Mx) \cdot \nabla \frac{|\rho^n|^2}{2} \, dx$$

$$= -\int_{\Omega_R} \operatorname{div} (u^n + b - Mx) \frac{|\rho^n|^2}{2} \, dx + \int_{\partial \Omega_R} \nu \cdot (u^n + b - Mx) \frac{|\rho^n|^2}{2} \, dx = 0. \tag{5.26}$$

This holds since $\operatorname{div}(u^n + b - Mx) = 0$, $u|_{\partial \Omega_R} = 0$, $b|_\Gamma = Mx$, $b|_{\partial B_R} = 0$ and $Mx \cdot \nu|_{\partial B_R} = 0$. From (5.25) we get

$$\|\rho^n(t)\|^2_{L^2(\Omega_R)} = \|\rho^n(0)\|^2_{L^2(\Omega_R)} \tag{5.27}$$

for all $t \in [0, T]$.

Next, we derive a norm estimate for $\nabla \rho^n$. For that purpose we apply ∂_i to equation (5.22) to obtain

$$\partial_t(\partial_i \rho^n) + (u^n + b - Mx) \cdot \nabla(\partial_i \rho^n) + (\partial_i u^n + \partial_i b - M_{\cdot i}) \cdot \nabla \rho^n = 0 \quad \text{in } [0, T] \times \Omega_R \tag{5.28}$$

for $i = 1, 2, 3$. Similarly as in (5.26), we can show that

$$\int_{\Omega_R} [(u^n + b - Mx) \cdot \nabla(\partial_i \rho^n)] \cdot \partial_i \rho^n \, dx = 0.$$

Thus, testing (5.28) with $\partial_i \rho^n$ and integrating by parts yield

$$\frac{d}{dt} \int_{\Omega_R} \frac{|\partial_i \rho^n|^2}{2} \, dx = -\int_{\Omega_R} [(\partial_i u^n + \partial_i b - M_{\cdot i}) \cdot \nabla \rho^n] \cdot \partial_i \rho^n \, dx.$$

Hence,

$$\frac{d}{dt} \int_{\Omega_R} \frac{|\partial_i \rho^n|^2}{2} \, dx \leq \left(\|\nabla u^n\|_{L^\infty(\Omega_R)} + \|\nabla b\|_{L^\infty(\Omega_R)} + |M| \right) \|\nabla \rho^n\|_{L^2(\Omega_R)}^2. \tag{5.29}$$

Integrating with respect to t, Gronwall's inequality and the Sobolev embedding $W^{1,6}(\Omega_R) \hookrightarrow L^\infty(\Omega_R)$ yield

$$\|\rho^n(t)\|_{H^1(\Omega_R)}^2 \leq \|\rho_0^n\|_{H^1(\Omega)}^2 \exp \left(C \int_0^t \left(\|u^n(s)\|_{W^{2,6}(\Omega_R)} + \|b\|_{W^{2,6}(\Omega_R)} + |M| \right) ds \right) \tag{5.30}$$

for some constant $C > 0$. Note that the constant $C > 0$ is essentially the embedding constant of $W^{1,6}(\Omega_R) \hookrightarrow L^\infty(\Omega_R)$ which is independent of $R > 0$, see Proposition 1.14.

We now derive a norm estimate for $\nabla^2 \rho^n$. Applying ∂_{ij} to equation (5.22), we obtain

$$\partial_t(\partial_{ij}\rho^n) + (u^n + b - Mx) \cdot \nabla(\partial_{ij}\rho^n) + (\partial_i u^n + \partial_i b - M_{\cdot i}) \cdot \nabla \partial_j \rho^n$$
$$+ (\partial_j u^n + \partial_j b - M_{\cdot j}) \cdot \nabla \partial_i \rho^n + (\partial_{ij} u^n + \partial_{ij} b) \cdot \nabla \rho^n = 0 \quad \text{in } [0,T] \times \Omega_R$$

for $i, j = 1, 2, 3$. Similarly as above, testing this equation with $\partial_{ij}\rho^n$ and integrating by parts yield

$$\frac{d}{dt} \int_{\Omega_R} \frac{|\partial_{ij}\rho^n|^2}{2} \, dx = - \int_{\Omega_R} [(\partial_i u^n + \partial_i b - M_{\cdot i}) \cdot \nabla(\partial_j \rho^n)] \cdot \partial_{ij}\rho^n \, dx$$
$$- \int_{\Omega_R} [(\partial_j u^n + \partial_j b - M_{\cdot j}) \cdot \nabla(\partial_i \rho^n)] \cdot \partial_{ij}\rho^n \, dx - \int_{\Omega_R} [(\partial_{ij} u^n + \partial_{ij} b) \cdot \nabla \rho^n] \cdot \partial_{ij}\rho^n \, dx.$$

Thus,

$$\frac{d}{dt} \int_{\Omega_R} \frac{|\partial_{ij}\rho^n|^2}{2} \, dx \leq 2 \left(\|\nabla u^n\|_{L^\infty(\Omega_R)} + \|\nabla b\|_{L^\infty(\Omega_R)} + |M| \right) \|\nabla^2 \rho^n\|_{L^2(\Omega_R)}^2$$
$$+ \left(\|\nabla^2 u^n\|_{L^6(\Omega_R)} + \|\nabla^2 b\|_{L^6(\Omega_R)} \right) \|\nabla \rho^n\|_{L^3(\Omega_R)} \|\nabla^2 \rho\|_{L^2(\Omega_R)}.$$

By using the embedding $H^1(\Omega_R) \hookrightarrow L^3(\Omega_R)$ we see that

$$\frac{d}{dt} \int_{\Omega_R} \frac{|\partial_{ij}\rho^n|^2}{2} \, dx \leq 2 \left(\|\nabla u^n\|_{L^\infty(\Omega_R)} + \|\nabla b\|_{L^\infty(\Omega_R)} + |M| \right) \|\nabla^2 \rho^n\|_{L^2(\Omega_R)}^2$$
$$+ \left(\|\nabla^2 u^n\|_{L^6(\Omega_R)} + \|\nabla^2 b\|_{L^6(\Omega_R)} \right) \|\nabla \rho^n\|_{H^1(\Omega_R)}^2.$$

Combining this estimate with (5.29), integrating with respect to t and using Gronwall's inequality yield

$$\|\nabla \rho^n(t)\|_{H^1(\Omega_R)}^2 \leq \|\nabla \rho_0^n\|_{H^1(\Omega_R)}^2 \exp \left(C \int_0^t \left(\|\nabla u^n(s)\|_{L^\infty(\Omega_R)} + \|u^n\|_{W^{2,6}(\Omega_R)} \right. \right.$$
$$\left. \left. + \|\nabla b\|_{L^\infty(\Omega_R)} + \|b\|_{W^{2,6}(\Omega_R)} + |M| \right) ds \right) \tag{5.31}$$

for a constant $C > 0$. Thus, by combining the estimates (5.27) and (5.31) and by applying the Sobolev embedding $W^{1,6}(\Omega_R) \hookrightarrow L^\infty(\Omega_R)$, we obtain

$$\|\rho^n(t)\|_{H^2(\Omega_R)} \le \|\rho_0^n\|_{H^2(\Omega_R)} \exp\left(C \int_0^t \left(\|u^n(s)\|_{W^{2,6}(\Omega_R)} + \|b\|_{W^{2,6}(\Omega_R)} + |M| \right) ds \right) \quad (5.32)$$

for all $t \in [0,T]$ and for some constant $C > 0$ independent of $R > 0$, see Proposition 1.14.

Step 4: Showing higher regularity for ρ. As a consequence of (5.32) and (5.21) and by recalling that $\rho_t^n = -(u^n + b - Mx) \cdot \nabla \rho^n$ we deduce that

$$\rho^n \subset L^\infty((0,T); H^2(\Omega_R)) \quad \text{is uniformly bounded,}$$
$$\rho_t^n \subset L^\infty((0,T); H^1(\Omega_R)) \quad \text{is uniformly bounded.}$$

Hence, by using the uniqueness of the solution ρ to the weak formulation (5.17), it follows – after extracting a suitable subsequence – that

$$\rho^n \stackrel{*}{\rightharpoonup} \rho \quad \text{weakly* in } L^\infty((0,T); H^2(\Omega_R)),$$

by that we mean

$$\int_0^T \int_{\Omega_R} \rho^n \cdot g \, dx \, dt \to \int_0^T \int_{\Omega_R} \rho \cdot g \, dx \, dt$$

for all $g \in L^1((0,T); H^2(\Omega_R))$. Moreover, by using classical compactness arguments, see Proposition 1.89, it follows – after extracting a suitable subsequence – that

$$\rho^n \to \rho \quad \text{strongly in } C([0,T]; H^1(\Omega_R)).$$

By using (5.21) and the fact that $\rho^n \to \rho$ strongly in $C([0,T]; H^1(\Omega_R))$ as $n \to \infty$, we immediately obtain (5.18) and (5.19) as a consequence of (5.27) and (5.30), respectively. Moreover, by Lemma 1.91 we know that ρ is weakly continuous in $[0,T]$ with values in $H^2(\Omega_R)$ and Lemma 1.92 yields that for fixed $t \in [0,T]$ we have $\rho^n(t) \rightharpoonup \rho(t)$ weakly in $H^2(\Omega_R)$ as $n \to \infty$. Hence, by using this weak convergence, estimate (5.20) follows directly from estimate (5.32).

It remains to prove the strong time-continuity of ρ in $H^2(\Omega_R)$. From (5.20) it follows that

$$\limsup_{t \to 0+} \|\rho(t)\|_{H^2(\Omega_R)} \le \|\rho_0\|_{H^2(\Omega_R)}.$$

This implies $\lim_{t \to 0+} \|\rho(t)\|_{H^2(\Omega_R)} = \|\rho_0\|_{H^2(\Omega_R)}$, and by the weak time-continuity we can even conclude that ρ is right-continuous at $t = 0$ with values in $H^2(\Omega_R)$. Then, it easily follows that ρ is strongly continuous in $[0,T]$ with values in $H^2(\Omega_R)$ since the transport equation (5.14) is invariant under reflections and translations in time. To make this more precise, let $t_0 \in (0,T]$ be arbitrary but fixed. Then it is easily seen that the function $\tilde{\rho} := \rho(t,x) := \rho(\pm t + t_0, x)$ is the unique solution to

$$\tilde{\rho}_t + (\tilde{u} + b - Mx) \cdot \nabla \tilde{\rho} = 0,$$

where $\tilde{u} := \tilde{u}(t,x) := \pm u(\pm t + t_0, x)$. By arguing as above, $\rho \in C([0,T]; H^2(\Omega_R))$ follows.

Moreover, by (5.14), we have $\rho_t = -(u + b - Mx) \cdot \nabla\rho$. Consequently, by using $u \in C([0,T]; H^1_0(\Omega_R)^3 \cap L^2_\sigma(\Omega_R))$ we conclude that $\rho_t \in C([0,T]; L^2(\Omega_R))$. Moreover, by using the Sobolev embedding $W^{1,6}(\Omega_R) \hookrightarrow L^\infty(\Omega_R)$ and $u \in L^2((0,T); W^{2,6}(\Omega_R)^3)$ we obtain $\rho_t \in L^2((0,T); H^1(\Omega_R))$.

The fact that ρ is a strong solution to (5.14) follows by integrating the weak formulation (5.17) by parts with respect to t and x. This concludes the proof. $\qquad\square$

To conclude this section we collect some technical norm estimates for the density ρ, which are needed later on.

Lemma 5.27. *Let ρ be the strong solution to (5.14) obtained in Lemma 5.26. Then there exists a constant $C > 0$, independent of $R > R_0$, such that*

$$\|\rho(t)Mx\|^2_{L^2(\Omega_R)} + \|\nabla\rho(t)(Mx)^\top\|^2_{L^2(\Omega_R)}$$

$$\leq \left(\|\rho_0 Mx\|^2_{L^2(\Omega_R)} + \|\nabla\rho_0(Mx)^\top\|^2_{L^2(\Omega_R)} + 2|M|^2 \int_0^t \|\rho(s)\|^2_{H^1(\Omega_R)} \, \mathrm{d}s \right)$$

$$\exp\left(C \int_0^t \left(\|u(s)\|^2_{W^{1,6}(\Omega_R)} + \|b\|^2_{W^{1,6}(\Omega_R)} + \|u(s)\|_{W^{2,6}(\Omega_R)} + \|b\|_{W^{2,6}(\Omega_R)} + |M| \right) \mathrm{d}s \right)$$

for all $t \in [0,T]$.

Proof. As in the proof of Lemma 5.26 let $\{\rho^n\}_{n\in\mathbb{N}}$ be the approximating sequence of solutions that solve the regularized problems (5.22). Since $\rho^n \to \rho$ strongly in $C([0,T]; H^1(\Omega_R))$ and $\rho^n_0 \to \rho_0$ strongly in $H^2(\Omega_R)$ as $n \to \infty$, it suffices to show the estimates for the approximating sequence $\{\rho^n\}_{n\in\mathbb{N}}$, only. To start with, we multiply (5.22) with $M_{i.}x$ and obtain .

$$\partial_t(\rho^n M_{i.}x) + (u^n + b - Mx) \cdot \nabla(\rho^n M_{i.}x) - M_{i.}(u^n + b - Mx)\rho^n = 0, \quad \text{in } [0,T] \times \Omega_R$$

for $i = 1,2,3$. Now we test this equation with $\rho^n M_{i.}x$ and integrate by parts to obtain

$$\frac{\mathrm{d}}{\mathrm{d}t} \int_{\Omega_R} \frac{|\rho^n M_{i.}x|^2}{2} \, \mathrm{d}x - \int_{\Omega_R} [M_{i.}(u^n + b - Mx)]\rho^n \cdot \rho^n M_{i.}x \, \mathrm{d}x. \tag{5.33}$$

Here we used that

$$\int_{\Omega_R} [(u^n + b - Mx) \cdot \nabla(\rho^n M_{i.}x)] \cdot (\rho^n M_{i.}x) \, \mathrm{d}x = \int_{\Omega_R} (u^n + b - Mx) \cdot \nabla \frac{|\rho^n M_{i.}x|^2}{2} \, \mathrm{d}x$$

$$= -\int_{\Omega_R} \mathrm{div}\,(u^n + b - Mx) \frac{|(\rho^n M_{i.}x)|^2}{2} \, \mathrm{d}x + \int_{\partial\Omega_R} \nu \cdot (u^n + b - Mx) \frac{|\rho^n M_{i.}x|^2}{2} \, \mathrm{d}x = 0.$$

This holds since $\mathrm{div}\,(u^n + b - Mx) = 0$, $u|_{\partial\Omega_R} = 0$, $b|_\Gamma = Mx$, $b|_{\partial B_R} = 0$ and $Mx \cdot \nu|_{\partial B_R} = 0$. Estimating the right-hand side in (5.33) by Young's inequality yields

$$\frac{\mathrm{d}}{\mathrm{d}t} \int_{\Omega_R} \frac{|\rho^n M_{i.}x|^2}{2} \, \mathrm{d}x \leq |M|^2 \|\rho^n\|^2_{L^2(\Omega_R)} + \left(\|u^n\|^2_{L^\infty(\Omega_R)} + \|b\|^2_{L^\infty(\Omega_R)} + |M| \right) \|\rho^n Mx\|^2_{L^2(\Omega_R)}.$$

Integrating with respect to t and applying Gronwall's inequality, we obtain

$$\|\rho^n(t)Mx\|^2_{L^2(\Omega_R)} \leq \left(\|\rho^n_0 Mx\|^2_{L^2(\Omega_R)} + 2|M|^2 \int_0^t \|\rho^n(s)\|^2_{L^2(\Omega_R)}\, ds \right)$$
$$\exp\left(2 \int_0^t \left(\|u^n(s)\|^2_{L^\infty(\Omega_R)} + \|b\|^2_{L^\infty(\Omega_R)} + |M| \right) ds \right). \tag{5.34}$$

Next, we apply ∂_i to equation (5.22) and multiply this equation with $M_j.x$ to obtain

$$\partial_t(\partial_i \rho^n M_j.x) + (u^n + b - Mx) \cdot \nabla(\partial_i \rho^n M_j.x)$$
$$- M_j.(u^n + b - Mx)\partial_i \rho^n + [(\partial_i u^n + \partial_i b - M_{.i}) \cdot \nabla \rho^n]M_j.x = 0 \quad \text{in } [0,T] \times \Omega_R$$

for $i, j = 1, 2, 3$. We test this equation with $\partial_i \rho^n M_j.x$ and integrate by parts. Similarly as above, some integrals vanish and we obtain

$$\frac{d}{dt} \int_{\Omega_R} \frac{|\partial_i \rho^n M_j.x|^2}{2}\, dx = \int_{\Omega_R} [M_j.(u^n + b - Mx)]\partial_i \rho^n \cdot \partial_i \rho^n M_j.x\, dx$$
$$- \int_{\Omega_R} [(\partial_i u^n + \partial_i b - M_{.i}) \cdot \nabla \rho^n]M_j.x \cdot \partial_i \rho^n M_j.x\, dx.$$

Thus, by Young's inequality, we obtain

$$\frac{d}{dt} \int_{\Omega_R} \frac{|\partial_i \rho^n M_j.x|^2}{2}\, dx \leq |M|^2 \|\nabla \rho^n\|^2_{L^2(\Omega_R)}$$
$$+ \left(\|u^n\|^2_{L^\infty(\Omega_R)} + \|b\|^2_{L^\infty(\Omega_R)} + \|\nabla u^n\|_{L^\infty(\Omega_R)} + \|\nabla b\|_{L^\infty(\Omega_R)} + 2|M| \right) \|Mx \cdot \nabla \rho^n\|^2_{L^2(\Omega_R)}.$$

Now integrating with respect to t and applying Gronwall's inequality, we obtain

$$\|Mx \cdot \nabla \rho^n(t)\|^2_{L^2(\Omega_R)} \leq \left(\|Mx \cdot \nabla \rho^n_0\|^2_{L^2(\Omega_R)} + 2|M|^2 \int_0^t \|\nabla \rho^n(s)\|^2_{L^2(\Omega_R)}\, ds \right)$$
$$\exp\left(4 \int_0^t \left(\|u^n(s)\|^2_{L^\infty(\Omega_R)} + \|b\|^2_{L^\infty(\Omega_R)} + \|\nabla u^n(s)\|_{L^\infty(\Omega_R)} + \|\nabla b\|_{L^\infty(\Omega_R)} + |M| \right) ds \right).$$

Combining this estimate with (5.34), applying the Sobolev embedding $W^{1,6}(\Omega_R) \hookrightarrow L^\infty(\Omega_R)$, see Proposition 1.14, and letting $n \to \infty$ yield the assertion. $\qquad \square$

5.3.3 Proof of elliptic L^6-estimate

Consider the modified stationary Stokes problem

$$\begin{cases} -\Delta u - Mx \cdot \nabla u + Mu + \nabla p &= g \quad \text{in } \Omega_R, \\ \operatorname{div} u &= 0 \quad \text{in } \Omega_R, \\ u &= 0 \quad \text{on } \partial\Omega_R. \end{cases} \tag{5.35}$$

We prove the elliptic L^6-estimate for the solution u stated in Proposition 5.17. Before that we prove the following L^2-estimate by using energy methods.

Lemma 5.28. *Let $R > R_0$ and $g \in L^2(\Omega_R)^3$. Moreover, let $(u, \mathrm{p}) \in H^2(\Omega_R)^3 \cap H^1_{0,\sigma}(\Omega_R) \times H^1(\Omega_R)$ be a solution to* (5.35). *Then there exists some constant $C := C(|M|) > 0$, independent of R, such that*

$$\|\nabla^2 u\|_{L^2(\Omega_R)} + \|Mx \cdot \nabla u - Mu\|_{L^2(\Omega_R)} + \|\nabla \mathrm{p}\|_{L^2(\Omega_R)} \leq C\left(\|g\|_{L^2(\Omega_R)} + \|\nabla u\|_{L^2(\Omega_R)}\right).$$

Proof. By Lemma 5.23 we know that $Mx \cdot \nabla u - Mu \in L^2_\sigma(\Omega_R)$. Multiplying equation (5.35) with $-A_{\Omega_R} u - Mx \cdot \nabla u + Mu$ yields

$$\|A_{B_R} u\|^2_{L^2(\Omega_R)} + \|Mx \cdot \nabla u - Mu\|^2_{L^2(\Omega_R)}$$
$$= 2\int_{\Omega_R} A_{\Omega_R} u \cdot (-Mx \cdot \nabla u + Mu)\, \mathrm{d}x - \int_{\Omega_R} (A_{\Omega_R} u + Mx \cdot \nabla u - Mu) \cdot g\, \mathrm{d}x.$$

Integrating by parts and using Lemma 5.23 yield

$$\int_{\Omega_R} A_{\Omega_R} u \cdot (-Mx \cdot \nabla u + Mu)\, \mathrm{d}x$$
$$= \int_\Gamma \left(-(\nu \cdot \nabla u) \cdot (Mx \cdot \nabla u) + \frac{1}{2}|\nabla u|^2 Mx \cdot \nu\right) \mathrm{d}\sigma - \int_{\Omega_R} \nabla Mu : \nabla u\, \mathrm{d}x.$$

By applying Young's inequality we conclude that

$$\|A_{B_R} u\|^2_{L^2(\Omega_R)} + \|Mx \cdot \nabla u + Mu\|^2_{L^2(\Omega_R)}$$
$$\leq 4\int_\Gamma \left(-(\nu \cdot \nabla u) \cdot (Mx \cdot \nabla u) + \frac{1}{2}|\nabla u|^2 Mx \cdot \nu\right) \mathrm{d}\sigma - 4\int_{\Omega_R} \nabla Mu : \nabla u\, \mathrm{d}x + 2\|g\|^2_{L^2(B_R)}.$$

For the boundary term we use the trace inequality in Corollary 1.11 to get

$$\int_\Gamma \left(-(\nu \cdot \nabla u) \cdot (Mx \cdot \nabla u) + \frac{1}{2}|\nabla u|^2 Mx \cdot \nu\right) \mathrm{d}\sigma \leq C\|\nabla u\|^2_{L^2(\Omega_R)} + \varepsilon\|\nabla^2 u\|^2_{L^2(\Omega_R)}$$

for arbitrary $\varepsilon > 0$ and some constant $C := C(|M|, \varepsilon) > 0$. Using Proposition 2.4 and the estimates above yields

$$\|\nabla^2 u\|_{L^2(\Omega_R)} + \|Mx \cdot \nabla u - Mu\|_{L^2(\Omega_R)} \leq C\left(\|g\|_{L^2(\Omega_R)} + \|\nabla u\|_{L^2(\Omega_R)}\right) + \varepsilon\|\nabla^2 u\|^2_{L^2(\Omega_R)}$$

for $\varepsilon > 0$ arbitrarily chosen and for some constant $C := C(|M|, \varepsilon) > 0$, independent of R. By choosing $\varepsilon > 0$ sufficiently small we obtain

$$\|\nabla^2 u\|_{L^2(\Omega_R)} + \|Mx \cdot \nabla u - Mu\|_{L^2(\Omega_R)} \leq C\left(\|g\|_{L^2(\Omega_R)} + \|\nabla u\|_{L^2(\Omega_R)}\right).$$

The estimate for the pressure follows now from the relation $\nabla \mathrm{p} = \Delta u + Mx \cdot \nabla u - Mu + g$. $\square$

We are now in position to give a proof of Proposition 5.17.

Proof of Proposition 5.17. The main idea is to use a localization or more precisely a cut-off technique in order to reduce the problem to a problem in B_R and a problem close to the boundary Γ. The construction used here is similar to the one outlined in Section 1.6.1 and to the one used in the proof of Proposition 2.7. We choose a cut-off function $\varphi_1 \in C_c^\infty(\mathbb{R}^3)$ such that $0 \le \varphi_1 \le 1$ and

$$\varphi_1(x) := \begin{cases} 1, & |x| \le R_0 - 3, \\ 0, & |x| \ge R_0 - 2. \end{cases}$$

Then, set $D := \Omega_{R_0}$, $D_1 := \{x \in \mathbb{R}^3 : R_0 - 4 < |x| < R_0 - 1\}$ and $\varphi_2 := 1 - \varphi_1$. Note that the bounded domain D is in particular independent of R. Now, for $i = 1, 2$, we define

$$\begin{aligned} u^{(i)} &:= \varphi_i u - \mathbb{B}_{D_1}((\nabla \varphi_i) \cdot u), \\ \mathrm{p}^{(i)} &:= \varphi_i \mathrm{p}. \end{aligned}$$

It is clear that $u = u^{(1)} + u^{(2)}$ and $\mathrm{p} = \mathrm{p}^{(1)} + \mathrm{p}^{(2)}$. Moreover, by Proposition 1.87 it is clear that $u^{(1)} \in W^{2,6}(D)^3 \cap W_0^{1,6}(D)^3 \cap L_\sigma^6(D)$ and $u^{(2)} \in W^{2,6}(B_R)^3 \cap W_0^{1,6}(B_R)^3 \cap L_\sigma^6(B_R)$. Then easy calculations show that $(u^{(1)}, \mathrm{p}^{(1)})$ solves the modified stationary Stokes problem

$$\begin{cases} -\Delta u^{(1)} - Mx \cdot \nabla u^{(1)} + Mu^{(1)} + \nabla \mathrm{p}^{(1)} &= \varphi_1 g + G^{(1)} \quad \text{in } D, \\ \operatorname{div} u^{(1)} &= 0 \quad \text{in } D, \\ u^{(1)} &= 0 \quad \text{on } \partial D, \end{cases}$$

in D, and $(u^{(2)}, \mathrm{p}^{(2)})$ solves the modified stationary Stokes problem

$$\begin{cases} -\Delta u^{(2)} - Mx \cdot \nabla u^{(2)} + Mu^{(2)} + \nabla \mathrm{p}^{(2)} &= \varphi_2 g + G^{(2)} \quad \text{in } B_R, \\ \operatorname{div} u^{(2)} &= 0 \quad \text{in } B_R, \\ u^{(2)} &= 0 \quad \text{on } \partial B_R, \end{cases}$$

in the ball B_R. Here the "error terms" $G^{(i)}$, $i = 1, 2$, are given by

$$\begin{aligned} G^{(i)} := &-2(\nabla \varphi_i) \cdot \nabla u - (\Delta \varphi_i) u - (Mx \cdot \nabla \varphi_i) u + (\nabla \varphi_i) \mathrm{p} \\ &- (\Delta + Mx \cdot \nabla - M) \mathbb{B}_{D_1}((\nabla \varphi_i) \cdot u). \end{aligned}$$

Note that $\operatorname{supp} G^{(i)} \subset D_1 \subset D$. By Proposition 2.3 there exists a constant $C := C(|M|) > 0$, depending on R_0 but independent of the actual value of $R > R_0$, such that

$$\|\nabla^2 u^{(1)}\|_{L^6(D)} + \|Mx \cdot \nabla u^{(1)} - Mu^{(1)}\|_{L^6(D)} \le C \left(\|g\|_{L^6(\Omega_R)} + \|G^{(1)}\|_{L^6(D)} \right). \tag{5.36}$$

Next we consider the function $u^{(2)}$. By Theorem 2.12 we have

$$\|\nabla^2 u^{(2)}\|_{L^6(B_R)} + \|Mx \cdot \nabla u^{(2)} - Mu^{(2)}\|_{L^6(B_R)} \le C \left(\|g\|_{L^6(\Omega_R)} + \|G^{(2)}\|_{L^6(D)} \right) \tag{5.37}$$

for some constant $C > 0$ independent of R. Now it only remains to estimate the "error terms" $G^{(i)}$, $i = 1, 2$, in the L^6-norm over D. Without loss of generality we may assume that $\mathrm{p} \in L_0^p(D)$, see Definition 2.8. Thus by using the Sobolev embedding $H^1(D) \hookrightarrow L^6(D)$, Poincaré's inequality (can be applied since $\mathrm{p} \in L_0^p(D)$) and equation (5.35), we obtain

$$\begin{aligned} \|\mathrm{p}\|_{L^6(D)} \le C \|\nabla \mathrm{p}\|_{L^2(D)} &\le C \left(\|\Delta u + Mx \cdot \nabla u - Mu\|_{L^2(D)} + \|g\|_{L^2(D)} \right) \\ &\le C \left(\|u\|_{H^2(D)} + \|g\|_{L^2(D)} \right) \end{aligned}$$

for a constant $C > 0$, depending only on R_0 but independent of the actual value of $R > R_0$.

Then by using standard properties of $\mathbb{B}_{D_1}$, see Proposition 1.87, and again the Sobolev embedding $H^1(D) \hookrightarrow L^6(D)$, we can easily conclude that

$$\|G^{(1)}\|_{L^6(D)} + \|G^{(2)}\|_{L^6(D)} \leq C\Big(\|g\|_{L^2(D)} + \|u\|_{H^2(D)}\Big)$$

holds for some constant $C := C(|M|) > 0$. By using the L^2-estimate stated in Lemma 5.28 and Poincaré's inequality for u on D, which can be applied here because $D = \Omega \cap B_{R_0}$ and $u|_\Gamma = 0$, we deduce that

$$\|u\|_{H^2(D)} \leq C\Big(\|g\|_{L^2(D)} + \|\nabla u\|_{L^2(D)}\Big)$$

for some constant $C := C(|M|) > 0$. Thus, the embedding $L^6(D) \hookrightarrow L^2(D)$ implies

$$\|G^{(1)}\|_{L^6(D)} + \|G^{(2)}\|_{L^6(D)} \leq C\Big(\|g\|_{L^6(D)} + \|\nabla u\|_{L^2(D)}\Big)$$

for a constant $C := C(|M|) > 0$, independent of the actual value of $R > R_0$. Now by combining this estimate with (5.36) and (5.37) we obtain

$$\|\nabla^2 u\|_{L^6(\Omega_R)} + \|Mx \cdot \nabla u - Mu\|_{L^6(\Omega_R)} \leq C\Big(\|g\|_{L^6(\Omega_R)} + \|\nabla u\|_{L^2(\Omega_R)}\Big).$$

The estimate for the pressure follows now from the relation $\nabla p = \Delta u + Mx \cdot \nabla u - Mu + g$. $\quad\square$

5.4 Solving the approximating problem

We consider the approximating problem (5.9)–(5.10) and give a proof of Proposition 5.9. Here we follow the approach of Kim [Kim87]. We first need the following lemma.

Lemma 5.29. *Let $T > 0$. For $n \in \mathbb{N}$ fixed, let $\rho^n \in C^1([0,T] \times \overline{\Omega}_R)$ be such that $\rho^n(t,x) \geq \alpha$ for all $(t,x) \in [0,T] \times \overline{\Omega}_R$ and for some $\alpha > 0$. Then the matrix $(A^n_{ik}(t))^n_{k,i=1}$, defined in (5.12), is invertible for each $t \in [0,T]$ and $((A^n_{ik}(\cdot))^n_{k,i=1})^{-1} \in C^1([0,T]; \mathbb{R}^{n \times n})$.*

Proof. Assume there exists some $t_0 \in [0,T]$ such that $(A^n_{ik}(t_0))^n_{k,i=1}$ is singular. Without loss of generality, we may assume that there exist constants $c_i \in \mathbb{R}$, $i = 2, \ldots, n$ such that

$$A^n_{1k}(t_0) = \sum_{i=2}^n c_i A^n_{ik}(t_0),$$

for all $k = 1, \ldots, n$. From that we deduce that for all $k = 1, \ldots, n$ we have

$$\int_{\Omega_R} \rho^n(t_0, x) \psi_k(x) \cdot \Psi(x) \, dx = 0,$$

where $\Psi := \psi_1 - \sum_{i=2}^n c_i \psi_i$. Since $\rho^n(t_0, x) \geq \alpha > 0$ for all $x \in \overline{\Omega}_R$, this yields $\Psi \equiv 0$, which is a contradiction because $\{\psi_k\}_{k \in \mathbb{N}} \subset L^2_\sigma(\Omega_R)$ is an orthonormal basis. Hence, the first assertion follows and the second follows from the fact that taking inverse is an analytic function. $\quad\square$

Recall that in (5.7) we defined the parameter

$$N_0 := \|F\|_{H^1(\Omega)} + \|a\|_{H^2(\Omega)} + \|Mx \cdot \nabla a\|_{L^2(\Omega)} + \|\rho_0 - 1\|_{H^2(\Omega)}$$
$$+ \|(\rho_0 - 1)Mx\|_{H^1(\Omega)} + \underline{\rho} + \bar{\rho} + |M|,$$

to express the dependence of constants on the given data. In the following we shall always estimate the functions b and f by using Remark 5.6.

Now we are in position to give a proof of Proposition 5.9.

Proof of Proposition 5.9. We start with showing the a priori estimate stated in Proposition 5.9. For that we assume that we have a solution $u^n \in C^1([0,\infty), \mathbb{P}^n_{\Omega_R} L^2(\Omega_R)^3)$ of (5.10) for some given function $\rho^n \in C^1([0,\infty) \times \overline{\Omega}_R)$ such that $\underline{\rho} \le \rho^n(t,x) \le \bar{\rho}$ for all $(t,x) \in [0,\infty) \times \overline{\Omega}_R$. Note that we do not assume that ρ^n solves the transport equation (5.10). Moreover, let us note that $\mathbb{P}^n_{\Omega_R} L^2(\Omega_R)^3 \subset \mathcal{D}(A_{\Omega_R}) \cap H^3(\Omega_R)^3$, see Remark 5.7. We multiply equation (5.10) with u^n_t and integrate by parts to obtain

$$\frac{\mathrm{d}}{\mathrm{d}t} \int_{\Omega_R} \frac{|\nabla u^n|^2}{2} \, \mathrm{d}x + \int_{\Omega_R} \rho^n |u^n_t|^2 \, \mathrm{d}x = - \int_{\Omega_R} (\rho^n(u^n + b) \cdot \nabla u^n) \cdot u^n_t \, \mathrm{d}x$$
$$- \int_{\Omega_R} (\rho^n u^n \cdot \nabla b) \cdot u^n_t \, \mathrm{d}x + \int_{\Omega_R} \rho^n (Mx \cdot \nabla u^n - Mu^n) \cdot u^n_t \, \mathrm{d}x \qquad (5.38)$$
$$+ \int_{\Omega_R} (\rho^n f + \Delta b) \cdot u^n_t \, \mathrm{d}x.$$

Here all the boundary terms vanish because $u^n \in H^1_{0,\sigma}(\Omega_R)$. Analogously, by testing with $-\varepsilon A_{\Omega_R} u^n$ for $0 < \varepsilon < 1$, we get

$$\varepsilon \int_{\Omega_R} |A_{\Omega_R} u^n|^2 \, \mathrm{d}x = \varepsilon \int_{\Omega_R} \rho^n u^n_t \cdot A_{\Omega_R} u^n \, \mathrm{d}x + \varepsilon \int_{\Omega_R} (\rho^n(u^n + b) \cdot \nabla u^n) \cdot A_{\Omega_R} u^n \, \mathrm{d}x$$
$$+ \varepsilon \int_{\Omega_R} (\rho^n u^n \cdot \nabla b) \cdot A_{\Omega_R} \, \mathrm{d}x - \varepsilon \int_{\Omega_R} \rho^n (Mx \cdot \nabla u^n - Mu^n) \cdot A_{\Omega_R} u^n \, \mathrm{d}x \qquad (5.39)$$
$$- \varepsilon \int_{\Omega_R} (\rho^n f + \Delta b) \cdot A_{\Omega_R} u^n \, \mathrm{d}x.$$

In the following we shall always estimate the L^∞-norm of ρ^n and $1/\rho^n$ by a constant just depending on N_0. This is always done without further reference. The terms on the right hand side of (5.38) are estimated as follows: By using Young's inequality we obtain

$$- \int_{\Omega_R} (\rho^n(u^n + b) \cdot \nabla u^n) \cdot u^n_t \, \mathrm{d}x \le \frac{1}{10} \int_{\Omega_R} \rho^n |u^n_t|^2 \, \mathrm{d}x + C \int_{\Omega_R} |(u^n + b) \cdot \nabla u^n|^2 \, \mathrm{d}x,$$

$$- \int_{\Omega_R} (\rho^n u^n \cdot \nabla b) \cdot u^n_t \, \mathrm{d}x \le \frac{1}{10} \int_{\Omega_R} \rho^n |u^n_t|^2 \, \mathrm{d}x + C \int_{\Omega_R} |u^n \cdot \nabla b|^2 \, \mathrm{d}x,$$

$$\int_{\Omega_R} \rho^n (Mx \cdot \nabla u^n - Mu^n) \cdot u^n_t \, \mathrm{d}x \le \frac{1}{10} \int_{\Omega_R} \rho^n |u^n_t|^2 \, \mathrm{d}x + C \int_{\Omega_R} |\nabla u^n|^2 \, \mathrm{d}x,$$

$$\int_{\Omega_R} (\rho^n f + \Delta b) \cdot u^n_t \, \mathrm{d}x \le \frac{1}{10} \int_{\Omega_R} \rho^n |u^n_t|^2 \, \mathrm{d}x + C \int_{\Omega_R} |f|^2 \, \mathrm{d}x + C \int_{\Omega_R} |\Delta b|^2 \, \mathrm{d}x,$$

for a $C := C(N_0, R) > 0$. Analogously, for the terms on the right-hand side of (5.39) we get

$$\int_{\Omega_R} \rho^n u_t^n \cdot A_{\Omega_R} u^n \, dx \leq \frac{1}{10} \int_{\Omega_R} |A_{\Omega_R} u^n|^2 \, dx + C \int_{\Omega_R} \rho^n |u_t^n|^2 \, dx$$

$$\int_{\Omega_R} (\rho^n (u^n + b) \cdot \nabla u^n) \cdot A_{\Omega_R} u^n \, dx \leq \frac{1}{10} \int_{\Omega_R} |A_{\Omega_R} u^n|^2 \, dx + C \int_{\Omega_R} |(u^n + b) \cdot \nabla u^n|^2 \, dx,$$

$$\int_{\Omega_R} (\rho^n u^n \cdot \nabla b) \cdot A_{\Omega_R} u^n \, dx \leq \frac{1}{10} \int_{\Omega_R} |A_{\Omega_R} u^n|^2 \, dx + C \int_{\Omega_R} |u^n \cdot \nabla b|^2 \, dx,$$

$$- \int_{\Omega_R} \rho^n (Mx \cdot \nabla u^n - Mu^n) \cdot A_{\Omega_R} u^n \, dx \leq \frac{1}{10} \int_{\Omega_R} |A_{\Omega_R} u^n|^2 \, dx + C \int_{\Omega_R} |\nabla u^n|^2 \, dx,$$

$$- \int_{\Omega_R} (\rho^n f + \Delta b) \cdot A_{\Omega_R} u^n \, dx \leq \frac{1}{10} \int_{\Omega_R} |A_{\Omega_R} u^n|^2 \, dx + C \int_{\Omega_R} |f|^2 \, dx + C \int_{\Omega_R} |\Delta b|^2 \, dx,$$

for a constant $C := C(N_0, R) > 0$. From (5.38) and the estimates above it follows that

$$\frac{d}{dt} \int_{\Omega_R} \frac{|\nabla u^n|^2}{2} \, dx + \int_{\Omega_R} \rho^n \frac{|u_t^n|^2}{2} \, dx \leq C\|u^n\|_{L^6(\Omega_R)}^2 \|\nabla u^n\|_{L^3(\Omega_R)}^2 + C\|b\|_{L^\infty(\Omega_R)}^2 \|\nabla u^n\|_{L^2(\Omega_R)}^2$$

$$+ C\|\nabla b\|_{L^3(\Omega_R)}^2 \|u^n\|_{L^6(\Omega_R)}^2 + C\|\nabla u^n\|_{L^2(\Omega_R)}^2 + C\|f\|_{L^2(\Omega_R)}^2 + C\|\Delta b\|_{L^2(\Omega_R)}^2$$

for a constant $C := C(N_0, R) > 0$. We use the Sobolev inequality, the second inequality in Proposition 2.4 and Young's inequality to see that

$$\|u^n\|_{L^6(\Omega_R)}^2 \|\nabla u^n\|_{L^3(\Omega_R)}^2 \leq \delta\|A_{\Omega_R} u^n\|_{L^2(\Omega_R)}^2 + C\left(\|\nabla u^n\|_{L^2(\Omega_R)}^6 + \|\nabla u^n\|_{L^2(\Omega_R)}^4\right)$$

for arbitrary δ and some $C := C(\delta) > 0$. Hence, we conclude that

$$\frac{d}{dt} \int_{\Omega_R} \frac{|\nabla u^n|^2}{2} \, dx + \int_{\Omega_R} \rho^n \frac{|u_t^n|^2}{2} \, dx$$
$$\leq C\left(\|\nabla u^n\|_{L^2(\Omega_R)}^6 + \|\nabla u^n\|_{L^2(\Omega_R)}^4 + \|\nabla u^n\|_{L^2(\Omega_R)}^2 + 1\right) + \delta\|A_{\Omega_R} u^n\|_{L^2(\Omega_R)}^2 \tag{5.40}$$

for arbitrary $\delta > 0$ and a constant $C := C(N_0, R, \delta) > 0$. Analogously, we get

$$\varepsilon \int_{\Omega_R} \frac{|A_{\Omega_R} u^n|^2}{2} \, dx - \varepsilon C \int_{\Omega_R} \rho^n |u_t^n|^2 \, dx$$
$$\leq \varepsilon \widetilde{C}\left(\|\nabla u^n\|_{L^2(\Omega_R)}^6 + \|\nabla u^n\|_{L^2(\Omega_R)}^4 + \|\nabla u^n\|_{L^2(\Omega_R)}^2 + 1\right) + \delta\|A_{\Omega_R} u^n\|_{L^2(\Omega_R)}^2 \tag{5.41}$$

for arbitrary $0 < \varepsilon, \delta < 1$ and constants $C := C(N_0, R), \widetilde{C} := \widetilde{C}(N_0, R, \delta) > 0$. By adding up (5.40) and (5.41) and by choosing first $\varepsilon > 0$ sufficiently small and then $\delta > 0$ sufficiently small, we conclude that there exists a $C := C(N_0, R) > 0$, independent of n, such that

$$\int_{\Omega_R} \frac{|\nabla u^n(t)|^2}{2} \, dx + \int_0^t \int_{\Omega_R} \rho^n \frac{|u_t^n|^2}{4} \, dx \, ds + \int_0^t \int_{\Omega_R} \frac{|A_{\Omega_R} u^n|^2}{4} \, dx \, ds$$
$$\leq \int_{\Omega_R} \frac{|\nabla u^n(0)|^2}{2} \, dx + C \int_0^t \left(\|\nabla u^n\|_{L^2(\Omega_R)}^6 + \|\nabla u^n\|_{L^2(\Omega_R)}^4 + \|\nabla u^n\|_{L^2(\Omega_R)}^2 + 1\right) ds. \tag{5.42}$$

By using the non-linear Gronwall inequality, see Lemma 1.24, we conclude that that there exists a time $T_0 := T_0(N_0, R) > 0$, independent of n, such that

$$\|\nabla u^n(t)\|_{L^2(\Omega_R)}^2 \leq C\|\nabla a^{n,R}\|_{L^2(\Omega_R)}^2$$

for all $t \in [0, T_0]$ and some constant $C := C(N_0, R, T_0) > 0$ independent of n. Lemma 5.19 directly yields that

$$\|\nabla u^n(t)\|_{L^2(\Omega_R)} \leq L \tag{5.43}$$

for all $t \in [0, T_0]$ and a suitable constant $L := L(N_0, R, T_0) > 0$ independent of n. Hence, the desired a priori estimate follows.

In the following the time T_0 and the constant L from estimate (5.43) are fixed. Moreover, we choose $\widetilde{R}$ such that $\|a^{n,R}\|_{L^2(\Omega)} \leq \widetilde{R}$ and $\widetilde{R} \geq C(\Omega_R) \cdot L$, where $C(\Omega_R)$ denotes the Poincaré constant of the bounded domain Ω_R. Then set

$$K^n := \left\{ u^n \in C([0, T_0]; \mathbb{P}_{\Omega_R}^n L^2(\Omega_R)^3) : \sup_{t \in [0, T_0]} \|u^n(t)\|_{L^2(\Omega_R)} \leq \widetilde{R} \right\}.$$

Take some $u^n \in K^n$, e.g. we can start with $u^n = a^{n,R}$, which is even defined on the whole time-interval $[0, \infty)$. Recall that $u^n \in \mathbb{P}_{\Omega_R}^n L^2(\Omega_R)^3 \subset H^3(\Omega_R)^3$. Then by Lemma 5.24 there exists some $\rho^n \in C^1([0, T_0] \times \overline{\Omega}_R)$ that solves (5.9). Now we insert this ρ^n into the ordinary differential equation (5.11). Note that

$$(A_{ik}^n(\cdot))_{i,k=1}^n, \ (C_{ik}^n(\cdot))_{i,k=1}^n \in C^1([0, T_0]; \mathbb{R}^{n \times n}), \ (B_{ikl}^n(\cdot))_{i,k,l=1}^n \in C^1([0, T_0]; \mathbb{R}^{n \times n \times n})$$

and

$$(d_i^n(\cdot))_{i=1}^n \in C^1([0, T_0]; \mathbb{R}^n).$$

Thus, by Lemma 5.29 and by the classical Picard-Lindelöf theorem, we conclude that there exists some maximal time $T_{\max} := T_{\max}(n)$ with $0 < T_{\max} \leq T_0$ and a solution $(\alpha_{nk}(\cdot))_{k=1}^n \in C^1([0, T_{\max}); \mathbb{R}^n)$ of the system (5.11). Set

$$\bar{u}^n(t, x) := \sum_{k=1}^n \alpha_{nk}(t) \psi_k(x).$$

It is clear that $\bar{u}^n \in C^1([0, T_{\max}); \mathbb{P}_{\Omega_R}^n L^2(\Omega_R)^3)$ and that $\bar{u}^n$ solves (5.10) on $[0, T_{\max})$. By the a priori estimate (5.43) and by Poincaré's inequality, which is applicable because $u^n \in H_{0,\sigma}^1(\Omega)$, it is clear that the L^2-norm of $\bar{u}^n$ remains bounded on the whole interval $[0, T_{\max})$. Hence, by classical arguments it is clear that $T_{\max} = T_0$ and that $\bar{u}^n \in C^1([0, T_0]; \mathbb{P}_{\Omega_R}^n L^2(\Omega_R)^3)$. Moreover, by using the a priori estimate (5.43) and again Poincaré's inequality it follows that $\bar{u}^n \in K^n$. Following this procedure we can define a mapping $F : K^n \to K^n$ by setting $F(u^n) = \bar{u}^n$. By (5.42) and (5.43) we conclude that

$$\|\bar{u}^n(t) - \bar{u}^n(s)\|_{L^2(\Omega_R)} \leq \int_s^t \|\bar{u}_t^n(\tau)\|_{L^2(\Omega_R)} \, d\tau \leq C(t - s)^{\frac{1}{2}}$$

for some constant $C > 0$, independent of n, and all $0 \leq s < t \leq T_0$. This shows that image of F is equicontinuous. The Arzelà-Ascoli theorem yields that the image is even compact. By Schauder's fixed point theorem the function F has a fixed point $u^n \in C([0, T_0]; \mathbb{P}_{\Omega_R}^n L^2(\Omega_R)^3)$ which together with $\rho^n \in C^1([0, T_0] \times \overline{\Omega}_R)$ solves (5.9)–(5.10) on the interval $[0, T_0]$. If we insert ρ^n again into (5.11), then the right-hand-side of this ordinary differential equation is continuously differentiable in t. Classical arguments for ordinary differential equations yield that even $u^n \in C^2([0, T_0]; \mathbb{P}^n_{\Omega_R} L^2(\Omega_R)^3)$. $\qquad\square$

5.5 A priori estimates

In this section will shall give the proofs of the *a priori estimates* stated in Proposition 5.13, Proposition 5.14 and Proposition 5.16. These estimates are crucial ingredients for the proof of Theorem 5.4. The presentation in this section is divided into two parts: First we derive estimates for the approximating solution $(\rho^n, u^n) := (\rho^{n,R}, u^{n,R})$ on Ω_R uniformly in n. In this step all constants may depend on R. Then, in the second step we derive estimates for the strong solution $(\rho, u) := (\rho^R, u^R)$ on Ω_R uniformly in the radius $R > R_0$. The reason why we proceed in two steps, rather than showing that all constants are independent of n and R directly in one step, is due to the fact that in order to achieve the independence of n and R, respectively, different strategies and arguments are needed. The differences will be pointed out later within the text.

Let us recall that in (5.7) we defined the parameter

$$N_0 := \|F\|_{H^1(\Omega)} + \|a\|_{H^2(\Omega)} + \|Mx \cdot \nabla a\|_{L^2(\Omega)} + \|\rho_0 - 1\|_{H^2(\Omega)}$$
$$+ \|(\rho_0 - 1)Mx\|_{H^1(\Omega)} + \underline{\rho} + \bar{\rho} + |M|$$

in order to express the dependence of constants on the given data. In the following, we make use of the estimates stated in Remark 5.6 without further reference. Moreover, Lemma 5.18 and Lemma 5.19 allow to estimate $a^{n,R}$ and a^R by a constant just depending on N_0. Similarly, by using Remark 5.21, we can estimate the norms of $\rho_0^{n,R}$ and ρ_0^R by a constant just depending on N_0. This is always done without further reference.

5.5.1 Uniform estimates in n

Here we prove uniform estimates for the approximating solution $(\rho^n, u^n) := (\rho^{n,R}, u^{n,R})$ on Ω_R which are stated in Proposition 5.13. We shall prove these estimates by combining different arguments used e.g. by Heywood [Hey80], Kim [Kim87], Boldrini, Rojas-Medar and Fernández-Cara [BRMFC03], Choe and Kim [CK03] and Cho and Kim [CK04]. Here all constants need to be independent of n, but they may depend on R.

In the following let $R > R_0$ and let $T_0 > 0$ denote the existence time from Proposition 5.9.

Lemma 5.30. *There exists a constant $C := C(N_0, R, T_0) > 0$ such that the approximating solution $(\rho^n, u^n) := (\rho^{n,R}, u^{n,R})$ from Proposition 5.9 satisfies*

$$\underline{\rho} \leq \rho^n(t, x) \leq \bar{\rho} \qquad \text{for all } x \in \overline{\Omega}_R,$$

$$\|u^n(t)\|^2_{L^2(\Omega_R)} + \|\nabla u^n(t)\|^2_{L^2(\Omega_R)} + \int_0^t \|\nabla^2 u^n(s)\|^2_{L^2(\Omega_R)} \, ds \leq C,$$

for all $t \in [0, T_0]$.

Proof. By Lemma 5.24 on linear transport equations, it follows that $\underline{\rho} \leq \rho^n(t,x) \leq \bar{\rho}$ for all $(t,x) \in [0,T_0] \times \overline{\Omega}_R$. In the proof of Proposition 5.9, see estimates (5.42) and (5.43), we have already shown that there exists a constant $C := C(N_0, R, T_0) > 0$ such that

$$\|\nabla u^n(t)\|^2_{L^2(\Omega_R)} + \int_0^t \|A_{\Omega_R} u^n(s)\|^2_{L^2(\Omega_R)} \, \mathrm{d}s \leq C \tag{5.44}$$

for all $t \in [0, T_0]$. By applying the elliptic estimate in Proposition 2.3 we conclude

$$\|\nabla^2 u^n(t)\|_{L^2(\Omega_R)} \leq C \|A_{\Omega_R} u^n(t)\|_{L^2(\Omega_R)}$$

for a constant $C := C(R) > 0$. Moreover, Poincaré's inequality yields

$$\|u^n(t)\|_{L^2(\Omega_R)} \leq C \|\nabla u^n(t)\|_{L^2(\Omega_R)}$$

for all $t \in [0, T_0]$ and some $C := C(R) > 0$. By combing these estimates with (5.44) the assertion follows. $\qquad\square$

Remark 5.31. The point-wise estimate

$$\underline{\rho} \leq \rho^n(t,x) \leq \bar{\rho} \qquad \text{for all } (t,x) \in [0,T_0] \times \overline{\Omega}_R$$

allows to estimate ρ^n and $1/\rho^n$ in the L^∞-norm by a constant $C := C(N_0) > 0$. This is done in the following without further reference.

Lemma 5.32. *There exists a constant* $C := C(N_0, R) > 0$ *such that the approximating solution* $(\rho^n, u^n) := (\rho^{n,R}, u^{n,R})$ *from Proposition 5.9 satisfies*

$$\|u_t^n(0)\|_{L^2(\Omega_R)} \leq C.$$

Proof. First note that

$$\mathbb{P}^n_{\Omega_R}\big(\rho_0^{n,R} u_t^n(0) + \rho_0^{n,R} u(0) \cdot \nabla u^n(0) + \rho_0^{n,R} b \cdot \nabla u^n(0) + \rho_0^{n,R} u^n(0) \cdot \nabla b$$

$$- \rho_0^{n,R}(Mx \cdot \nabla u^n(0) - Mu^n(0)) - \Delta u^n(0)\big) = \mathbb{P}^n_{\Omega_R}(\rho_0^{n,R} f + \Delta b) \quad \text{in } \Omega_R.$$

We multiply this equation with $u_t^n(0)$ and then we integrate over Ω_R. Then by applying Hölder's and Young's inequality we obtain

$$\left\| \sqrt{\rho_0^{n,R}} u_t^n(0) \right\|^2_{L^2(\Omega_R)} \leq C \left\| a^{n,R} \right\|^2_{L^\infty(\Omega_R)} \left\| \nabla a^{n,R} \right\|^2_{L^2(\Omega_R)} + C \left\| \nabla a^{n,R} \right\|^2_{L^2(\Omega_R)} + C \left\| a^{n,R} \right\|^2_{L^2(\Omega_R)}$$

$$+ C \left\| Mx \cdot \nabla a^{n,R} \right\|^2_{L^2(\Omega_R)} + C \left\| \mathbb{P}_{\Omega_R} \Delta a^{n,R} \right\|^2_{L^2(\Omega_R)} + C$$

for some constant $C := C(N_0, R) > 0$. The right-hand side of the previous inequality is finite by the assumptions on the initial data and it can be controlled uniformly in n. To see this, we apply the embedding $H^2(\Omega_R) \hookrightarrow L^\infty(\Omega_R)$ and Proposition 2.3 to obtain

$$\|a^{n,R}\|_{L^\infty(\Omega_R)} \leq C \|a^{n,R}\|_{H^2(\Omega_R)} \leq C \|A_{\Omega_R} a^{n,R}\|_{L^2(\Omega_R)}$$

for some constant $C := C(R) > 0$. The assertion follows now from Lemma 5.19. $\qquad\square$

Lemma 5.33. *There exists a constant $C := C(N_0, R, T_0) > 0$ such that the approximating solution $(\rho^n, u^n) := (\rho^{n,R}, u^{n,R})$ from Proposition 5.9 satisfies*

$$\|u_t^n(t)\|_{L^2(\Omega_R)}^2 + \int_0^t \|\nabla u_t^n(s)\|_{L^2(\Omega_R)}^2 \, ds \le C$$

for all $t \in [0, T_0]$.

Proof. We differentiate equation (5.10) with respect to t which is possible due to the time-regularity of ρ^n and u^n stated in Proposition 5.9. Then we see that

$$\mathbb{P}_{\Omega_R}^n \, (\rho^n u_{tt}^n + \rho_t^n u_t^n + \rho^n u^n \cdot \nabla u_t^n + \rho_t^n u^n \cdot \nabla u^n + \rho^n u_t^n \cdot \nabla u^n + \rho^n b \cdot \nabla u_t^n$$
$$+ \rho_t^n b \cdot \nabla u^n + \rho^n u_t^n \cdot \nabla b + \rho_t^n u^n \cdot \nabla b - \rho^n M x \cdot \nabla u_t^n - \rho_t^n M x \cdot \nabla u^n \qquad (5.45)$$
$$+ \rho^n M u_t^n + \rho_t^n M u^n - \Delta u_t^n) = \mathbb{P}_{\Omega_R}^n (\rho_t^n f)$$

holds in $[0, T_0] \times \Omega_R$. We set $w^n := u^n + b - Mx$ for notational simplicity. Recall that $\rho_t^n = -(w^n \cdot \nabla)\rho^n$. We multiply equation (5.45) with u_t^n and integrate by parts. Here we use in particular the following computations:

$$\int_{\Omega_R} \rho^n u_{tt}^n \cdot u_t^n \, dx = \frac{d}{dt} \int_{\Omega_R} \rho^n \frac{|u_t^n|^2}{2} \, dx - \int_{\Omega_R} \rho_t^n \frac{|u_t^n|^2}{2} \, dx$$
$$= \frac{d}{dt} \int_{\Omega_R} \rho^n \frac{|u_t^n|^2}{2} \, dx + \int_{\Omega_R} (w^n \cdot \nabla \rho^n) \frac{|u_t^n|^2}{2} \, dx,$$

$$\int_{\Omega_R} (\rho^n u^n \cdot \nabla u_t^n) \cdot u_t^n \, dx = - \int_{\Omega_R} u_t^n \cdot (\rho^n u^n \cdot \nabla u_t^n) \, dx - \int_{\Omega_R} u_t^n \cdot \rho^n (\operatorname{div} u^n) u_t^n \, dx$$
$$- \int_{\Omega_R} (u^n \cdot \nabla \rho^n) |u_t^n|^2 \, dx + \int_{\partial \Omega_R} u_t^n \cdot \rho^n (\nu \cdot u^n) u_t^n \, d\sigma.$$

Note that the boundary integral above vanishes because $u_t^n \in H_{0,\sigma}^1(\Omega_R)$. Moreover, since $u^n \in H_{0,\sigma}^1(\Omega_R)$ we have $\operatorname{div} u^n = 0$. Therefore, the last inequality yields

$$\int_{\Omega_R} (\rho^n u^n \cdot \nabla u_t^n) \cdot u_t^n \, dx = - \int_{\Omega_R} (u^n \cdot \nabla \rho^n) \frac{|u_t^n|^2}{2} \, dx.$$

Analogously, by using $\operatorname{div}(Mx) = 0$ and $\operatorname{div} b = 0$ we see that

$$\int_{\Omega_R} (\rho^n Mx \cdot \nabla u_t^n) \cdot u_t^n \, dx = - \int_{\Omega_R} (Mx \cdot \nabla \rho^n) \frac{|u_t^n|^2}{2} \, dx,$$

$$\int_{\Omega_R} (\rho^n b \cdot \nabla u_t^n) \cdot u_t^n \, dx = - \int_{\Omega_R} (b \cdot \nabla \rho^n) \frac{|u_t^n|^2}{2} \, dx.$$

Moreover, by using that $M^\top = -M$ we see that

$$\int_{\Omega_R} \rho^n M u_t^n \cdot u_t^n \, dx = 0.$$

Therefore we deduce that

$$\int_{\Omega_R} \rho^n u_{tt}^n \cdot u_t^n \, dx$$

$$+ \int_{\Omega_R} (\rho^n u^n \cdot \nabla u_t^n + \rho^n b \cdot \nabla u_t^n - \rho^n Mx \cdot \nabla u_t^n + \rho^n Mu_t^n) \cdot u_t^n \, dx = \frac{d}{dt} \int_{\Omega_R} \rho^n \frac{|u_t^n|^2}{2} \, dx.$$

By taking everything together we obtain

$$\frac{d}{dt} \int_{\Omega_R} \rho^n \frac{|u_t^n|^2}{2} \, dx + \int_{\Omega_R} |\nabla u_t^n|^2 \, dx$$

$$= \int_{\Omega_R} (w^n \cdot \nabla) \rho^n (u_t^n + u^n \cdot \nabla u^n + b \cdot \nabla u^n + u^n \cdot \nabla b - Mx \cdot \nabla u^n + Mu^n - f) \cdot u_t^n \, dx$$

$$- \int_{\Omega_R} \rho^n (u_t^n \cdot \nabla u^n + u_t^n \cdot \nabla b) \cdot u_t^n \, dx.$$

In the next step, we integrate again by parts to see that

$$\int_{\Omega_R} (w^n \cdot \nabla \rho) u_t^n \cdot u_t^n \, dx = \int_{\Omega_R} \operatorname{div}(\rho^n w^n) \cdot |u_t^n|^2 \, dx$$

$$= -2 \int_{\Omega_R} \rho^n w^n \cdot (\nabla u_t^n) u_t^n \, dx + \int_{\partial\Omega_R} \rho(\nu \cdot w^n) |u_t^n|^2 \, d\sigma$$

$$= -2 \int_{\Omega_R} \rho^n w^n \cdot (\nabla u_t^n) u_t^n \, dx.$$

Here we used that $\operatorname{div} w^n = 0$ and that $\nu \cdot w|_{\partial\Omega_R} = 0$, see Lemma 5.22. The other terms involving $w^n \cdot \nabla \rho^n$ are treated analogously and therefore the detailed calculations are omitted here. This shows that we can remove the gradient in front of ρ^n using integration by parts. This has the advantage that we can now estimate ρ^n in the L^∞-norm by some constant just depending on N_0. Therefore, we conclude that there exists a constant $C := C(N_0, R) > 0$ such that

$$\frac{d}{dt} \int_{\Omega_R} \rho^n \frac{|u_t^n|^2}{2} \, dx + \int_{\Omega_R} |\nabla u_t^n|^2 \, dx$$

$$\leq C \int_{\Omega_R} |w^n| \Big(|u_t^n||\nabla u_t^n| + |\nabla u^n|^2 |u_t^n| + |\nabla^2 u^n||u_t^n| + |u^n||\nabla u^n||\nabla u_t^n|$$

$$+ |\nabla u^n||u_t^n| + |\nabla u^n||\nabla u_t^n| + |u^n||u_t^n| + |u^n||\nabla u_t^n| + |\nabla f||u_t^n| + |f||\nabla u_t^n| \Big) \, dx$$

$$+ \int_{\Omega_R} \Big(|u_t^n|^2 |\nabla u^n| + |u_t^n|^2 \Big) \, dx$$

$$=: \sum_{j=1}^{12} I_j.$$

In the following we shall estimate the terms I_i, $i = 1, \ldots, 12$. For that purpose we shall make use of the Sobolev inequality

$$\|u_t^n\|_{L^6(\Omega_R)} \leq C \|\nabla u_t^n\|_{L^2(\Omega_R)}.$$

Moreover, the interpolation inequality and the Sobolev embedding $H^1(\Omega_R) \hookrightarrow L^6(\Omega_R)$ yield

$$\|\nabla u^n\|_{L^3(\Omega_R)}^2 \leq \|\nabla u^n\|_{L^2(\Omega_R)}\|\nabla u^n\|_{L^6(\Omega_R)} \leq \|\nabla u^n\|_{L^2(\Omega_R)}\|\nabla u^n\|_{H^1(\Omega_R)}.$$

Let $2 \leq p \leq 6$. By applying Remark 5.6, the Sobolev embedding (Proposition 1.14) and Lemma 5.30 we see that

$$\|w^n(t)\|_{L^p(\Omega_R)} \leq C$$

for all $t \in [0, T_0]$ and some constant $C := C(N_0, R, T_0)$. Now let $\varepsilon > 0$ be arbitrarily chosen and denote by $\widetilde{C} := \widetilde{C}(N_0, R, T_0, \varepsilon) > 0$ and $C := C(N_0, R, T_0) > 0$ generic constants which may change from line to line. By applying Lemma 5.30 and Young's inequality we obtain

$$I_1 \leq C\|w^n\|_{L^6(\Omega_R)}\|u_t^n\|_{L^3(\Omega_R)}\|\nabla u_t^n\|_{L^2(\Omega_R)}$$
$$\leq \widetilde{C}\|u_t^n\|_{L^2(\Omega_R)}^2 + \varepsilon\|\nabla u_t^n\|_{L^2(\Omega_R)}^2$$

$$I_2 \leq C\|w^n\|_{L^6(\Omega_R)}\|\nabla u^n\|_{L^6(\Omega_R)}\|\nabla u^n\|_{L^2(\Omega_R)}\|u_t^n\|_{L^6(\Omega_R)}$$
$$\leq \widetilde{C}\|\nabla u^n\|_{H^1(\Omega_R)}^2 + \varepsilon\|\nabla u_t^n\|_{L^2(\Omega_R)}^2$$

$$I_3 \leq C\|w^n\|_{L^6(\Omega_R)}\|\nabla^2 u^n\|_{L^2(\Omega_R)}\|u_t^n\|_{L^3(\Omega_R)}$$
$$\leq C\|\nabla^2 u^n\|_{L^2(\Omega_R)}^2 + \widetilde{C}\|u_t^n\|_{L^2(\Omega_R)}^2 + \varepsilon\|\nabla u_t^n\|_{L^2(\Omega_R)}^2$$

$$I_4 \leq C\|w^n\|_{L^6(\Omega_R)}\|u^n\|_{L^6(\Omega_R)}\|\nabla u^n\|_{L^6(\Omega_R)}\|\nabla u_t^n\|_{L^2(\Omega_R)}$$
$$\leq \widetilde{C}\|\nabla u^n\|_{H^1(\Omega_R)}^2 + \varepsilon\|\nabla u_t^n\|_{L^2(\Omega_R)}^2$$

$$I_5 \leq C\|w^n\|_{L^6(\Omega_R)}\|\nabla u^n\|_{L^2(\Omega_R)}\|u_t^n\|_{L^3(\Omega_R)}$$
$$\leq C\|\nabla u^n\|_{L^2(\Omega_R)}^2 + \widetilde{C}\|u_t^n\|_{H^1(\Omega_R)}^2 + \varepsilon\|\nabla u_t^n\|_{L^2(\Omega_R)}^2$$

$$I_6 \leq C\|w^n\|_{L^6(\Omega_R)}\|\nabla u^n\|_{L^3(\Omega_R)}\|\nabla u_t^n\|_{L^2(\Omega_R)}$$
$$\leq \widetilde{C}\|\nabla u^n\|_{H^1(\Omega_R)}^2 + \varepsilon\|\nabla u_t^n\|_{L^2(\Omega_R)}^2$$

$$I_7 \leq C\|w^n\|_{L^6(\Omega_R)}\|u^n\|_{L^3(\Omega_R)}\|u_t^n\|_{L^2(\Omega_R)}$$
$$\leq C\|u^n\|_{L^2(\Omega_R)}^2 + C\|\nabla u^n\|_{L^2(\Omega_R)}^2 + C\|u_t^n\|_{L^2(\Omega_R)}^2$$

$$I_8 \leq C\|w^n\|_{L^6(\Omega_R)}\|u^n\|_{L^3(\Omega_R)}\|\nabla u_t^n\|_{L^2(\Omega_R)}$$
$$\leq \widetilde{C}\|u^n\|_{L^2(\Omega_R)}^2 + \widetilde{C}\|\nabla u^n\|_{L^2(\Omega_R)}^2 + \varepsilon\|\nabla u_t^n\|_{L^2(\Omega_R)}^2$$

$$I_9 \leq C\|w^n\|_{L^3(\Omega_R)}\|\nabla f\|_{L^2(\Omega)}\|u_t^n\|_{L^6(\Omega_R)} \leq \widetilde{C} + \varepsilon\|\nabla u_t^n\|_{L^2(\Omega_R)}^2$$

$$I_{10} \leq C\|w^n\|_{L^3(\Omega_R)}\|f\|_{L^6(\Omega_R)}\|\nabla u_t^n\|_{L^2(\Omega_R)} \leq \widetilde{C} + \varepsilon\|\nabla u_t^n\|_{L^2(\Omega_R)}^2$$

$$I_{11} + I_{12} \leq C\|u_t^n\|_{L^6(\Omega_R)}\|u_t^n\|_{L^3(\Omega_R)}\|\nabla u^n\|_{L^2(\Omega_R)} + C\|u_t^n\|_{L^2(\Omega_R)}^2$$
$$\leq \widetilde{C}\|u_t^n\|_{L^2(\Omega_R)}^2 + \varepsilon\|\nabla u_t^n\|_{L^2(\Omega_R)}^2.$$

Summing up and by choosing $\varepsilon > 0$ sufficiently, we conclude that

$$\frac{d}{dt} \int_{\Omega_R} \rho^n \frac{|u_t^n|^2}{2}\, dx + \int_{\Omega_R} \frac{|\nabla u_t^n|^2}{2}\, dx \leq C + C\|\nabla u^n\|_{H^1(\Omega_R)}^2 + C\|u_t^n\|_{L^2(\Omega_R)}^2$$

for some constant $C := C(N_0, R, T_0) > 0$. Thus, by integrating with respect to t and by using Lemma 5.30, we obtain

$$\|u_t^n(t)\|_{L^2(\Omega_R)}^2 + \int_0^t \|\nabla u_t^n(s)\|_{L^2(\Omega_R)}^2\, ds$$

$$\leq C + C\|u_t^n(0)\|_{L^2(\Omega_R)}^2 + C \int_0^t \|u_t^n(s)\|_{L^2(\Omega_R)}^2\, ds$$

for some $C := C(N_0, R, T_0) > 0$. Then, Gronwall's inequality and Lemma 5.32 yield

$$\|u_t^n(t)\|_{L^2(\Omega_R)}^2 \leq C$$

for all $t \in [0, T_0]$ and some $C := C(N_0, R, T_0) > 0$. Hence, the assertion directly follows. $\quad\square$

Lemma 5.34. *There exists a constant $C := C(N_0, R, T_0) > 0$ such that the approximating solution $(\rho^n, u^n) := (\rho^{n,R}, u^{n,R})$ from Proposition 5.9 satisfies*

$$\|\nabla^2 u^n(t)\|_{L^2(\Omega_R)}^2 \leq C$$

for all $t \in [0, T_0]$.

Proof. The idea of the proof is to use elliptic theory for the Stokes system. Set

$$g^n := -\mathbb{P}_{\Omega_R}^n (\rho^n u_t^n + \rho^n u^n \cdot \nabla u^n + \rho^n b \cdot \nabla u^n + \rho^n u^n \cdot b - \rho^n (Mx \cdot \nabla u^n - Mu^n) - \rho^n f - \Delta b).$$

Note that $\mathbb{P}_{\Omega_R}^n \Delta u^n = A_{\Omega_R} u^n$ because $u^n \in \operatorname{span}\{\psi_1, \ldots, \psi_n\}$, where $\{\psi_k\}_{k \in \mathbb{N}}$ is a basis of eigenfunctions of the Stokes operator A_{Ω_R}.

For fixed $t \in [0, T_0]$ we choose $\mathrm{p}^n(t) \in \widehat{W}^{1,2}(\Omega_R)$ such that $\nabla \mathrm{p}^n(t) = (\mathrm{Id} - \mathbb{P}_{\Omega_R})\Delta u^n(t)$. Thus, by using (5.10) we see that for every fixed $t \in [0, T_0]$ the pair of functions $(u^n(t), \mathrm{p}^n(t))$ solves

$$\begin{cases} -\Delta u^n(t) + \nabla \mathrm{p}^n(t) &= g^n(t) \quad \text{in } \Omega_R, \\ \operatorname{div} u^n(t) &= 0 \quad\ \ \text{in } \Omega_R, \\ u^n(t) &= 0 \quad\ \ \text{on } \partial\Omega_R. \end{cases}$$

The elliptic estimate in Proposition 2.3 then yields

$$\|\nabla^2 u^n(t)\|_{L^2(\Omega_R)} \leq C\|g^n\|_{L^2(\Omega_R)} \tag{5.46}$$

for some $C := C(N_0, R) > 0$. It remains to estimate the L^2-norm of g^n. It is easily seen that

$$\|g^n(t)\|_{L^2(\Omega_R)}$$

$$\leq C\Big(\|u_t^n(t)\|_{L^2(\Omega_R)} + \|u^n(t)\|_{L^6(\Omega_R)}\|\nabla u^n(t)\|_{L^3(\Omega_R)} + \|\nabla u^n(t)\|_{L^2(\Omega_R)} + \|u^n(t)\|_{L^2(\Omega_R)} + 1\Big)$$

holds for all $t \in [0, T_0]$ and some $C := C(N_0, R) > 0$. Next, we apply the Sobolev inequality, the second estimate in Proposition 2.4 and Young's inequality to obtain

$$\|u^n\|^2_{L^6(\Omega_R)} \|\nabla u^n\|^2_{L^3(\Omega_R)} \leq \varepsilon \|A_{\Omega_R} u^n\|^2_{L^2(\Omega_R)} + C\left(\|\nabla u^n\|^6_{L^2(\Omega_R)} + \|\nabla u^n\|^4_{L^2(\Omega_R)} \right)$$

for arbitrary $\varepsilon > 0$ and a $C := C(\varepsilon) > 0$. Then, by applying Lemma 5.30 and Lemma 5.33, we conclude that for arbitrary $\varepsilon > 0$ there exists some $C := C(N_0, R, T_0, \varepsilon) > 0$ such that

$$\|g^n(t)\|_{L^2(\Omega_R)} \leq C + \varepsilon \|\nabla^2 u^n(t)\|_{L^2(\Omega_R)}$$

for all $t \in [0, T_0]$. The assertion follows now from (5.46) and the estimate above if we choose $\varepsilon > 0$ sufficiently small. $\qquad \square$

To conclude this subsection, we give the $W^{2,6}$-estimate for the weak solution $(\rho, u) := (\rho^R, u^R)$ stated in Proposition 5.14. Recall that by Proposition 5.11 the (weak) solution (ρ, u) has the following regularity:

$$\underline{\rho} \leq \rho(t, x) \leq \bar{\rho} \quad \text{almost everywhere in } (0, T_0) \times \Omega_R,$$

$$u \in L^\infty((0, T_0); H^2(\Omega_R)^3 \cap H^1_{0,\sigma}(\Omega_R)), \tag{5.47}$$

$$u_t \in L^\infty((0, T_0); L^2(\Omega_R)^3) \cap L^2((0, T_0); H^1_{0,\sigma}(\Omega_R)).$$

By Definition 5.10 of weak solutions we know that

$$-\int_0^T \int_{\Omega_R} \rho u \cdot \partial_t \Phi \, dx \, dt - \sum_{i=1}^3 \int_0^T \int_{\Omega_R} \rho u_i u \cdot \partial_i \Phi \, dx \, dt + \int_0^T \int_{\Omega_R} \nabla u \cdot \nabla \Phi \, dx \, dt$$

$$+ \int_0^T \int_{\Omega_R} [(\rho b \cdot \nabla) u + (\rho u \cdot \nabla) b] \cdot \Phi \, dx \, dt - \int_0^T \int_{\Omega_R} \rho (Mx \cdot \nabla u - Mu) \cdot \Phi \, dx \, dt \tag{5.48}$$

$$= \int_0^T \int_{\Omega_R} (\rho f + \Delta b) \cdot \Phi \, dx \, dt + \int_{\Omega_R} \rho_0(x) a^R(x) \cdot \Phi(0, x) \, dx,$$

and

$$-\int_0^T \int_{\Omega_R} \rho \cdot \partial_t \Psi \, dx \, dt - \int_0^T \int_{\Omega_R} \rho \cdot [(u + b - Mx) \cdot \nabla \Psi] \, dx \, dt$$

$$= \int_{\Omega_R} \rho_0^R(x) \cdot \Psi(0, x) \, dx \tag{5.49}$$

for all $\Phi \in C^1([0, T]; H^1_{0,\sigma}(\Omega_R))$ and all $\Psi \in C^1([0, T]; H^1(\Omega_R))$ with $\Phi(T, x) = \Psi(T, x) = 0$ almost everywhere in Ω_R. The regularity for u stated in (5.47) allows to conclude that

$$\int_0^{T_0} \int_{\Omega_R} \rho u_t \cdot \Phi \, dx \, dt + \int_0^{T_0} \int_{\Omega_R} (\rho u \cdot \nabla u) \cdot \Phi \, dx \, dt - \int_0^{T_0} \int_{\Omega_R} \Delta u \cdot \Phi \, dx \, dt$$

$$+ \int_0^{T_0} \int_{\Omega_R} ((\rho b \cdot \nabla) u + (\rho u \cdot \nabla) b) \cdot \Phi \, dx \, dt - \int_0^{T_0} \int_{\Omega_R} \rho (Mx \cdot \nabla u - Mu) \cdot \Phi \, dx \, dt \tag{5.50}$$

$$= \int_0^{T_0} \int_{\Omega_R} (\rho f + \Delta b) \cdot \Phi \, dx \, dt$$

holds for all $\Phi \in L^2((0,T); L^2_\sigma(\Omega_R))$. This equality simply follows from (5.48) by doing integration by parts with respect to t and x, by using the equation (5.49) and by using that $C^1([0,T]; H^1_{0,\sigma}(\Omega_R))$ is dense in $L^2((0,T); L^2_\sigma(\Omega_R))$.

Lemma 5.35. Let $(\rho, u) := (\rho^R, u^R)$ be the weak solution from Proposition 5.11 on the interval $(0,T_0)$. Then there exist a constant $C := C(N_0, R, T_0) > 0$ such that

$$\int_0^t \|u(s)\|^2_{W^{2,6}(\Omega_R)}\, ds \leq C$$

for all $t \in [0,T_0]$.

Proof. The main tool in this proof is the elliptic estimate for the stationary Stokes system stated in Proposition 2.3. Set

$$g := -\rho u_t - \rho u \cdot \nabla u - \rho b \cdot \nabla u - \rho u \cdot b + \rho(Mx \cdot \nabla u - Mu) + \rho f + \Delta b.$$

Equation (5.50) states that

$$-A_{\Omega_R} u(t) = \mathbb{P}_{\Omega_R} g(t)$$

holds in $L^2_\sigma(\Omega_R)$ for almost all $t \in (0,T_0)$. If we choose $\mathrm{p}(t) \in \widehat{W}^{1,2}(\Omega_R)$ such that $\nabla \mathrm{p}(t) = (\mathrm{Id} - \mathbb{P}_{\Omega_R})\Delta u(t) + (\mathrm{Id} - \mathbb{P}_{\Omega_R})g(t)$ for almost all $t \in (0,T_0)$, then the pair of functions $(u(t), \mathrm{p}(t))$ solves the stationary Stokes problem

$$\begin{cases} -\Delta u(t) + \nabla \mathrm{p}(t) &=& g(t) & \text{in } \Omega, \\ \mathrm{div}\, u(t) &=& 0 & \text{in } \Omega, \\ u(t) &=& 0 & \text{on } \partial\Omega, \end{cases}$$

for almost all $t \in (0,T_0)$. It is not difficult to see that for almost all $t \in (0,T_0)$ the function g is even in $L^6(\Omega_R)^3$. More precisely, by using the Soboev embedding $H^1(\Omega_R) \hookrightarrow L^6(\Omega_R)$ and by the estimates stated in Proposition 5.11 we can conclude that

$$\int_0^t \|g(s)\|^2_{L^6(\Omega_R)}\, ds$$

$$\leq C \int_0^t \left(\|u_t\|^2_{L^6(\Omega_R)} + \|u\|^2_{L^6(\Omega_R)} \|\nabla u\|^2_{L^\infty(\Omega_R)} + \|\nabla u\|^2_{L^6(\Omega_R)} + \|u\|^2_{L^6(\Omega_R)} + 1 \right)$$

$$\leq C + C \int_0^t \|\nabla u(s)\|^2_{L^\infty(\Omega_R)}\, ds$$

for all $t \in [0,T_0]$. Next, we apply the Gagliardo-Nirenberg inequality, see Proposition 1.17, with $p = \infty$, $r = 6$, $q = 2$, $j = 0$ and $m = 1$ to obtain

$$\int_0^t \|\nabla u(s)\|^2_{L^\infty(\Omega_R)}\, ds \leq C \int_0^t \left(\|\nabla^2 u(s)\|^{\frac{3}{2}}_{L^6(\Omega_R)} \|\nabla u(s)\|^{\frac{1}{2}}_{L^2(\Omega_R)} + \|\nabla u(s)\|_{L^2(\Omega)} \right) ds$$

for some $C > 0$. By using Young's inequality and the estimate for $\|\nabla u(s)\|_{L^2(\Omega_R)}$ in Lemma

5.30 we see that for arbitrary $\varepsilon > 0$ there exists some $C := C(N_0, R, T_0, \varepsilon) > 0$ such that

$$\int_0^t \|g(s)\|_{L^6(\Omega_R)}^2 \, ds \leq C + \varepsilon \int_0^t \|\nabla^2 u^R(s)\|_{L^6(\Omega_R)}^2 \, ds \qquad (5.51)$$

for all $t \in [0, T_0]$. Thus, we apply the elliptic estimate in Proposition 2.3 to conclude that

$$\int_0^t \|u(s)\|_{W^{2,6}(\Omega_R)}^2 \, ds \leq C \int_0^t \|g(s)\|_{L^6(\Omega_R)}^2 \, ds \qquad (5.52)$$

for all $t \in [0, T_0]$ and some constant $C := C(R) > 0$. If we choose $\varepsilon > 0$ in (5.51) sufficiently small, then the assertion follows from (5.52) and (5.51). $\qquad \square$

5.5.2 Uniform estimates in R

In this section we derive the key estimates for the purposes of this chapter. We show that all necessary norm estimates for the strong solution $(\rho, u) := (\rho^R, u^R)$ on the bounded domain Ω_R hold with constants independent of $R > R_0$. For that purpose we use arguments similar to the ones used by Galdi and Silvestre in [GS02, GS05]. However, dealing with non-constant density causes additional difficulties which were not faced by Galdi and Silvestre. In the following let $R > R_0$, and let $T_{\max}^R > 0$ denote the *maximal existence time* of the strong solution (ρ, u) on Ω_R obtained in Proposition 5.15. We have the following information on (ρ, u):

$$\begin{aligned}
&\underline{\rho} \leq \rho(t, x) \leq \bar{\rho} \quad \text{for all } (t, x) \in [0, T] \times \Omega, \\
&\rho \in C([0, T] \times \Omega_R), \\
&\rho \in C([0, T]; H^2(\Omega_R)), \\
&\rho_t \in C([0, T]; H^1(\Omega_R)), \\
&u \in C([0, T]; H^2(\Omega_R)^3 \cap H^1_{0,\sigma}(\Omega_R)) \cap L^2((0, T); W^{2,6}(\Omega_R)^3), \\
&u_t \in C([0, T]; L^2(\Omega_R)^3) \cap L^2((0, T); H^1_{0,\sigma}(\Omega_R)),
\end{aligned} \qquad (5.53)$$

for all $0 < T < T_{\max}^R$. This regularity and the fact that (ρ, u) is a strong solution to (5.8) yield that for all $0 \leq t < T_{\max}^R$ we have

$$\rho_t(t) + u(t) \cdot \nabla \rho(t) + b \cdot \nabla \rho(t) - Mx \cdot \nabla \rho(t) = 0 \qquad (5.54)$$

in $L^2(\Omega_R)$, and

$$\begin{aligned}
\mathbb{P}_{\Omega_R} \big(\rho(t) u_t(t) + \rho u(t) \cdot \nabla u(t) + \rho(t) b \cdot \nabla u(t) + \rho(t) u(t) \cdot \nabla b \\
-\rho(t) Mx \cdot \nabla u(t) + \rho(t) Mu(t) \big) - A_{\Omega_R} u(t) = \mathbb{P}_{\Omega_R} \big(\rho(t) f + \Delta b \big)
\end{aligned} \qquad (5.55)$$

in $L^2_\sigma(\Omega)$. Moreover, the initial condition $(\rho(0), u(0)) = (\rho_0^R, a^R)$ is satisfied.

Remark 5.36. The point-wise estimate

$$\underline{\rho} \leq \rho(t, x) \leq \bar{\rho}, \qquad \text{for all } (t, x) \in [0, T_{\max}^R) \times \Omega_R$$

allows to estimate ρ and $1/\rho$ in the L^∞-norm by a constant $C := C(N_0) > 0$. This is done in the following without further reference.

So far we do not know whether the maximal existence time $T_{\max}^R$ goes to zero as $R \to \infty$. Therefore, we also need to make sure that we have a time-interval $[0, T_0) \subset [0, T_{\max}^R)$ for some $T_0 > 0$ and all $R > R_0$. In a first step we show that there exists some time $T_0 > 0$ independent of R such that all estimates hold on $[0, T]$ for every $T < \min\{T_0, T_{\max}^R\}$ uniformly in R. Then, by using the uniform norm estimates and a classical blow-up argument we show that $T_0 \leq T_{\max}^R$ for all $R > R_0$.

Lemma 5.37. *For every* $0 < T < T_{\max}^R$, *there exists a constant* $C := C(N_0, T) > 0$, *independent of* $R > R_0$, *such that the strong solution* $(\rho, u) := (\rho^R, u^R)$ *from Proposition 5.15 satisfies*

$$\|u(t)\|_{L^2(\Omega_R)}^2 + \int_0^t \|\nabla u(s)\|_{L^2(\Omega_R)}^2 \, ds \leq C$$

for all $t \in [0, T]$.

Proof. Let $0 < T < T_{\max}^R$ be arbitrary but fixed. We multiply equation (5.55) with u and then we integrate by parts with respect to x to obtain

$$\frac{d}{dt} \int_{\Omega_R} \rho \frac{|u|^2}{2} \, dx + \int_{\Omega_R} |\nabla u|^2 \, dx = -\int_{\Omega_R} (\rho u \cdot \nabla b) \cdot u \, dx + \int_{\Omega_R} (\rho f + \Delta b) \cdot u \, dx. \qquad (5.56)$$

Here we used in particular the following computations: Since $u \in H_{0,\sigma}^1(\Omega_R)$ we have

$$\int_{\Omega_R} \Delta u \cdot u \, dx = -\int_{\Omega_R} |\nabla u|^2 \, dx.$$

Moreover, we see that

$$\int_{\Omega_R} \rho u_t \cdot u \, dx = \frac{d}{dt} \int_{\Omega_R} \rho \frac{|u|^2}{2} \, dx - \int_{\Omega_R} \rho_t \frac{|u|^2}{2} \, dx$$

$$= \frac{d}{dt} \int_{\Omega_R} \rho \frac{|u|^2}{2} \, dx + \int_{\Omega_R} [(u + b - Mx) \cdot \nabla \rho] \frac{|u|^2}{2} \, dx,$$

$$\int_{\Omega_R} (\rho u \cdot \nabla u) \cdot u \, dx = -\int_{\Omega_R} u \cdot (\rho u \cdot \nabla u) \, dx - \int_{\Omega_R} u \cdot \rho(\operatorname{div} u)u \, dx$$

$$- \int_{\Omega_R} (u \cdot \nabla \rho)|u|^2 \, dx + \int_{\partial \Omega_R} u \cdot \rho(\nu \cdot u)u \, d\sigma.$$

From the last inequality we conclude that

$$\int_{\Omega_R} (\rho u \cdot \nabla u) \cdot u \, dx = -\int_{\Omega_R} (u \cdot \nabla \rho) \frac{|u|^2}{2} \, dx,$$

because $u \in H_{0,\sigma}^1(\Omega)$ and therefore $\operatorname{div} u = 0$ and the boundary term vanishes. Analogously, by using $\operatorname{div}(Mx) = 0$ and $\operatorname{div} b = 0$ we see that

$$\int_{\Omega_R} (\rho Mx \cdot \nabla u) \cdot u \, dx = -\int_{\Omega_R} (Mx \cdot \nabla \rho) \frac{|u|^2}{2} \, dx,$$

$$\int_{\Omega_R} (\rho b \cdot \nabla u) \cdot u \, dx = -\int_{\Omega_R} (b \cdot \nabla \rho) \frac{|u|^2}{2} \, dx.$$

Moreover, by using that $M^\top = -M$ we obtain

$$\int_{\Omega_R} \rho M u \cdot u \, \mathrm{d}x = 0.$$

Summing up, equality (5.56) follows. Estimating the right-hand side of (5.56) yields

$$\int_{\Omega_R} (\rho f + \Delta b) \cdot u \, \mathrm{d}x \leq \int_{\Omega_R} \rho |u|^2 \, \mathrm{d}x + C \int_{\Omega_R} |f|^2 \, \mathrm{d}x + C \int_{\Omega_R} |\Delta b|^2 \, \mathrm{d}x,$$

$$-\int_{\Omega_R} (\rho u \cdot \nabla b) \cdot u \, \mathrm{d}x \leq C \int_{\Omega_R} \rho |u|^2 \, \mathrm{d}x$$

for some $C := C(N_0) > 0$. Thus, we have

$$\frac{\mathrm{d}}{\mathrm{d}t} \int_{\Omega_R} \rho \frac{|u|^2}{2} \, \mathrm{d}x + \int_{\Omega_R} |\nabla u|^2 \, \mathrm{d}x \leq C \int_{\Omega_R} \rho |u|^2 \, \mathrm{d}x + C \int_{\Omega_R} |f|^2 \, \mathrm{d}x + C \int_{\Omega_R} |\Delta b|^2 \, \mathrm{d}x. \quad (5.57)$$

Next, we integrate (5.57) with respect to t and then we apply Gronwall's inequaility to obtain

$$\int_{\Omega_R} \rho(t) |u(t)|^2 \, \mathrm{d}x + \int_0^t \int_{\Omega_R} |\nabla u(s)|^2 \, \mathrm{d}x \, \mathrm{d}s \leq C e^{Ct} \int_{\Omega_R} \rho_0^R |a^R|^2 \, \mathrm{d}x$$

for all $t \in [0, T]$ and some constant $C := C(N_0) > 0$. Now the assertion follows from Lemma 5.18 and from the fact that ρ is bounded from below by $\underline{\rho} > 0$. $\qquad \square$

Lemma 5.38. *There exist a time $T_0 := T_0(N_0) > 0$, independent of $R > R_0$, and a constant $C := C(N_0, T_0) > 0$, independent of $R > R_0$, such that the strong solution $(\rho, u) := (\rho^R, u^R)$ from Proposition 5.15 satisfies*

$$\|\nabla u(t)\|_{L^2(\Omega_R)}^2 + \int_0^t \|\nabla^2 u(s)\|_{L^2(\Omega_R)}^2 \, \mathrm{d}s \leq C$$

for all $t \in [0, T]$ and all $T < \min\{T_0, T_{\max}^R\}$.

Proof. Let $T < \min\{T_0, T_{\max}^R\}$ and note that $u_t - Mx \cdot \nabla u + Mu \in L^2((0, T); L_\sigma^2(\Omega_R))$, see Lemma 5.23. Multiply equation (5.55) with $u_t - Mx \cdot \nabla u + Mu$ and integrate in $(0, t) \times \Omega$ for some $t \in [0, T]$. Then, integrate by parts with respect to x and apply Lemma 5.23 to obtain

$$\int_{\Omega_R} \frac{|\nabla u(t)|^2}{2} \, \mathrm{d}x - \int_{\Omega_R} \frac{|\nabla u(0)|^2}{2} \, \mathrm{d}x + \int_0^t \int_{\Omega_R} \rho |u_t - Mx \cdot \nabla u + Mu|^2 \, \mathrm{d}x \, \mathrm{d}s$$

$$= -\int_0^t \int_{\Omega_R} (\rho(u + b) \cdot \nabla u) \cdot (u_t - Mx \cdot \nabla u + Mu) \, \mathrm{d}x \, \mathrm{d}s$$

$$- \int_0^t \int_{\Omega_R} (\rho u \cdot \nabla b) \cdot (u_t - Mx \cdot \nabla u + Mu) \, \mathrm{d}x \, \mathrm{d}s$$

$$+ \int_0^t \int_\Gamma \left(-(\nu \cdot \nabla u) \cdot (Mx \cdot \nabla u) + \frac{1}{2} |\nabla u|^2 Mx \cdot \nu \right) \mathrm{d}\sigma \, \mathrm{d}s$$

$$- \int_0^t \int_{\Omega_R} \nabla u : \nabla M u \, \mathrm{d}x \, \mathrm{d}s + \int_0^t \int_{\Omega_R} (\rho f + \Delta b) \cdot (u_t - Mx \cdot \nabla u + Mu) \, \mathrm{d}x \, \mathrm{d}s.$$

The terms on the right hand side are estimated as follows: By Young's inequality

$$-\int_{\Omega_R} (\rho(u+b) \cdot \nabla u) \cdot (u_t - Mx \cdot \nabla u + Mx) \, dx$$

$$\leq \frac{1}{10} \int_{\Omega_R} \rho |u_t - Mx \cdot \nabla u + Mx|^2 \, dx + C \int_{\Omega_R} |(u+b) \cdot \nabla u|^2 \, dx,$$

$$-\int_{\Omega_R} (\rho u \cdot \nabla b) \cdot (u_t - Mx \cdot \nabla u + Mx) \, dx$$

$$\leq \frac{1}{10} \int_{\Omega_R} \rho |u_t - Mx \cdot \nabla u + Mx|^2 \, dx + C \int_{\Omega} |u \cdot \nabla b|^2 \, dx$$

$$\int_{\Omega_R} (\rho f + \Delta b) \cdot (u_t - Mx \cdot \nabla u + Mx) \, dx$$

$$\leq \frac{1}{10} \int_{\Omega_R} \rho |u_t - Mx \cdot \nabla u + Mx)|^2 \, dx + C \int_{\Omega_R} |f|^2 \, dx + C \int_{\Omega_R} |\Delta b|^2 \, dx,$$

for some $C := C(\underline{\rho}, \bar{\rho}) > 0$. For the boundary term we use Corollary 1.11 to get

$$\int_\Gamma \left| -(\nu \cdot \nabla u) \cdot (Mx \cdot \nabla u) + \frac{1}{2} |\nabla u|^2 Mx \cdot \nu \right| \, d\sigma \leq C \|\nabla u\|^2_{L^2(\Omega)} + \varepsilon \|\nabla^2 u\|^2_{L^2(\Omega)}$$

for arbitrary $\varepsilon > 0$ and some $C := C(|M|, \varepsilon) > 0$. Combining the above estimates yields

$$\int_{\Omega_R} \frac{|\nabla u(t)|^2}{2} \, dx - \int_{\Omega_R} \frac{|\nabla u(0)|^2}{2} \, dx + \int_0^t \int_{\Omega_R} \rho |u_t - Mx \cdot \nabla u + Mu|^2 \, dx \, ds$$

$$\leq \frac{1}{2} \int_0^t \int_{\Omega_R} \rho |u_t - Mx \cdot \nabla u + Mx|^2 \, dx \, ds + C \int_0^t \int_{\Omega_R} |(u+b) \cdot \nabla u|^2 \, dx \, ds$$

$$+ C \int_0^t \int_{\Omega_R} |u \cdot \nabla b|^2 \, dx$$

$$+ C \int_0^t \int_{\Omega_R} |\nabla u|^2 \, dx \, ds + C \int_0^t \int_{\Omega_R} |f|^2 \, dx \, ds + C \int_0^t \int_{\Omega_R} |\Delta b|^2 \, dx \, ds$$

$$+ \int_0^t \int_\Gamma \left| -(\nu \cdot \nabla u) \cdot (Mx \cdot \nabla u) + \frac{1}{2} |\nabla u|^2 Mx \cdot \nu \right| \, d\sigma \, ds$$

$$\leq \frac{1}{2} \int_0^t \int_{\Omega_R} \rho |u_t - Mx \cdot \nabla u + Mx|^2 \, dx \, ds + \varepsilon \int_0^t \|\nabla^2 u\|^2_{L^2(\Omega_R)} \, ds$$

$$+ C \int_0^t \left(\|u\|^2_{L^6(\Omega_R)} \|\nabla u\|^2_{L^3(\Omega_R)} + \|\nabla u\|^2_{L^2(\Omega_R)} \|u\|^2_{L^2(\Omega_R)} + 1 \right) \, ds$$

for $\varepsilon > 0$ arbitrarily chosen and some constant $C := C(N_0, \varepsilon) > 0$. Next, we use the interpolation inequality and the Sobolev embedding $H^1(\Omega_R) \hookrightarrow L^6(\Omega_R)$ to conclude

$$\|\nabla u\|^2_{L^3(\Omega_R)} \leq \|\nabla u\|_{L^2(\Omega_R)} \|\nabla u\|_{L^6(\Omega_R)} \leq C \|\nabla u\|_{L^2(\Omega_R)} \|\nabla u\|_{H^1(\Omega_R)}$$

for some constant $C > 0$ independent of R, see Proposition 1.14. Moreover, by Sobolev inequality, see Proposition 1.15, there exists a constant $C > 0$, independent of R, such that

$$\|u\|_{L^6(\Omega_R)} \leq C \|\nabla u\|_{L^2(\Omega_R)}.$$

Thus, by applying Lemma 5.37 and Young's inequality we conclude that for arbitrary $\varepsilon > 0$ there exists a constant $C := C(N_0, \varepsilon) > 0$ such that

$$\int_{\Omega_R} \frac{|\nabla u(t)|^2}{2} \, dx - \int_{\Omega_R} \frac{|\nabla u(0)|^2}{2} \, dx + \int_0^t \int_{\Omega_R} \rho |u_t - Mx \cdot \nabla u + Mx|^2 \, dx \, ds$$

$$\leq C \int_0^t \left(\|\nabla u\|_{L^2(\Omega_R)}^6 + \|\nabla u\|_{L^2(\Omega_R)}^4 + \|\nabla u\|_{L^2(\Omega_R)}^2 + 1 \right) ds + \varepsilon \int_0^t \|\nabla^2 u\|_{L^2(\Omega_R)}^2 \, ds$$

$$\tag{5.58}$$

for all $t \in [0, T]$.

Next we derive an estimate for the last term on the right-hand side of (5.58). Set

$$g := -\rho u_t - \rho u \cdot \nabla u - \rho b \cdot \nabla u - \rho u \cdot \nabla b + \rho(Mx \cdot \nabla u - Mu) + \rho f + \Delta b.$$

For fixed $t \in [0, T]$, the regularity of (ρ, u) stated in (5.53) allows to choose $\mathrm{p}(t) \in \widehat{W}^{1,2}(\Omega_R)$ such that $\nabla \mathrm{p}(t) = (\mathrm{Id} - \mathbb{P}_{\Omega_R})\Delta u(t) + (\mathrm{Id} - \mathbb{P}_{\Omega_R})g(t)$. Then, by using equation (5.55), we see that for fixed $t \in [0, T]$ the pair of functions $(u(t), \mathrm{p}(t))$ solves the stationary Stokes problem

$$\begin{cases} -\Delta u(t) + \nabla \mathrm{p}(t) &= g(t) \quad \text{in } \Omega_R, \\ \operatorname{div} u(t) &= 0 \quad \text{in } \Omega_R, \\ u(t) &= 0 \quad \text{on } \partial\Omega_R. \end{cases}$$

The elliptic estimate in Proposition 2.4 yields

$$\int_0^t \|\nabla^2 u(s)\|_{L^2(\Omega_R)}^2 \, ds \leq C \int_0^t \left(\|\mathbb{P}_{\Omega_R} g(s)\|_{L^2(\Omega_R)}^2 + \|\nabla u(s)\|_{L^2(\Omega_R)}^2 \right) ds \tag{5.59}$$

for some constant $C > 0$ independent of R. Then, by using as above the interpolation inequality, the Sobolev embedding and Lemma 5.37, we obtain

$$\|g\|_{L^2(\Omega_R)} \leq C \|\sqrt{\rho}(u_t - Mx \cdot \nabla u + Mu)\|_{L^2(\Omega_R)} + C\|u\|_{L^6(\Omega_R)} \|\nabla u\|_{L^3(\Omega_R)}$$
$$+ C\|\nabla u\|_{L^2} + \|u\|_{L^2(\Omega_R)} + C$$
$$\leq C \|\sqrt{\rho}(u_t - Mx \cdot \nabla u + Mu)\|_{L^2(\Omega_R)} + C\|\nabla u\|_{L^2(\Omega_R)}^{\frac{3}{2}} \|\nabla^2 u\|_{L^2(\Omega_R)}^{\frac{1}{2}}$$
$$+ C\|\nabla u\|_{L^2(\Omega_R)}^2 + C\|\nabla u\|_{L^2(\Omega_R)} + C$$

for a constant $C := C(N_0) > 0$. By Young's inequality wee see that for arbitrary $\delta > 0$ there exists a constant $C := C(N_0, \delta) > 0$ such that

$$\|g\|_{L^2(\Omega_R)} \leq C \|\sqrt{\rho}(u_t - Mx \cdot \nabla u + Mu)\|_{L^2(\Omega_R)} + C\|\nabla u\|_{L^2(\Omega_R)}^3 + \delta\|\nabla^2 u\|_{L^2(\Omega_R)}$$
$$+ C\|\nabla u\|_{L^2(\Omega_R)}^2 + C\|\nabla u\|_{L^2(\Omega_R)} + C. \tag{5.60}$$

By choosing $\delta > 0$ sufficiently small it follows from (5.59) and (5.60) that there exists a constant $C := C(N_0) > 0$ such that

$$\int_0^t \|\nabla^2 u(s)\|_{L^2(\Omega_R)}^2 \, ds \leq C \int_0^t \left(\|\sqrt{\rho}(u_t - Mx \cdot \nabla u + Mu)\|_{L^2(\Omega_R)}^2 + \|\nabla u(s)\|_{L^2(\Omega_R)}^6 \right.$$
$$\left. + \|\nabla u(s)\|_{L^2(\Omega_R)}^4 + \|\nabla u(s)\|_{L^2(\Omega_R)}^2 + 1 \right) ds. \tag{5.61}$$

Now, if we combine (5.58) and (5.61), and if we choose $\varepsilon > 0$ sufficiently small, then it follows that

$$\|\nabla u(t)\|_{L^2(\Omega_R)}^2 + \int_0^t \|\sqrt{\rho}(u_t(s) - Mx \cdot \nabla u(s) + Mu(s))\|_{L^2(\Omega_R)}^2 \, ds$$

$$\leq C\|\nabla u(0)\|_{L^2(\Omega_R)}^2 + C\int_0^t \left(1 + \|\nabla u(s)\|_{L^2(\Omega_R)}^2 + \|\nabla u(s)\|_{L^2(\Omega_R)}^4 + \|\nabla u(s)\|_{L^2(\Omega_R)}^6\right) \, ds$$

for some constant $C := C(N_0)$. In particular,

$$\|\nabla u(t)\|_{L^2(\Omega_R)}^2 \leq C\|\nabla u(0)\|_{L^2(\Omega_R)}^2$$
$$+ C\int_0^t \left(1 + \|\nabla u(s)\|_{L^2(\Omega_R)}^2 + \|\nabla u(s)\|_{L^2(\Omega_R)}^4 + \|\nabla u(s)\|_{L^2(\Omega_R)}^6\right) \, ds.$$

By using the non-linear version of Gronwall's inequality, see Lemma 1.24, there exists a $T_0 := T_0(N_0) > 0$ and a constant $C := C(N_0, T_0) > 0$ such that

$$\|\nabla u(t)\|_{L^2(\Omega_R)}^2 \leq C$$

for all $t \in [0, T]$ and all $T < \min\{T_0, T_{\max}^R\}$. Here T_0 and C are in particular independent of R. Hence, we conclude that

$$\|\nabla u(t)\|_{L^2(\Omega_R)}^2 + \int_0^t \|\sqrt{\rho}(u_t(s) - Mx \cdot \nabla u(s) + Mu(s))\|_{L^2(\Omega_R)}^2 \, ds \leq C,$$

and the estimate for $\nabla^2 u$ follows now from (5.61). This concludes the proof. $\qquad\square$

At this point, let us briefly compare the proof of Lemma 5.38 and the proof of the a priori estimate for ∇u^n stated in Proposition 5.9. It turns out that in order to control the L^2-norm of ∇u with constant independent of R, one needs to test the equation with the element $u_t - Mx \cdot \nabla u + Mu$, as it is done in the proof of Lemma 5.38. However, in order to control L^2-norm of ∇u^n uniformly in n one cannot test with the element $u_t^n - Mx \cdot \nabla u^n + Mu^n$. This is due to the fact that u^n solves the equation (5.10) in the subspace span $\{\psi_1, \dots, \psi_n\}$. However, $u_t^n - Mx \cdot \nabla u^n + Mu^n$ is in general not contained in this finite dimensional subspace. Therefore (5.10) cannot be tested with the element $u_t^n - Mx \cdot \nabla u^n + Mu^n$. This explains why we have to derive the a priori estimates uniformly in n and uniformly in R separately.

As a direct consequence of the proof of Lemma 5.38 we obtain the following corollary.

Corollary 5.39. *There exist a time $T_0 := T_0(N_0) > 0$, independent of $R > R_0$, and a constant $C := C(N_0, T_0) > 0$, independent of $R > R_0$, such that the strong solution $(\rho, u) := (\rho^R, u^R)$ from Proposition 5.15 satisfies*

$$\int_0^t \|Mx \cdot \nabla u(s) - Mu(s)\|_{L^2(\Omega_R)}^2 \, ds \leq C \int_0^t \|u_t(s)\|_{L^2(\Omega_R)}^2 \, ds + C$$

for all $t \in [0, T]$ and all $T < \min\{T_0, T_{\max}^R\}$.

Proof. In the proof of Lemma 5.38 it was shown that there exists a $T_0 := T_0(N_0) > 0$ and constant $C := C(N_0, T_0) > 0$ such that

$$\int_0^t \|\sqrt{\rho}(u_t(s) - Mx \cdot \nabla u(s) + Mu(s))\|_{L^2(\Omega_R)}^2 \, ds \leq C$$

for all $t \in [0, T]$ and all $T < \min\{T_0, T_{\max}^R\}$. Since ρ is bounded from below by $\underline{\rho} > 0$ the assertion directly follows. $\square$

Lemma 5.40. *There exists a constant $C := C(N_0) > 0$, independent of R, such that the strong solution $(\rho, u) := (\rho^R, u^R)$ from Proposition 5.15 satisfies*

$$\|u_t(0)\|_{L^2(\Omega_R)} \leq C.$$

Proof. We multiply equation (5.55) for $t = 0$ with $u_t(0)$ and then integrate over Ω_R. By Hölder's and Young's inequality, we obtain

$$\left\|\sqrt{\rho_0^R} u_t(0)\right\|_{L^2(\Omega_R)}^2 \leq C \left\|a^R\right\|_{L^\infty(\Omega_R)}^2 \left\|\nabla a^R\right\|_{L^2(\Omega_R)}^2 + C \left\|\nabla a^R\right\|_{L^2(\Omega_R)}^2 + C \left\|a^R\right\|_{L^2(\Omega_R)}^2$$

$$+ C \left\|Mx \cdot \nabla a^R\right\|_{L^2(\Omega_R)}^2 + C \left\|\mathbb{P}_{\Omega_R} \Delta a^R\right\|_{L^2(\Omega_R)}^2 + C$$

for some constant $C := C(N_0) > 0$. The assertion now follows by applying the embedding $H^2(\Omega_R) \hookrightarrow L^\infty(\Omega)$ and Lemma 5.18. $\square$

Lemma 5.41. *There exist a time $T_0 := T_0(N_0) > 0$, independent of $R > R_0$, and a constant $C := C(N_0, T_0) > 0$, independent of R, such that the strong solution $(\rho, u) := (\rho^R, u^R)$ from Proposition 5.15 satisfies*

$$\|u_t(t)\|_{L^2(\Omega_R)}^2 + \int_0^t \|\nabla u_t(s)\|_{L^2(\Omega_R)}^2 \, ds \leq C,$$

$$\|\nabla^2 u(t)\|_{L^2(\Omega_R)}^2 + \|Mx \cdot \nabla u(t) - Mu(t)\|_{L^2(\Omega_R)}^2 \leq C,$$

$$\int_0^t \|\nabla^2 u(s)\|_{L^6(\Omega_R)}^2 \, ds + \int_0^t \|Mx \cdot \nabla u(t) - Mu(t)\|_{L^6(\Omega_R)}^2 \, ds \leq C,$$

$$\|\rho(t) - 1\|_{H^2(\Omega_R)}^2 + \|\rho_t(t)\|_{L^2(\Omega_R)}^2 + \|(\rho(t) - 1)Mx\|_{H^1(\Omega_R)}^2 \leq C,$$

for all $t \in [0, T]$ and all $T < \min\{T_0, T_{\max}^R\}$.

Proof. Introduce functions $\Psi_R(t)$ and $\Phi_R(t)$ defined by

$$\Psi_R(t) := \sup_{0 \leq s \leq t} \left(\|\rho(s) - 1\|_{H^2(\Omega_R)}^2 + \|\rho_t(s)\|_{L^2(\Omega_R)}^2 + \|(\rho(s) - 1)Mx\|_{H^1(\Omega_R)}^2 + 1\right),$$

$$\Phi_R(t) := \sup_{0 \leq s \leq t} \left(\|u_t(s)\|_{L^2(\Omega_R)}^2 + \|\nabla^2 u(s)\|_{L^2(\Omega_R)}^2 + \|Mx \cdot \nabla u(t) - Mu(t)\|_{L^2(\Omega_R)}^2 + 1\right)$$

$$+ \int_0^t \left(\|\nabla u_t(s)\|_{L^2(\Omega_R)}^2 + \|\nabla^2 u(s)\|_{L^6(\Omega_R)}^2 + \|Mx \cdot \nabla u(s) - Mu(s)\|_{L^6(\Omega_R)}^2\right) ds.$$

In the following we control $\Psi_R(t)$ and $\Phi_R(t)$, respectively. It turns out that in order control $\Psi_R(t)$ one needs the function $\Phi_R(t)$ and in order to control $\Phi_R(t)$ one needs the function $\Psi_R(t)$. Our proof is based on a continuation method. The proof is divided into 3 steps. In the following, let $T^* := T^*(N_0) > 0$ denote the time independent of R such that the estimates from Lemma 5.38 hold for all $t \in [0,T]$ and all $T < \min\{T^*, T_{\max}^R\}$.

Step 1: Estimate for $\Psi_R(t)$. First of all let us not that if ρ solves the transport equation (5.14) in the strong sense for initial value ρ_0, then $\rho - 1$ solves the transport equation (5.14) in the strong sense for initial value $\rho_0 - 1$. Therefore, by applying Lemma 5.26 and Lemma 5.27 we conclude that

$$\|\rho(t) - 1\|^2_{H^2(\Omega_R)} \leq \|\rho_0 - 1\|^2_{H^2(\Omega_R)} \exp\left(C \int_0^t \left(\|u(s)\|_{W^{2,6}(\Omega_R)} + \|b\|_{W^{2,6}(\Omega_R)} + |M|\right) ds\right)$$

and

$$\|(\rho(t) - 1)Mx\|^2_{L^2(\Omega_R)} + \|\nabla\rho(t)(Mx)^\top\|^2_{L^2(\Omega_R)}$$

$$\leq \left(\|(\rho_0 - 1)Mx\|^2_{L^2(\Omega_R)} + \|\nabla\rho_0(Mx)^\top\|^2_{L^2(\Omega_R)} + 2|M|^2 \int_0^t \|\rho(s)\|^2_{H^1(\Omega_R)} ds\right)$$

$$\exp\left(C \int_0^t \left(\|u(s)\|^2_{W^{1,6}(\Omega_R)} + \|b\|^2_{W^{1,6}(\Omega_R)} + \|u(s)\|_{W^{2,6}(\Omega_R)} + \|b\|_{W^{2,6}(\Omega_R)} + |M|\right) ds\right)$$

for all $t \in [0,T]$ and all $T < \min\{T^*, T_{\max}^R\}$. By the Sobolev embedding $H^2(\Omega_R) \hookrightarrow W^{1,6}(\Omega_R)$, by Lemma 5.37 and Lemma 5.38 and by Hölder's inequality we see that there exists a constant $C := C(N_0, T^*) > 0$ such that

$$\int_0^t \|u(s)\|^2_{W^{1,6}(\Omega_R)} ds + \int_0^t \|u(s)\|_{W^{2,6}(\Omega_R)} ds$$

$$\leq C \int_0^t \|\nabla^2 u(s)\|^2_{L^2(\Omega_R)} ds + C \int_0^t \|\nabla^2 u(s)\|_{L^6(\Omega_R)} ds + Ct \leq Ct^{\frac{1}{2}}\Phi_R(t)$$

for all $t \in [0,T]$ and all $T < \min\{T^*, T_{\max}^R\}$. Therefore,

$$\|\rho(t) - 1\|^2_{H^2(\Omega_R)} + \|(\rho(t) - 1)Mx\|^2_{H^1(\Omega_R)} \leq C + C \exp\left(C\Phi_R(t)t^{\frac{1}{2}}\right) \tag{5.62}$$

for all $t \in [0,T]$ and all $T < \min\{T^*, T_{\max}^R\}$ and some constant $C := C(N_0, T^*) > 0$. Recall $\rho_t = -(u + b - Mx) \cdot \nabla\rho$. Thus, we see that

$$\|\rho_t(t)\|_{L^2(\Omega_R)} \leq \|u\|_{L^3(\Omega_R)}\|\nabla\rho(t)\|_{L^6(\Omega_R)} + \|b\|_{L^\infty(\Omega_R)}\|\nabla\rho(t)\|_{L^2(\Omega_R)} + \|\nabla\rho(t)(Mx)^\top\|_{L^2(\Omega_R)}.$$

Taking the Sobolev embedding $H^1(\Omega_R) \hookrightarrow L^p(\Omega_R)$ for $2 < p \leq 6$ into account and using Lemma 5.37 and Lemma 5.38, we obtain

$$\|\rho_t(t)\|_{L^2(\Omega_R)} \leq C \left(\|\rho(t) - 1\|^2_{H^2(\Omega_R)} + \|(\rho(t) - 1)Mx\|^2_{L^2(\Omega_R)}\right) \tag{5.63}$$

for all $t \in [0,T]$ and all $T < \min\{T^*, T_{\max}^R\}$ and some constant $C := C(N_0, T^*) > 0$. Combining (5.62) and (5.63) yields

$$\Psi_R(T) \leq L + L \exp\left(L\Phi_R(T)T^{\frac{1}{2}}\right) \tag{5.64}$$

for all $T < \min\{T^*, T^R_{\max}\}$ and a constant $L := L(N_0, T^*) > 0$, which is now fixed for the rest of the proof.

Step 2: Estimate for $\Phi_R(t)$. We start with estimating the L^2-norm of u_t. In contrast to the proof of Lemma 5.33, it is not directly possible to differentiate (5.55) with respect to time since the solution (ρ, u) does not have sufficient time-regularity. Therefore, we shall work with difference-quotients instead. For that purpose we define for all $h \in \mathbb{R}$ (sufficiently small) and for some function $v(t, x)$ the operation τ_h by

$$\tau_h v(t, x) := \frac{v(t+h, x) - v(t, x)}{h}.$$

We apply τ_h to (5.55), then we multiply with $\tau_h u$ and finally we integrate over Ω_R to obtain

$$\int_{\Omega_R} \big(\rho \partial_t \tau_h u + \tau_h \rho u_t + \rho u \cdot \nabla \tau_h u + \tau_h \rho u \cdot \nabla u + \rho \tau_h u \cdot \nabla u + \rho b \cdot \nabla \tau_h u + \tau_h \rho b \cdot \nabla u$$

$$+ \rho \tau_h u \cdot \nabla b + \tau_h \rho u \cdot \nabla b - \rho M x \cdot \nabla \tau_h u - \tau_h \rho M x \cdot \nabla u + \rho M \tau_h u \qquad (5.65)$$

$$+ \tau_h \rho M u - \Delta \tau_h u \big) \cdot \tau_h u \, dx = \int_{\Omega_R} \tau_h \rho f \cdot \tau_h u \, dx$$

in $[h, T^R_{\max} - h)$. Note that (5.65) includes the term $\partial_t \tau_h u$ which is well-defined, however the regularity of u stated in (5.53) is not high enough to pass to the limit $h \to 0$ at this point. Therefore, we first integrate (5.65) with respect to time in the interval $(|h|, t)$ for some $t \in [|h|, T^R_{\max} - |h|)$ and then we do integration by parts with respect to x. Here we essentially use the same computations as in the proof of Lemma 5.33 and therefore we omit the details. We end up with

$$\int_{\Omega_R} \rho(t) \frac{|\tau_h u(t)|^2}{2} \, dx - \int_{\Omega_R} \rho(|h|) \frac{|\tau_h u(|h|)|^2}{2} \, dx + \int_{|h|}^t \int_{\Omega_R} |\nabla \tau_h u|^2 \, dx \, ds$$

$$= -\int_{|h|}^t \int_{\Omega_R} \tau_h \rho (u_t + u \cdot \nabla u + b \cdot \nabla u + u \cdot \nabla b - Mx \cdot \nabla u + Mu) \cdot \tau_h u \, dx \, ds$$

$$- \int_{|h|}^t \int_{\Omega_R} \rho(\tau_h u \cdot \nabla u + \tau_h u \cdot \nabla b) \cdot \tau_h u \, dx \, ds + \int_{|h|}^t \int_{\Omega_R} \tau_h \rho f \cdot \tau_h u \, dx \, ds.$$

Now the regularity of all terms is high enough in order to pass to the limit $h \to 0$. Lebesgue's theorem implies

$$\int_{\Omega_R} \rho(t) \frac{|u_t(t)|^2}{2} \, dx - \int_{\Omega_R} \rho(0) \frac{|u_t(0)|^2}{2} \, dx + \int_0^t \int_{\Omega_R} |\nabla u_t|^2 \, dx \, ds$$

$$= -\int_0^t \int_{\Omega_R} \rho_t (u_t + u \cdot \nabla u + b \cdot \nabla u + u \cdot \nabla b - Mx \cdot \nabla u + Mu) \cdot u_t \, dx \, ds$$

$$- \int_0^t \int_{\Omega_R} \rho(u_t \cdot \nabla u + u_t \cdot \nabla b) \cdot u_t \, dx \, ds + \int_0^t \int_{\Omega_R} \rho_t f \cdot u_t \, dx \, ds =: I_1 + I_2 + I_3.$$

Since $u_t \in L^2((0, T^R_{\max}); H^1_{0,\sigma}(\Omega_R))$, the Sobolev inequality, see Proposition 1.15, implies that there exists a constant $C > 0$, independent of $R > 0$, such that

$$\|u_t(t)\|_{L^6(\Omega_R)} \le C \|\nabla u_t(t)\|_{L^2(\Omega_R)} \qquad (5.66)$$

for almost all $t \in (0, T_{\max}^R)$. Moreover, the interpolation inequality yields

$$\|u_t\|_{L^3(\Omega_R)} \le \|u_t\|_{L^2(\Omega_R)}^{\frac{1}{2}} \|u_t\|_{L^6(\Omega_R)}^{\frac{1}{2}},$$

$$\|Mx \cdot \nabla u - Mu\|_{L^3(\Omega_R)} \le \|Mx \cdot \nabla u - Mu\|_{L^2(\Omega_R)}^{\frac{1}{2}} \|Mx \cdot \nabla u - Mu\|_{L^6(\Omega_R)}^{\frac{1}{2}}.$$

Thus, we obtain

$$
\begin{aligned}
|I_1| \le &\int_0^t \|\rho_t\|_{L^2(\Omega_R)} \|u_t\|_{L^6(\Omega_R)} \Big(\|u_t\|_{L^3(\Omega_R)} + \|u \cdot \nabla u\|_{L^3(\Omega_R)} \\
&+ \|b \cdot \nabla u\|_{L^3(\Omega_R)} + \|u \cdot \nabla b\|_{L^3(\Omega_R)} + \|Mx \cdot \nabla u - Mu\|_{L^3(\Omega_R)} \Big) \, \mathrm{d}s \\
\le &\, C \int_0^t \|\rho_t\|_{L^2(\Omega_R)} \|\nabla u_t\|_{L^2(\Omega_R)} \Big(\|u_t\|_{L^2(\Omega_R)}^{\frac{1}{2}} \|\nabla u_t\|_{L^2(\Omega_R)}^{\frac{1}{2}} \\
&+ \|u\|_{L^6(\Omega_R)} \|\nabla u\|_{L^6(\Omega_R)} + \|b\|_{L^6(\Omega_R)} \|\nabla u\|_{L^6(\Omega_R)} + \|u\|_{L^6(\Omega_R)} \|\nabla b\|_{L^6(\Omega_R)} \\
&+ \|Mx \cdot \nabla u - Mu\|_{L^2(\Omega_R)}^{\frac{1}{2}} \|Mx \cdot \nabla u - Mu\|_{L^6(\Omega_R)}^{\frac{1}{2}} \Big) \, \mathrm{d}s
\end{aligned}
$$

for some constant $C > 0$. By Young's inequality and Corollary 5.39 we conclude that for arbitrary $\varepsilon > 0$ there exists a constant $C := C(N_0, T^*, \varepsilon) > 0$, independent of R, such that

$$
\begin{aligned}
&\int_0^t \|\rho_t\|_{L^2(\Omega_R)} \|\nabla u_t\|_{L^2(\Omega_R)} \|Mx \cdot \nabla u - Mu\|_{L^2(\Omega_R)}^{\frac{1}{2}} \|Mx \cdot \nabla u - Mu\|_{L^6(\Omega_R)}^{\frac{1}{2}} \, \mathrm{d}s \\
&\qquad \le \varepsilon \Psi_R(t) \int_0^t \|\nabla u_t\|_{L^2(\Omega_R)}^2 \, \mathrm{d}s + \varepsilon \int_0^t \|Mx \cdot \nabla u - Mu\|_{L^6(\Omega_R)}^2 \, \mathrm{d}s \\
&\qquad\qquad + C \int_0^t \|u_t\|_{L^2(\Omega_R)}^2 \, \mathrm{d}s + C.
\end{aligned}
$$

Now we use again Young's inequality, Lemma 5.37 and Lemma 5.38 to conclude that for $\varepsilon > 0$ arbitrarily chosen there exists a constant $C := C(N_0, T^*, \varepsilon) > 0$ such that

$$
\begin{aligned}
|I_1| \le &\, \varepsilon \Psi_R(t) \int_0^t \|\nabla u_t\|_{L^2(\Omega_R)}^2 \, \mathrm{d}s + \varepsilon \int_0^t \|Mx \cdot \nabla u - Mu\|_{L^6(\Omega_R)}^2 \, \mathrm{d}s \\
&+ C \int_0^t \|u_t\|_{L^2(\Omega_R)}^2 \, \mathrm{d}s + C.
\end{aligned}
$$

It is easy to see that there exists a constant $C := C(N_0) > 0$ such that

$$|I_2| \le C \int_0^t \|u_t\|_{L^4(\Omega_R)}^2 \|\nabla u\|_{L^2(\Omega_R)} \, \mathrm{d}s + C \int_0^t \|u_t\|_{L^2(\Omega_R)}^2 \, \mathrm{d}s.$$

We now use the interpolation inequality and the Sobolev inequality (5.66) to deduce that

$$\|u_t\|_{L^4(\Omega_R)}^2 \le \|u_t\|_{L^2(\Omega_R)}^{\frac{1}{2}} \|u_t\|_{L^6(\Omega_R)}^{\frac{3}{2}} \le C \|u_t\|_{L^2(\Omega_R)}^{\frac{1}{2}} \|\nabla u_t\|_{L^2(\Omega_R)}^{\frac{3}{2}}$$

for some constant $C > 0$. Using this estimate together with Lemma 5.38 and Young's inequality yield that for arbitrary $\varepsilon > 0$ there exists some $C := C(N_0, T^*, \varepsilon) > 0$ such that

$$|I_2| \le C \int_0^t \|u_t\|_{L^2(\Omega_R)}^2 \, \mathrm{d}s + \varepsilon \int_0^t \|\nabla u_t\|_{L^2(\Omega_R)}^2 \, \mathrm{d}s.$$

The term I_3 is easily estimated by

$$|I_3| \leq \int_0^t \|\rho_t\|_{L^2(\Omega_R)} \|f\|_{L^3(\Omega_R)} \|u_t\|_{L^6(\Omega_R)} \leq C + \varepsilon \Psi_R(t) \int_0^t \|\nabla u_t\|_{L^2(\Omega_R)}^2 \, ds$$

for some constant $C := C(N_0, T^*, \varepsilon) > 0$. Then, combining the above estimates, we see that for arbitrary $\varepsilon > 0$ there exists a constant $C := C(N_0, T^*, \varepsilon) > 0$ such that

$$\int_{\Omega_R} \rho \frac{|u_t(t)|^2}{2} \, dx + \int_0^t \int_{\Omega_R} |\nabla u_t|^2 \, dx \, ds$$

$$\leq \varepsilon \Psi_R(t) \int_0^t \|\nabla u_t(s)\|_{L^2(\Omega_R)}^2 \, ds + \varepsilon \int_0^t \|Mx \cdot \nabla u(s) - Mu(s)\|_{L^6(\Omega_R)}^2 \, ds \quad (5.67)$$

$$+ C \int_0^t \|u_t(s)\|_{L^2(\Omega_R)}^2 \, ds + C\Psi_R(t)$$

for all $t \in [0, T]$ and all $T < \min\{T^*, T_{\max}^R\}$.

Next we derive an estimate for $\nabla^2 u$ in the L^2-norm and in the L^6-norm. Set

$$g := -\rho u_t - \rho u \cdot \nabla u - \rho b \cdot \nabla u - \rho u \cdot \nabla b + (\rho - 1)(Mx \cdot \nabla u - Mu) + \rho f + \Delta b.$$

Note that by the regularity of the solution (ρ, u) stated in (5.53) it directly follows that $g \in C([0, T]; L^2(\Omega_R)^3)$ for all $T < T_{\max}^R$ and that $g \in L^2((0, T_{\max}^R); L^6(\Omega_R)^3)$.

For $t \in [0, T_{\max}^R)$, choose $\mathrm{p}(t) \in \widehat{W}^{1,2}(\Omega_R)$ such that $\nabla \mathrm{p}(t) = (\mathrm{Id} - \mathbb{P}_{\Omega_R})\Delta u(t) + (\mathrm{Id} - \mathbb{P}_{\Omega_R})g(t)$. Note that for almost all $t \in (0, T_{\max}^R)$ we even have $\mathrm{p}(t) \in \widehat{W}^{1,6}(\Omega_R)$. Equation (5.55) yields that for fixed $t \in (0, T_{\max}^R)$ the pair of functions $(u(t), \mathrm{p}(t))$ solves the modified stationary Stokes problem

$$\begin{cases} -\Delta u(t) - Mx \cdot \nabla u(t) + Mu(t) + \nabla \mathrm{p}(t) &= g(t) \quad \text{in } \Omega, \\ \operatorname{div} u(t) &= 0 \quad \text{in } \Omega, \\ u(t) &= 0 \quad \text{on } \partial\Omega. \end{cases}$$

Thus, we can apply the elliptic estimates in Proposition 5.17 and Lemma 5.28 to obtain

$$\|\nabla^2 u(t)\|_{L^2(\Omega_R)}^2 + \|Mx \cdot \nabla u(t) - Mu(t)\|_{L^2(\Omega_R)}^2$$

$$+ \int_0^t \|\nabla^2 u(s)\|_{L^6(\Omega_R)}^2 \, ds + \int_0^t \|Mx \cdot \nabla u(s) - Mu(s)\|_{L^6(\Omega_R)}^2 \, ds \quad (5.68)$$

$$\leq C\|g(t)\|_{L^2(\Omega_R)}^2 + C\|\nabla u\|_{L^2(\Omega_R)}^2 + C \int_0^t \|g(s)\|_{L^6(\Omega_R)}^2 \, ds + C \int_0^t \|\nabla u(s)\|_{L^2(\Omega_R)}^2 \, ds$$

for some constant $C := C(|M|) > 0$ independent of R. Next let us estimate the L^2-norm and the L^6-norm of g, respectively. By Hölder's inequality and the estimates for b and f in Remark 5.6 we obtain

$$\|g\|_{L^2(\Omega_R)} \leq \|\rho\|_{L^\infty(\Omega_R)} \Big(\|u_t\|_{L^2(\Omega_R)} + \|u\|_{L^6(\Omega_R)} \|\nabla u\|_{L^3(\Omega_R)} + C\|\nabla u\|_{L^2(\Omega_R)} + C\|u\|_{L^2(\Omega_R)} \Big)$$

$$+ \|(\rho - 1)Mx\|_{L^6(\Omega_R)} \|\nabla u\|_{L^3(\Omega_R)} + \|\rho - 1\|_{L^\infty(\Omega_R)} \|u\|_{L^2(\Omega_R)} + C$$

and

$$\|g\|_{L^6(\Omega_R)} \le \|\rho\|_{L^\infty(\Omega_R)} \Big(\|u_t\|_{L^6(\Omega_R)} + \|u\|_{L^6(\Omega_R)} \|\nabla u\|_{L^\infty(\Omega_R)} + C\|\nabla u\|_{L^6(\Omega_R)} + C\|u\|_{L^6(\Omega_R)} \Big)$$
$$+ \|(\rho - 1)Mx\|_{L^6(\Omega_R)} \|\nabla u\|_{L^\infty(\Omega_R)} + \|\rho - 1\|_{L^\infty(\Omega_R)} \|u\|_{L^6(\Omega_R)} + C$$

for some $C := C(|M|) > 0$. In the following we use the Sobolev inequality, see Propoisition 1.15, and the Sobolev embedding $H^1(\Omega_R) \hookrightarrow L^6(\Omega_R)$, see Proposition 1.14, to obtain

$$\|u\|_{L^6(\Omega_R)} \le C\|\nabla u\|_{L^2(\Omega_R)}, \qquad \|\nabla u\|_{L^6(\Omega_R)} \le C\|u\|_{H^2(\Omega_R)},$$

and

$$\|(\rho - 1)Mx\|_{L^6(\Omega_R)} \le C\|(\rho - 1)Mx\|_{H^1(\Omega_R)}$$

for a constant $C > 0$ independent of R. Next, we apply the Gagliardo-Nirenberg inequality, see Proposition 1.17, with $p = \infty$, $r = 6$, $q = 2$, $j = 0$, $m = 1$ and $\theta = 3/4$ to obtain

$$\int_0^t \|\nabla u(s)\|_{L^\infty(\Omega_R)}^2 \, ds \le C \int_0^t \Big(\|\nabla^2 u(s)\|_{L^6(\Omega_R)}^{\frac{3}{2}} \|\nabla u(s)\|_{L^2(\Omega_R)}^{\frac{1}{2}} + \|\nabla u(s)\|_{L^2(\Omega)} \Big) \, ds$$

for some constant $C > 0$ independent[2] of R. Moreover, the interpolation inequality yields

$$\|\nabla u\|_{L^3(\Omega_R)}^2 \le \|\nabla u\|_{L^2(\Omega_R)} \|\nabla u\|_{L^6(\Omega_R)}.$$

By using these estimates together with Young's inequality, Lemma 5.37 and Lemma 5.38 we obtain that for arbitrary $\varepsilon > 0$ there exists a constant $C := C(N_0, T^*, \varepsilon) > 0$ such that

$$\|g(t)\|_{L^2(\Omega_R)}^2 + \int_0^t \|g(s)\|_{L^6(\Omega_R)}^2 \, ds$$
$$\le C\|u_t\|_{L^2(\Omega_R)}^2 + \varepsilon\|\nabla^2 u\|_{L^2(\Omega_R)} + C\Psi_R^4(t) + C \int_0^t \|\nabla u_t\|_{L^2(\Omega_R)}^2 \, ds + \varepsilon \int_0^t \|\nabla^2 u\|_{L^6(\Omega_R)}^2 \, ds$$

for all $t \in [0, T]$ and all $T < \min\{T^*, T_{\max}^R\}$. Then, by choosing $\varepsilon > 0$ sufficiently small, we can see from (5.68) that there exists a constant $C := C(N_0, T^*) > 0$ such that

$$\|\nabla^2 u(t)\|_{L^2(\Omega_R)}^2 + \|Mx \cdot \nabla u(t) - Mu(t)\|_{L^2(\Omega_R)}^2$$
$$+ \int_0^t \|\nabla^2 u(s)\|_{L^6(\Omega_R)}^2 \, ds + \int_0^t \|Mx \cdot \nabla u(s) - Mu(s)\|_{L^6(\Omega_R)}^2 \, ds \qquad (5.69)$$
$$\le C\|u_t(t)\|_{L^2(\Omega_R)}^2 + C\Psi_R^4(t) + C \int_0^t \|\nabla u_t(s)\|_{L^2(\Omega_R)}^2 \, ds$$

for all $t \in [0, T]$ and all $T < \min\{T^*, T_{\max}^R\}$. Finally, by combining (5.67) and (5.69), we conclude that for arbitrary $\varepsilon > 0$ there exists a constant $C := C(N_0, T^*, \varepsilon) > 0$ such that

$$\Phi_R(T) \le \varepsilon\Psi_R(T)\Phi_R(T) + C \int_0^T \Phi_R(s) \, ds + C\Psi_R^4(T), \qquad T < \min\{T^*, T_{\max}^R\}. \qquad (5.70)$$

---

[2]This can be seen as follows: There exists some radius $\delta := \delta(R_0) > 0$ and a sequence of points $\{x_n\}_{n \in \mathbb{N}} \subset \Omega_R$ such that $\Omega_R = \bigcup_{n \in \mathbb{N}} B(x_n, \delta) \cup \Omega_{R_0}$. Then we apply the Gagliardo-Nirenberg inequality on each ball separately. Therefore, the constant depends only on R_0 and δ, but is independent of $R > R_0$.

Step 3: A continuation argument. First, by the regularity of u stated in (5.53) we conclude that the map $t \mapsto \Phi_R(t)$ is a continuous on $[0, T]$ for all $T < T_{\max}^R$. In particular, this means that for any $T < T_{\max}^R$ the function $\Phi_R(\cdot)$ is uniformly bounded on $[0, T]$ by some constant which might depend on R. Therefore, we can find a time $0 < T^R < \min\{T^*, T_{\max}^R\}$, depending on R, such that

$$\Psi_R(t) \leq L + L \exp\left(L\Phi_R(t) t^{\frac{1}{2}}\right) \leq 3L$$

for all $t \in [0, T^R]$, where $L := L(N_0, T^*) > 0$ is the constant from (5.64). From (5.70) it now follows that

$$\Phi_R(t) \leq \varepsilon 3L\Phi_R(t) + C \int_0^t \Phi_R(s)\,\mathrm{d}s + C(3L)^4 \tag{5.71}$$

for all $t \in [0, T^R]$. By choosing $\varepsilon < 1/(6L)$ we see that

$$\Phi_R(t) \leq C_0 \int_0^t \Phi_R(s)\,\mathrm{d}s + C_0(3L)^4 \tag{5.72}$$

for all $t \in [0, T^R]$ and some constant $C_0 := C_0(N_0, T^*) > 0$. By applying Gronwall's inequality we obtain a constant $K := K(N_0, L, T^*)$ independent of R such that

$$\Phi_R(t) \leq K$$

for all $t \in [0, T^R]$. This constant K is now fixed for the rest of the proof. Next, set

$$T_0 := \min\left\{\left(\frac{\log 2}{2LK}\right)^2, T^*\right\}.$$

Note that the choice of T_0 is independent of R. Moreover, we introduce the interval

$$I^R := \left\{0 < t < \min\{T_0, T_{\max}^R\} : \Phi_R(t) \leq 2K\right\}.$$

This interval is not empty, since we showed that there exists a time T^R such that $t \in I^R$ for all $t \in [0, T^R]$. Next, we take some $t \in I^R$. By the choice of T_0 we know

$$\Psi_R(t) \leq L + L \exp\left(L\Phi_R(t) t^{\frac{1}{2}}\right) \leq L + L \exp\left(L\Phi_R(T_0) T_0^{\frac{1}{2}}\right) \leq 3L,$$

where $L := L(N_0, T^*) > 0$ is the constant from (5.64). Then estimate (5.70) with $\varepsilon < 1/(6L)$ yields that

$$\Phi_R(t) \leq C_0 \int_0^t \Phi_R(s)\,\mathrm{d}s + C_0(3L)^4,$$

where C_0 is the same constant as in (5.72). So, again by applying Gronwall's inequality we see that

$$\Phi_R(t) \leq K.$$

The continuity of $t \mapsto \Phi_R(t)$ yields that there exists a $\delta > 0$ such that $t + \delta \in I^R$. Hence, it follows that $I^R = (0, \min\{T_0, T_{\max}^R\})$. This concludes the proof. $\qquad\square$

So far all estimates are stated for time intervals $[0, T]$ with $T < \min\{T_0, T_{\max}^R\}$, where T_0 is independent of R. This formulation is necessary as we do not have any information about $T_{\max}^R$ so far. However, now it is easy to conclude that $T_0 < T_{\max}^R$ holds: From general theory it is clear that if $T_{\max}^R$ is finite, then the solution blows up at time $T_{\max}^R$. However, the estimates derived above show that the solution does not blow up on the interval $[0, \min\{T_0, T_{\max}^R\})$. Hence, it is clear that $T_0 < T_{\max}^R$. We can also make this argument more precise.

Lemma 5.42. *Let $T_0 > 0$ be independent of R, such that the assertions in Lemma 5.38 and Lemma 5.41 hold. Then, $T_0 < T_{\max}^R$ for all $R > R_0$.*

Proof. Let $(\rho, u) := (\rho^R, u^R)$ be the strong solution obtained in Proposition 5.15 with maximal existence time $T_{\max}^R$. Let us assume that $T_{\max}^R \leq T_0$. Then, from Lemma 5.37, Lemma 5.38 and Lemma 5.41 it follows that there exists a constant $N_0^* := N_0^*(N_0, T_0) > 0$ such that

$$\sup_{t \in [0, T_{\max}^R)} \left(\|\rho(t) - 1\|_{H^2(\Omega_R)} + \|u(t)\|_{H^2(\Omega_R)} \right) \leq N_0^*.$$

Note that the local existence time T in Proposition 5.15 depends on the data (i.e. on N_0) and on the actual value of R. In the following the constant N_0^* allows to control the initial data $(\rho(t_0), u(t_0))$ for any initial time $t_0 \in [0, T_{\max}^R)$. Thus, we conclude that for every $t_0 \in [0, T_{\max}^R)$ we obtain a time $T := T(N_0^*, R) > 0$ and strong solution $(\tilde{\rho}, \tilde{u})$ on $[t_0, t_0 + T]$ satisfying (5.8) with initial condition $(\tilde{\rho}, \tilde{u})|_{t=t_0} = (\rho(t_0), u(t_0))$. In particular, we can set $t_0 := T_{\max}^R - T/2$. Then, we obtain a solution $(\tilde{\rho}, \tilde{u})$ on the interval $[T_{\max}^R - T/2, T_{\max}^R + T/2]$ satisfying (5.8) with initial condition $(\tilde{\rho}, \tilde{u})|_{t=t_0} = (\rho(t_0), u(t_0))$. Since the strong solution (ρ, u) is unique, we conclude that (ρ, u) even exists on the interval $[0, T_{\max}^R + T/2]$. This is a contradiction. Hence, $T_0 < T_{\max}^R$. $\qquad\square$

By taking all the results of Section 5.5.2 together, we immediately obtain Proposition 5.16.

5.6 Conclusion of the proof

We finally put all the pieces together and give the proofs of Proposition 5.11, Proposition 5.15 and Theorem 4.4.

Proof of Proposition 5.11

Let $R > R_0$ and let $(\rho^n, u^n) := (\rho^{n,R}, u^{n,R})$ be the approximating solution from Proposition 5.9. By Proposition 5.13 we know that there exists a time $T_0 > 0$ such that

$$u^n \subset L^\infty((0, T_0); H^2(\Omega_R)^3 \cap H_{0,\sigma}^1(\Omega_R)) \quad \text{is uniformly bounded,}$$
$$u_t^n \subset L^\infty((0, T_0); L_\sigma^2(\Omega_R)) \cap L^2((0, T_0); H_{0,\sigma}^1(\Omega_R)) \quad \text{is uniformly bounded,} \qquad (5.73)$$
$$\rho^n \subset L^\infty((0, T_0); L^\infty(\Omega_R)) \quad \text{is uniformly bounded.}$$

Hence, it is clear – after extracting a subsequence if necessary – that there exists a function $u := u^R$ such that

$$u^n \overset{*}{\rightharpoonup} u \quad \text{weakly* in } L^\infty((0, T_0); H^2(\Omega_R)^3 \cap H^1_{0,\sigma}(\Omega_R)),$$

by this we mean

$$\int_0^{T_0} \int_{\Omega_R} u^n \cdot g \, \mathrm{d}x \, \mathrm{d}t \to \int_0^{T_0} \int_{\Omega_R} u \cdot g \, \mathrm{d}x \, \mathrm{d}t$$

for all $g \in L^1((0, T_0); H^2(\Omega_R)^3 \cap H^1_{0,\sigma}(\Omega_R))$. Moreover, we conclude that there exists some $v \in L^2((0, T_0); H^1_{0,\sigma}(\Omega_R))$ such that – after extracting a subsequence if necessary – we have

$$u_t^n \overset{*}{\rightharpoonup} v \quad \text{weakly* in } L^\infty((0, T_0); L^2_\sigma(\Omega_R)),$$

by this we mean

$$\int_0^{T_0} \int_{\Omega_R} u_t^n \cdot g \, \mathrm{d}x \, \mathrm{d}t \to \int_0^{T_0} \int_{\Omega_R} v \cdot g \, \mathrm{d}x \, \mathrm{d}t$$

for all $g \in L^1((0, T); L^2_\sigma(\Omega_R))$. Further, we have

$$u_t^n \rightharpoonup v \quad \text{weakly in } L^2((0, T_0); H^1_{0,\sigma}(\Omega_R)).$$

It is easy to see that v is the time-derivative of u in distributional sense and therefore we write $v = u_t$.

By classical compactness arguments, see Proposition 1.89, we may even suppose – after extracting a subsequence – that

$$u^n \to u \quad \text{strongly in } C([0, T_0]; H^1_{0,\sigma}(\Omega_R)).$$

The estimates in Proposition 5.11 follow directly from Proposition 5.13 and the convergence of u^n to u and u_t^n to u_t in the sense stated above. Moreover, also by the convergence stated above, and also by the convergence of $a^{n,R}$ to a^R stated in Lemma 5.19, it is clear that the initial condition $u(0) = a^R$ is satisfied. Similarly, we may suppose – after extracting a subsequence – that there exists a function $\rho := \rho^R$ such that

$$\rho^n \overset{*}{\rightharpoonup} \rho \quad \text{weakly* in } L^\infty((0, T_0); L^\infty(\Omega_R)).$$

Moreover, from (5.73) we see that for all $2 \le p \le 6$

$$\rho^n u^n \subset L^\infty((0, T_0); L^p(\Omega_R)^3) \quad \text{is uniformly bounded.}$$

Now, by using $\rho_t^n = -\mathrm{div}\,(\rho^n u^n + \rho^n b - \rho^n Mx)$ we can conclude that

$$\rho_t^n \subset L^\infty((0, T_0); H^{-1}(\Omega_R)) \quad \text{is uniformly bounded.}$$

Then, by the Arzelà-Ascoli theorem, it is easy to deduce that $\rho^n \to \rho$ in $C([0, T_0]; H^{-1}(\Omega_R))$. Moreover, Lemma 1.91 yields that ρ is weakly continuous on $[0, T_0]$ with values in $L^2(\Omega_R)$, and by using Lemma 1.92 we see that ρ^n converges – after extracting a subsequence – to ρ in $C([0, T_0]; L^2(\Omega_R) - weak)$. In particular, we conclude that $\rho(0) = \rho_0^R$ almost everywhere in Ω_R.

Next, we show that (u, ρ) solves the transport equation in the sense of Definition 5.10 (ii). For that we need to show that

$$\int_0^T \int_{\Omega_R} \rho^n \cdot (u^n \cdot \nabla \Psi) \, dx \, dt \to \int_0^T \int_{\Omega_R} \rho \cdot (u \cdot \nabla \Psi) \, dx \, dt \qquad (5.74)$$

as $n \to \infty$, for all $\Psi \in C^\infty([0, T] \times \Omega)$ with $\Psi(T, x) = 0$ for all $x \in \Omega_R$. Note that

$$\int_0^T \int_{\Omega_R} [\rho^n \cdot (u^n \cdot \nabla \Psi) - \rho \cdot (u \cdot \nabla \Psi)] \, dx \, dt$$

$$= \int_0^T \int_{\Omega_R} (\rho^n - \rho) \cdot (u^n \cdot \nabla \Psi) \, dx \, dt + \int_0^T \int_{\Omega_R} \rho \cdot ([u^n - u] \cdot \nabla \Psi) \, dx \, dt =: I_1 + I_2.$$

Then, by using the convergence of ρ^n in $C([0, T_0]; H^{-1}(\Omega_R))$ and of u^n in $C([0, T_0]; H^1_{0,\sigma}(\Omega_R))$, we see that

$$|I_1| \leq C \int_0^T \|\rho^n - \rho\|_{H^{-1}(\Omega_R)} \|u^n \cdot \nabla \Psi\|_{H^1(\Omega_R)} \, dt \to 0,$$

and

$$|I_2| \leq C \int_0^T \|\rho\|_{H^{-1}(\Omega_R)} \|u^n \cdot \nabla \Psi - u \cdot \nabla \Psi\|_{H^1(\Omega_R)} \, dt \to 0,$$

as $n \to \infty$. Hence, (5.74) follows and we conclude (u, ρ) solves the transport equation in the sense of Definition 5.10 (ii). Lemma 5.25 then states that

$$\|\rho(t)\|_{L^2(\Omega_R)} = \|\rho_0^R\|_{L^2(\Omega_R)} \qquad (5.75)$$

for all $t \in [0, T_0]$. Now fix $t \in [0, T_0]$ and let $\{t_k\}_{k \in \mathbb{N}} \subset [0, T_0]$ such that $t_k \to t$ as $k \to \infty$. Then we know that $\|\rho(t_k)\|_{L^2(\Omega_R)} = \|\rho_0^R\|_{L^2(\Omega_R)}$ for all $k \in \mathbb{N}$ and $\rho(t_k) \rightharpoonup \rho(t)$ weakly in $L^2(\Omega_R)$. From these two properties it follows that even $\rho(t_k) \to \rho(t)$ strongly in $L^2(\Omega_R)$. This shows that $\rho \in C([0, T_0]; L^2(\Omega_R))$. Now we want to show that $\rho^n \to \rho$ strongly in $C([0, T_0]; L^2(\Omega_R))$. Since we already know the convergence in $C([0, T_0]; L^2(\Omega_R) - weak)$ we only need to show that $\|\rho^n(t_n)\|_{L^2(\Omega_R)}$ converges to $\|\rho(t)\|_{L^2(\Omega_R)}$, if $\{t_n\}_{n \in \mathbb{N}} \subset [0, T_0]$ converges to $t \in [0, T_0]$. Note that (5.18) states $\|\rho^n(t)\|_{L^2(\Omega_R)} = \|\rho_0^{n,R}\|_{L^2(\Omega_R)}$. By using this equality together with (5.75) and the convergence $\|\rho_0^{n,R}\|_{L^2(\Omega_R)} \to \|\rho_0^R\|_{L^2(\Omega_R)}$ as $n \to \infty$, we easily see that

$$\|\rho^n(t_n)\|_{L^2(\Omega_R)} = \|\rho_0^{n,R}\|_{L^2(\Omega_R)} \to \|\rho_0^R\|_{L^2(\Omega_R)} = \|\rho(t)\|_{L^2(\Omega)}$$

as $n \to \infty$. Hence, we can conclude – after extracting a subsequence – that

$$\rho^n \to \rho \quad \text{strongly in } C([0, T_0]; L^2(\Omega_R)).$$

It remains to show that (ρ, u) solves (5.8) on $(0, T_0)$ in the weak sense, see Definition 5.10. In the following let $\Phi \in C^1([0, T_0]; \mathcal{D}(A_{\Omega_R}))$ with $\Phi(T_0, x) = 0$ for almost all $x \in \Omega_R$ be an arbitrary test function. Note that by the Sobolev embedding $L^\infty(\Omega_R) \hookrightarrow H^2(\Omega_R)$ it is clear that $\Phi \in L^\infty((0, T_0) \times \Omega_R)^3$. First we prove that

$$\int_0^{T_0} \int_{\Omega_R} \rho^n u^n \cdot \partial_t \Phi \, dx \, dt \to \int_0^{T_0} \int_{\Omega_R} \rho u \cdot \partial_t \Phi \, dx \, dt \qquad (5.76)$$

as $n \to \infty$. We observe that

$$\left| \int_0^{T_0} \int_{\Omega_R} (\rho^n u^n - \rho u) \cdot \partial_t \Phi \, \mathrm{d}x \, \mathrm{d}t \right|$$

$$\leq \left| \int_0^{T_0} \int_{\Omega_R} (\rho^n - \rho) u^n \cdot \partial_t \Phi \, \mathrm{d}x \, \mathrm{d}t \right| + \left| \int_0^{T_0} \int_{\Omega_R} \rho(u^n - u) \cdot \partial_t \Phi \, \mathrm{d}x \, \mathrm{d}t \right|. \tag{5.77}$$

For the first integral on the right-hand side of (5.77) we have

$$\left| \int_0^{T_0} \int_{\Omega_R} (\rho^n - \rho) u^n \cdot \partial_t \Phi \, \mathrm{d}x \, \mathrm{d}t \right| \leq C \int_0^{T_0} \|\rho^n(t) - \rho(t)\|_{L^2(\Omega_R)} \|u^n(t)\|_{L^2(\Omega_R)} \, \mathrm{d}x \, \mathrm{d}t \to 0$$

as $n \to \infty$ because $\rho^n \to \rho$ strongly in $C([0, T_0]; L^2(\Omega_R))$, and $u^n \subset L^\infty((0, T_0); L^2_\sigma(\Omega_R))$ is uniformly bounded. The second integral on the right-hand side of (5.77) converges also to 0 because $u_n \to u$ strongly in $C([0, T_0]; H^1_0(\Omega_R)^3)$, and $\rho \in L^\infty((0, T_0) \times \Omega_R)$. Thus, (5.76) is proved. Next we show that

$$\int_0^{T_0} \int_{\Omega_R} \rho^n u_i^n u^n \cdot \partial_i \Phi \, \mathrm{d}x \, \mathrm{d}t \to \int_0^{T_0} \int_{\Omega_R} \rho u_i u \cdot \partial_i \Phi \, \mathrm{d}x \, \mathrm{d}t \tag{5.78}$$

for $i = 1, 2, 3$ as $n \to \infty$. We observe that

$$\int_0^{T_0} \int_{\Omega_R} (\rho^n u_i^n u^n - \rho u_i u) \cdot \partial_i \Phi \, \mathrm{d}x \, \mathrm{d}t$$

$$= \int_0^{T_0} \int_{\Omega_R} (\rho^n - \rho) u_i^n u^n \cdot \partial_i \Phi \, \mathrm{d}x \, \mathrm{d}t + \int_0^{T_0} \int_{\Omega_R} \rho(u_i^n - u_i) u^n \cdot \partial_i \Phi \, \mathrm{d}x \, \mathrm{d}t$$

$$+ \int_0^{T_0} \int_{\Omega_R} \rho u_i (u^n - u) \cdot \partial_i \Phi \, \mathrm{d}x \, \mathrm{d}t =: I_1 + I_2 + I_3.$$

For I_1 we have

$$|I_1| \leq C(R) \|\rho^n - \rho\|_{L^\infty((0,T_0);L^2(\Omega_R))} \int_0^{T_0} \|u_i^n\|_{L^6(\Omega_R)} \|u^n\|_{L^3(\Omega_R)} \, \mathrm{d}t \to 0$$

as $n \to \infty$. This follows by applying the Sobolev embeddings $H^1(\Omega_R) \hookrightarrow L^6(\Omega_R)$ and $H^1(\Omega_R) \hookrightarrow L^3(\Omega_R)$ and by using that $\rho^n \to \rho$ strongly in $C([0, T_0]; L^2(\Omega_R))$ and that $u^n \subset L^\infty((0, T_0); H^1_0(\Omega_R)^3)$ is uniformly bounded. The integral I_2 also goes to zero as $n \to \infty$ which can be seen from the estimate

$$|I_2| \leq C \|\rho\|_{L^\infty((0,T_0);L^\infty(\Omega_R))} \int_0^{T_0} \|u_i^n - u_i\|_{L^6(\Omega_R)} \|u^n\|_{L^3(\Omega_R)} \, \mathrm{d}t$$

and from the facts that $u^n \to u$ strongly in $C([0, T_0]; H^1_{0,\sigma}(\Omega_R))$ and $\rho \in L^\infty((0, T_0) \times \Omega_R)$. Similarly, for I_3 we have

$$|I_3| \leq C \|\rho\|_{L^\infty((0,T_0);L^\infty(\Omega_R))} \int_0^{T_0} \|u\|_{L^6(\Omega_R)} \|u^n - u\|_{L^3(\Omega_R)} \, \mathrm{d}t \to 0$$

as $n \to \infty$. This shows that (5.78) holds. Similarly, we show that

$$\int_0^{T_0} \int_{\Omega_R} \rho^n (Mx \cdot \nabla u^n - Mu^n) \cdot \Phi \, \mathrm{d}x \, \mathrm{d}t \to \int_0^{T_0} \int_{\Omega_R} \rho(Mx \cdot \nabla u - Mu) \cdot \Phi \, \mathrm{d}x \, \mathrm{d}t$$

177

and

$$\int_0^{T_0} \int_{\Omega_R} \rho^n [(b \cdot \nabla) u^n + (u^n \cdot \nabla) b] \cdot \Phi \, dx \, dt \rightarrow \int_0^{T_0} \int_{\Omega_R} \rho [(b \cdot \nabla) u + (u \cdot \nabla) b] \cdot \Phi \, dx \, dt$$

as $n \to \infty$. By Proposition 5.9 we know that for any $\Phi \in \bigcup_{k \geq 1} C^1([0, T_0]; \mathbb{P}_{\Omega_R}^k L_\sigma^2(\Omega_R))$ with $\Phi(T_0, x) = 0$ for almost all $x \in \Omega_R$ we have

$$\int_0^{T_0} \int_{\Omega_R} (\rho^n u_t^n + \rho^n u^n \cdot \nabla u^n + \rho^n b \cdot \nabla u^n + \rho^n u^n \cdot \nabla b$$

$$-\rho^n Mx \cdot \nabla u^n + \rho^n M u^n - \Delta u^n - \rho^n f - \Delta b) \cdot \Phi \, dx \, dt = 0$$

for all sufficiently large n. Integration by parts with respect to t and x and by using that $\rho_t^n = -(u^n + b - Mx) \cdot \nabla \rho^n$ yields

$$-\int_0^{T_0} \int_{\Omega_R} \rho^n u^n \cdot \partial_t \Phi \, dx \, dt - \sum_{i=1}^3 \int_0^{T_0} \int_{\Omega_R} \rho^n u_i^n u^n \cdot \partial_i \Phi \, dx \, dt + \int_0^{T_0} \int_{\Omega_R} \nabla u^n \cdot \nabla \Phi \, dx \, dt$$

$$+ \int_0^{T_0} \int_{\Omega_R} [\rho^n b \cdot \nabla u^n + \rho^n u^n \cdot \nabla b] \cdot \Phi \, dx \, dt - \int_0^T \int_{\Omega_R} \rho^n (Mx \cdot \nabla u^n - M u^n) \cdot \Phi \, dx \, dt$$

$$= \int_0^{T_0} \int_{\Omega_R} (\rho^n f + \Delta b) \cdot \Phi \, dx \, dt + \int_{\Omega_R} \rho_0^{n,R}(x) a^R(x) \cdot \Phi(0, x) \, dx.$$

Now we let $n \to \infty$ and we use that

$$\bigcup_{k \geq 1} C^1([0, T_0]; \mathbb{P}_{\Omega_R}^k L_\sigma^2(\Omega_R)) \subset C^1([0, T_0]; H_{0,\sigma}^1(\Omega_R))$$

is dense to obtain that (ρ, u) satisfies the first equation in Definition 5.10 (ii). Above it was already shown that (ρ, u) also satisfies the transport equation in the sense of Definition 5.10 (ii), and hence, (ρ, u) is indeed a weak solution to (5.8) on $(0, T_0)$. $\square$

Next, we show that the weak solution (ρ, u) has even higher regularity and that it is a strong solution with the regularity stated in Proposition 5.15. For that we proceed in several steps. First we show the following lemma.

Lemma 5.43. *Let* $R > R_0$ *and let* $(\rho, u) := (\rho^R, u^R)$ *be the weak solution obtained in Proposition 5.11. Then* $\rho \in C([0, T_0]; H^2(\Omega_R))$, $\rho_t \in C([0, T_0]; L^2(\Omega_R)) \cap L^2((0, T_0); H^1(\Omega_R))$ *and* (ρ, u) *is a even strong solution to* (5.8) *on* $(0, T_0)$, *i.e.*

$$\rho_t(t) + u(t) \cdot \nabla \rho(t) + b \cdot \nabla \rho(t) - Mx \cdot \nabla \rho(t) = 0 \tag{5.79}$$

in $L^2(\Omega_R)$, *and*

$$\mathbb{P}_{\Omega_R} \left(\rho(t) u_t(t) + \rho u(t) \cdot \nabla u(t) + \rho(t) b \cdot \nabla u(t) + \rho(t) u(t) \cdot \nabla b \right.$$
$$\left. -\rho(t) Mx \cdot \nabla u(t) + \rho(t) M u(t) \right) - A_{\Omega_R} u(t) = \mathbb{P}_{\Omega_R} \left(\rho(t) f + \Delta b \right) \tag{5.80}$$

in $L_\sigma^2(\Omega_R)$ *for almost all* $t \in (0, T_0)$. *Moreover, the initial condition* $(\rho(0), u(0)) = (\rho_0^R, a^R)$ *is satisfied.*

Proof. Let $(\rho, u) := (\rho^R, u^R)$ be the weak solution obtained in Proposition 5.11. In particular we know that

$$u \in C([0, T_0]; H^1_{0,\sigma}(\Omega_R)) \cap L^\infty((0, T_0); H^2(\Omega_R)^3).$$

Moreover, Proposition 5.14 yields $u \in L^2((0, T_0); W^{2,6}(\Omega_R)^3)$. We apply Lemma 5.26 on linear transport equations and we use the uniqueness of weak solutions to transport equations to obtain that

$$\rho \in C([0, T_0]; H^2(\Omega_R)), \qquad \rho_t \in C([0, T_0]; L^2(\Omega_R)) \cap L^2((0, T_0); H^1(\Omega_R))$$

and that ρ satisfies (5.79). Moreover, by Definition 5.10 of weak solutions we know that

$$
-\int_0^{T_0} \int_{\Omega_R} \rho u \cdot \partial_t \Phi \, dx \, dt - \sum_{i=1}^3 \int_0^{T_0} \int_{\Omega_R} \rho u_i u \cdot \partial_i \Phi \, dx \, dt + \int_0^{T_0} \int_{\Omega_R} \nabla u \cdot \nabla \Phi \, dx \, dt
$$

$$
+ \int_0^{T_0} \int_{\Omega_R} [(\rho b \cdot \nabla) u + (\rho u \cdot \nabla) b] \cdot \Phi \, dx \, dt - \int_0^{T_0} \int_{\Omega_R} \rho(Mx \cdot \nabla u - Mu) \cdot \Phi \, dx \, dt \quad (5.81)
$$

$$
= \int_0^{T_0} \int_{\Omega_R} (\rho f + \Delta b) \cdot \Phi \, dx \, dt + \int_{\Omega_R} \rho_0(x) a^R(x) \cdot \Phi(0, x) \, dx,
$$

holds for all $\Phi \in C^1([0, T_0]; H^1_{0,\sigma}(\Omega_R))$ satisfying $\Phi(T_0, x) = 0$ almost everywhere in Ω_R. The regularity for u stated in Proposition 5.11 allows to conclude that

$$
\int_0^{T_0} \int_{\Omega_R} \rho u_t \cdot \Phi \, dx \, dt + \int_0^{T_0} \int_{\Omega_R} (\rho u \cdot \nabla u) \cdot \Phi \, dx \, dt - \int_0^{T_0} \int_{\Omega_R} \Delta u \cdot \Phi \, dx \, dt
$$

$$
+ \int_0^{T_0} \int_{\Omega_R} ((\rho b \cdot \nabla) u + (\rho u \cdot \nabla) b) \cdot \Phi \, dx \, dt - \int_0^{T_0} \int_{\Omega_R} \rho(Mx \cdot \nabla u - Mu) \cdot \Phi \, dx \, dt \quad (5.82)
$$

$$
= \int_0^{T_0} \int_{\Omega_R} (\rho f + \Delta b) \cdot \Phi \, dx \, dt
$$

for all $\Phi \in L^2((0, T_0); L^2_\sigma(\Omega_R))$. This equality simply follows from (5.81) by integrating by parts with respect to t and x, by using the equation (5.79) and by using density arguments. Equation (5.80) follows now directly from (5.82). That the initial conditions are satisfied was already shown in the proof of Proposition 5.11. $\qquad\square$

To show that u_t is a continuous function in time we need the following lemma, see also [CK04, Remark 6]).

Lemma 5.44. *Let $R > R_0$ and let $(\rho, u) := (\rho^R, u^R)$ be the weak solution obtained in Proposition 5.11 with the properties stated in Lemma 5.43. Then $(\rho u_t)_t \in L^2(0, T_0); H^{-1}_\sigma(\Omega))$ and for almost all $t \in (0, T_0)$ it holds that*

$$
\frac{d}{dt} \int_{\Omega_R} \rho \frac{|u_t|^2}{2} \, dx + \int_{\Omega_R} |\nabla u_t|^2 \, dx
$$

$$
= -\int_{\Omega_R} \rho_t (u_t + u \cdot \nabla u + b \cdot \nabla u + u \cdot \nabla b - Mx \cdot \nabla u + Mu - f) \cdot u_t \, dx \quad (5.83)
$$

$$
-\int_{\Omega_R} \rho(u_t \cdot \nabla u + u_t \cdot \nabla b) \cdot u_t \, dx.
$$

An analogous equality has been derived in the proof of Lemma 5.33 for the approximating solution (ρ^n, u^n). The main difference is that here (ρ, u) has less regularity than the approximating solution (ρ^n, u^n). Therefore, the left-hand side in (5.83) and some terms in the derivation have to be interpreted in distributional sense. However, the formal calculations are the same as in the proof of Lemma 5.33.

Proof. Let $w \in H^1_{0,\sigma}(\Omega)$. Then from (5.80) it follows that

$$
\int_{\Omega_R} \rho u_t \cdot w \, dx = - \int_{\Omega_R} (\rho u \cdot \nabla u + \rho b \cdot \nabla u + \rho u \cdot \nabla b - \rho M x \cdot \nabla u
$$
$$
+ \rho M u - \Delta u - \rho f - \Delta b) \cdot w \, dx
$$
$$
= - \int_{\Omega_R} \nabla u : \nabla w \, dx - \int_{\Omega_R} (\rho u \cdot \nabla u + \rho b \cdot \nabla u + \rho u \cdot \nabla b - \rho M x \cdot \nabla u
$$
$$
+ \rho M u - \rho f - \Delta b) \cdot w \, dx
$$

for almost all $t \in (0, T_0)$. Thus,

$$
\frac{d}{dt} \int_{\Omega_R} \rho u_t \cdot w \, dx = - \int_{\Omega_R} \rho_t (u \cdot \nabla u + b \cdot \nabla u + u \cdot \nabla b - M x \cdot \nabla u + M u - f) \cdot w \, dx
$$
$$
- \int_{\Omega_R} \rho (u_t \cdot \nabla u + u \cdot \nabla u_t + b \cdot \nabla u_t + u_t \cdot \nabla b \tag{5.84}
$$
$$
- M x \cdot \nabla u_t + M u_t) \cdot w \, dx - \int_{\Omega_R} \nabla u_t : \nabla w \, dx
$$

in the scalar distributional sense on $(0, T_0)$. Now by using the regularity of (ρ, u) stated in Proposition 5.11 and in Lemma 5.43 it is easy to see that the right-hand side in (5.84) is bounded above by $G(t) \|w\|_{H^1(\Omega_R)}$ for some function $G \in L^2(0, T_0)$. Hence, we can conclude that $(\rho u_t)_t \in L^2((0, T_0); H^{-1}_\sigma(\Omega_R))$ and that

$$
\int_{\Omega_R} (\rho u_t)_t \cdot w \, dx = \frac{d}{dt} \int_{\Omega_R} \rho u_t \cdot w \, dx
$$

for almost all $t \in (0, T_0)$ (see e.g. [Tem01, Chapter 3, Lemma 1.1]). Set $w = u_t$ in equation (5.84). Next we use integration by parts to obtain the following equalities (see the computations in the proof of Lemma 5.33 for more details):

$$
\int_{\Omega_R} (\rho u \cdot \nabla u_t) \cdot u_t \, dx = - \int_{\Omega_R} (u \cdot \nabla \rho) \frac{|u_t|^2}{2} \, dx,
$$
$$
\int_{\Omega_R} (\rho M x \cdot \nabla u_t) \cdot u_t \, dx = - \int_{\Omega_R} (M x \cdot \nabla \rho) \frac{|u_t|^2}{2} \, dx, \tag{5.85}
$$
$$
\int_{\Omega_R} (\rho b \cdot \nabla u_t) \cdot u_t \, dx = - \int_{\Omega_R} (b \cdot \nabla \rho) \frac{|u_t|^2}{2} \, dx.
$$

Moreover, since $M^\top = -M$, we conclude that

$$
\int_{\Omega_R} \rho M u_t \cdot u_t \, dx = 0.
$$

A formal calculation, which can be done in a rigorous way by using regularization or difference quotients, shows that

$$\frac{\mathrm{d}}{\mathrm{d}t} \int_{\Omega_R} \rho |u_t|^2 \,\mathrm{d}x = 2 \int_{\Omega} (\rho u_t)_t \cdot u_t \,\mathrm{d}x - \int_{\Omega_R} \rho_t |u_t| \,\mathrm{d}x$$

$$= 2 \int_{\Omega} (\rho u_t)_t \cdot u_t \,\mathrm{d}x + \int_{\Omega_R} (u + b - Mx) \cdot \nabla \rho |u_t|^2 \,\mathrm{d}x$$

for almost all $t \in (0, T_0)$. Now using this equality together with (5.84) for $w := u_t$ and the equalities in (5.85) yield the assertion. $\qquad\square$

Now we can prove the remaining parts of Proposition 5.15

Proof of Proposition 5.15

Let $R > R_0$ and let $(\rho, u) := (\rho^R, u^R)$ be the weak solution obtained in Proposition 5.11. Most of the assertions were already proved in Lemma 5.43. The uniqueness of the strong (ρ, u) is standard and the proof follows exactly by the same arguments used later in the proof of Theorem 5.4. Therefore it is omitted here. So it only remains to show the time-continuity of u with values in $H^2(\Omega_R)^3$ and the time-continuity of u_t with values in $L^2_\sigma(\Omega_R)$. By Proposition 5.11 we know that $u \in C([0, T_0]; H^1_{0,\sigma}(\Omega_R)) \cap L^\infty((0, T_0); H^2(\Omega_R)^3)$. Thus Lemma 1.91 yields that u is weakly continuous in $[0, T_0]$ with values in $H^2(\Omega_R)^3$ which is denoted by $u \in C([0, T_0], H^2(\Omega_R)^3 - weak)$, see Section 1.7 for details. By using equation (5.80) we then see that $u_t \in C([0, T_0]; L^2_\sigma(\Omega_R) - weak)$. From (5.83) it directly follows that

$$t \mapsto \| \sqrt{\rho(t)} u_t(t) \|^2_{L^2(\Omega_R)}$$

is continuous on $[0, T_0]$. Since we know $\rho \in C([0, T_0]; H^2(\Omega_R))$ and $\rho(t, x) \geq \underline{\rho} > 0$ for all $(t, x) \in [0, T_0] \times \Omega_R$, we conclude that the map

$$t \mapsto \| u_t(t) \|^2_{L^2(\Omega_R)}$$

is continuous on $[0, T_0]$. This together with the weak continuity yields $u_t \in C([0, T_0]; L^2_\sigma(\Omega_R))$. Next, we show the time-continuity of u with values in $H^2(\Omega_R)^3$. For that purpose set

$$g := -\rho u_t - \rho u \cdot \nabla u - \rho b \cdot \nabla u - \rho u \cdot b + \rho(Mx \cdot \nabla u - Mu) + \rho f + \Delta b.$$

By equation (5.80) we see that

$$-A_{\Omega_R} u(t) = \mathbb{P}_{\Omega_R} g(t)$$

holds almost everywhere in $(0, T_0)$. If we choose $\mathrm{p}(t) \in \widehat{W}^{1,2}(\Omega_R)$ such that $\nabla \mathrm{p}(t) = (\mathrm{Id} - \mathbb{P}_{\Omega_R}) \Delta u(t) + (\mathrm{Id} - \mathbb{P}_{\Omega_R}) g(t)$ for almost all $t \in (0, T_0)$, then for almost all $t \in (0, T_0)$ the pair of functions $(u(t), \mathrm{p}(t))$ solves the stationary Stokes problem

$$\begin{cases} -\Delta u(t) + \nabla \mathrm{p}(t) &= g(t) & \text{in } \Omega, \\ \operatorname{div} u(t) &= 0 & \text{in } \Omega, \\ u(t) &= 0 & \text{on } \partial\Omega. \end{cases}$$

Hence, by using the elliptic estimate stated in Proposition 2.3 we obtain

$$\|u(t) - u(s)\|_{H^2(\Omega_R)} \leq C\|g(t) - g(s)\|_{L^2(\Omega_R)}$$

for some constant $C := C(R) > 0$ and for almost all $t, s \in (0, T_0)$. This estimate yields $u \in C([0, T_0]; H^2(\Omega_R)^3)$, provided that g is continuous in time with values in $L^2(\Omega_R)^3$. To see that g is indeed continuous with values in $L^2(\Omega_R)^3$ we observe that

$$\|\rho(t)u_t(t) - \rho(s)u_t(s)\|_{L^2(\Omega_R)} \leq \|\rho(t)(u_t(t) - u_t(s))\|_{L^2(\Omega_R)} + \|u_t(s)(\rho(t) - \rho(s))\|_{L^2(\Omega_R)}$$
$$\leq \|\rho\|_{L^\infty((0,T_0);L^\infty(\Omega_R))}\|u_t(t) - u_t(s)\|_{L^2(\Omega_R)}$$
$$+ \|u_t\|_{L^\infty((0,T_0);L^2_\sigma(\Omega_R))}\|\rho(t) - \rho(s)\|_{L^\infty(\Omega_R)}.$$

We can make the right-hand side above smaller than any given $\varepsilon > 0$ if $|t - s| < \delta$ for some $\delta := \delta(\varepsilon) > 0$. This follows directly from the fact $u_t \in C([0, T_0]; L^2(\Omega_R))$, the Sobolev embedding $H^2(\Omega_R) \hookrightarrow L^\infty(\Omega_R)$ and since $\rho \in C([0, T_0]; H^2(\Omega_R))$. The other terms in g are treated similarly and therefore we can conclude that $g \in C([0, T_0]; L^2(\Omega_R)^3)$. $\qquad\square$

Finally, we give the proof of the main result, Theorem 5.4.

Proof of Theorem 5.4

Step 1: Existence: We choose a sequence $\{R_n\}_{n\in\mathbb{N}}$ such that $R_1 > R_0$ and $R_{n+1} > R_n$ for $n \in \mathbb{N}$. In the following we consider the sequence of increasing domains

$$\{\Omega_{R_n} := \Omega \cap B_{R_n} : n \in \mathbb{N}\}.$$

Note that $\Omega = \bigcup_{n\in\mathbb{N}} \Omega_{R_n}$. Let $(\rho^n, u^n) := (\rho^{R_n}, u^{R_n})$ denote the solution to (5.8) on Ω_{R_n}. By Proposition 5.16 the solutions $\{(\rho^n, u^n)\}_{n\in\mathbb{N}}$ are all defined on a common time interval $[0, T_0]$ and they all satisfy the uniform norm estimates stated in Proposition 5.16. Hence, similarly as in the proof of Proposition 5.11, it is clear that for fixed $m \in \mathbb{N}$ – after extracting a suitable subsequence denoted by $\{(\rho^n_m, u^n_m)\}_{n\in\mathbb{N}}$ – there exist functions u_m and ρ_m such that

$$u^n_m \xrightarrow{*} u_m \quad \text{weakly* in } L^\infty((0, T_0); H^2(\Omega_{R_m})^3 \cap H^1(\Omega_{R_m})^3 \cap L^2_\sigma(\Omega_{R_m})),$$
$$u^n_m \rightharpoonup u_m \quad \text{weakly in } L^2((0, T_0); W^{2,6}(\Omega_{R_m})^3),$$
$$\partial_t u^n_m \xrightarrow{*} \partial_t u_m \quad \text{weakly* in } L^\infty((0, T_0); L^2_\sigma(\Omega_{R_m})),$$
$$\partial_t u^n_m \rightharpoonup \partial_t u_m \quad \text{weakly in } L^2((0, T_0); H^1(\Omega_{R_m})^3 \cap L^2_\sigma(\Omega_{R_m})),$$
$$\rho^n_m \xrightarrow{*} \rho_m \quad \text{weakly* in } L^\infty((0, T_0) \times \Omega_{R_m}),$$
$$\rho^n_m - 1 \xrightarrow{*} \rho_m - 1 \quad \text{weakly* in } L^\infty((0, T_0); H^2(\Omega_{R_m})),$$
$$\partial_t \rho^n_m \xrightarrow{*} \partial_t \rho_m \quad \text{weakly* in } L^\infty((0, T_0); H^1(\Omega_{R_m})),$$

as $n \to \infty$ and such that $\{(\rho^n_{m+1}, u^n_{m+1})\}_{n\in\mathbb{N}} \subset \{(\rho^n_m, u^n_m)\}_{n\in\mathbb{N}}$ for $m \in \mathbb{N}$. Note that here $\partial_t u_m$ and $\partial_t \rho_m$ have to be interpreted again in distributional sense. Now we define a diagonal sequence, also denoted by $\{(\rho^n, u^n)\}_{n\in\mathbb{N}}$, by setting

$$\rho^n := \rho^n_n \quad \text{and} \quad u^n := u^n_n \quad \text{for } n \in \mathbb{N}.$$

Recall that in Section 1.5 we defined the Stokes operator with rotating effect by

$$\mathcal{D}(L_{\Omega,p}) \;:=\; \{u \in W^{2,p}(\Omega)^3 \cap W_0^{1,p}(\Omega)^3 \cap L_\sigma^p(\Omega) : Mx \cdot \nabla u \in L^p(\Omega)^3\},$$
$$L_{\Omega,p} u \;:=\; \mathbb{P}_{\Omega,p}\left(\Delta u + Mx \cdot \nabla u - Mu\right),$$

where $1 < p < \infty$.

By the uniform estimates stated in Proposition 5.16 we conclude that there exists a pair of functions (ρ, u) with regularity

$$u \in L^\infty((0,T_0); \mathcal{D}(L_{\Omega,2})) \cap L^2((0,T_0); \mathcal{D}(L_{\Omega,6})),$$
$$u_t \in L^\infty((0,T_0); L_\sigma^2(\Omega)) \cap L^2((0,T_0); H_{0,\sigma}^1(\Omega)),$$
$$\rho - 1 \in L^\infty((0,T_0); H^2(\Omega)), \tag{5.86}$$
$$\rho_t \in L^\infty((0,T_0); L^2(\Omega)),$$
$$\underline{\rho} \le \rho(t,x) \le \bar{\rho} \qquad \text{for almost all } (t,x) \in (0,T_0) \times \Omega.$$

such that

$$u^n \xrightarrow{*} u \quad \text{weakly* in } L^\infty((0,T_0); H^2(\Omega_{R_m})^3 \cap H^1(\Omega_{R_m})^3 \cap L_\sigma^2(\Omega_{R_m})),$$
$$u^n \rightharpoonup u \quad \text{weakly in } L^2((0,T_0); W^{2,6}(\Omega_{R_m})^3),$$
$$u_t^n \xrightarrow{*} u_t \quad \text{weakly* in } L^\infty((0,T_0); L_\sigma^2(\Omega_{R_m})),$$
$$u_t^n \rightharpoonup u_t \quad \text{weakly in } L^2((0,T_0); H^1(\Omega_{R_m})^3 \cap L_\sigma^2(\Omega_{R_m})),$$
$$\rho^n \xrightarrow{*} \rho \quad \text{weakly* in } L^\infty((0,T_0) \times \Omega_{R_m}),$$
$$\rho^n - 1 \xrightarrow{*} \rho - 1 \quad \text{weakly* in } L^\infty((0,T_0); H^2(\Omega_{R_m})),$$
$$\rho_t^n \xrightarrow{*} \rho_t \quad \text{weakly* in } L^\infty((0,T_0); L^2(\Omega_{R_m})),$$

for all $m \in \mathbb{N}$. As above, u_t and ρ_t have to be interpreted in distributional sense. Furthermore, by applying the classical compactness result in Proposition 1.89 we even obtain

$$u^n \to u \quad \text{strongly in } C([0,T_0]; H^1(\Omega_{R_m})^3 \cap L_\sigma^2(\Omega_{R_m})),$$
$$\rho^n - 1 \to \rho - 1 \quad \text{strongly in } C([0,T_0]; H^1(\Omega_{R_m})).$$

Note that the described convergence and the uniform estimates in Proposition 5.16 ensure that the regularity for (ρ, u) stated in (5.86) holds. Moreover, by the uniform estimates for (ρ^n, u^n) we obtain that there exists a $C > 0$, such that

$$\|u(t)\|_{H^1(\Omega_{R_m})} \le C, \qquad \|\rho(t) - 1\|_{H^1(\Omega_{R_m})} \le C$$

for all $m \in \mathbb{N}$ and all $t \in [0, T_0]$. Hence, we conclude that

$$\|u(t)\|_{H^1(\Omega)} = \sup_{m \in \mathbb{N}} \|u(t)\|_{H^1(\Omega_{R_m})} \le C, \qquad \|\rho(t) - 1\|_{H^1(\Omega)} = \sup_{m \in \mathbb{N}} \|\rho(t) - 1\|_{H^1(\Omega_{R_m})} \le C$$

for all $t \in [0, T_0]$. This means that for all fixed t we have $u(t) \in H_{0,\sigma}^1(\Omega)$ and $\rho(t) - 1 \in H^1(\Omega)$.

So in particular, it makes sense to write $u(0)$ and $\rho(0) - 1$ as functions in $H^1_{0,\sigma}(\Omega)$ and in $H^1(\Omega)$, respectively. Next let us check if the initial conditions are satisfied. For all $m \in \mathbb{N}$ and $n \geq m$ we have

$$\|u(0) - a\|_{H^1(\Omega_{R_m})} \leq \|u(0) - u^n(0)\|_{H^1(\Omega_{R_m})} + \|u^n(0) - a\|_{H^1(\Omega_{R_m})} \to 0$$

as $n \to \infty$ because $u^n \to u$ strongly in $C([0, T_0]; H^1_{0,\sigma}(\Omega_{R_m}))$ and $u^n(0) = a^{R_n} \to a$ strongly in $H^2(\Omega)^3 \cap H^1_{0,\sigma}(\Omega)$, see Lemma 5.18. As this holds for all $m \in \mathbb{N}$, we have $u(0) = a$ as functions in $H^1_{0,\sigma}(\Omega)$. Analogously, we can see that $\rho(0) = \rho$ holds. Next we show time continuity of u and $\rho - 1$ in $[0, T_0]$ with values in $H^1_{0,\sigma}(\Omega)$ and $L^2(\Omega)$, respectively. For that purpose let $t, s \in [0, T_0]$. By (5.86) we see that

$$\|u(t) - u(s)\|_{H^1(\Omega)} \leq \int_s^t \|u_t(\tau)\|_{H^1(\Omega)} \, d\tau \leq C(t - s)^{\frac{1}{2}}$$

and

$$\|\rho(t) - \rho(s)\|_{L^2(\Omega)} \leq \int_s^t \|\rho_t(\tau)\|_{L^2(\Omega)} \, d\tau \leq C(t - s)$$

for some constant $C > 0$. This shows that $u \in C([0, T_0]; H^1_{0,\sigma}(\Omega))$ and $\rho - 1 \in C([0, T_0]; L^2(\Omega))$.

To conclude Step 1 it only remains to show that (ρ, u) solves (5.5) in the strong sense. For that purpose let $\Psi_m \in C_c^\infty([0, T_0] \times \Omega_{R_m})$ and let $\Phi_m \in C_c^\infty([0, T_0] \times \Omega_{R_m})$ such that $\operatorname{div} \Phi_m(t, \cdot) = 0$ for all $t \in [0, T_0]$. We extend Ψ_m and Φ_m by zero to the whole exterior domain Ω and the extensions are still denoted by Ψ_m and Φ_m, respectively. Analogously, we also extend u^n, which is defined on Ω_{R_n}, to the whole Ω by zero and we denote the extension still by u^n. By Proposition 5.15 we know that for $n \geq m$ we have

$$\int_0^{T_0} \int_\Omega (\rho_t^n + u^n \cdot \nabla \rho^n + b \cdot \nabla \rho^n - Mx \cdot \nabla \rho^n) \cdot \Psi_m \, dx \, dt = 0, \tag{5.87}$$

and

$$\int_0^{T_0} \int_\Omega (\rho^n u_t^n + \rho^n u^n \cdot \nabla u^n + \rho^n b \cdot \nabla u^n + \rho^n u^n \cdot \nabla b$$
$$- \rho^n Mx \cdot \nabla u^n + \rho^n Mu^n - \Delta u^n) \cdot \Phi_m \, dx \, dt = \int_0^{T_0} \int_\Omega (\rho^n f + \Delta b) \cdot \Phi_m \, dx \, dt. \tag{5.88}$$

At first we shall prove that

$$\int_0^{T_0} \int_\Omega \rho^n u_t^n \cdot \Phi_m \, dx \, dt \to \int_0^{T_0} \int_{\Omega_R} \rho u_t \cdot \Phi_m \, dx \, dt \tag{5.89}$$

as $n \to \infty$. We observe that

$$\left| \int_0^{T_0} \int_\Omega (\rho^n u_t^n - \rho u_t) \cdot \Phi_m \, dx \, dt \right|$$
$$\leq \left| \int_0^{T_0} \int_\Omega (\rho^n - \rho) u_t^n \cdot \Phi \, dx \, dt \right| + \left| \int_0^{T_0} \int_\Omega \rho(u_t^n - u_t) \cdot \Phi_m \, dx \, dt \right| \to 0$$

as $n \to \infty$. The convergence of the first integral on the right-hand side is due to the strong convergence $\rho^n \to \rho$ in $C([0,T]; H^1(\Omega_{R_m}))$ and the uniform boundedness of $u^n_t \subset L^\infty([0,T_0]; L^2_\sigma(\Omega_{R_m}))$. The second integral converges to 0 because $u^n_t \overset{*}{\rightharpoonup} u$ weakly-$*$ in $L^\infty([0,T]; L^2_\sigma(\Omega_{R_m}))$ and $\rho \in L^\infty((0,T_0); H^2(\Omega_{R_m}))$. Thus, (5.89) is proved. Next, we show that

$$\int_0^{T_0} \int_\Omega (\rho^n u^n \cdot \nabla u^n) \cdot \Phi_m \, dx \, dt \to \int_0^{T_0} \int_\Omega (\rho u \cdot \nabla u) \cdot \Phi_m \, dx \, dt$$

as $n \to \infty$. For that purpose, observe that

$$\int_0^{T_0} \int_\Omega [(\rho^n u^n \cdot \nabla u^n) - (\rho u \cdot \nabla u)] \cdot \Phi_m \, dx \, dt = \int_0^{T_0} \int_\Omega [\rho^n - \rho](u^n \cdot \nabla u^n) \cdot \Phi_m \, dx \, dt$$

$$+ \int_0^{T_0} \int_{\Omega_R} \rho([u^n - u] \cdot \nabla u^n) \cdot \Phi_m \, dx \, dt + \int_0^{T_0} \int_\Omega \rho(u \cdot \nabla[u^n - u]) \cdot \Phi_m \, dx \, dt =: I_1 + I_2 + I_3.$$

For I_1 we have

$$|I_1| \le C \|\rho_n - \rho\|_{L^\infty((0,T_0); L^2(\Omega_{R_m}))} \int_0^{T_0} \|u^n\|_{L^6(\Omega_{R_m})} \|\nabla u^n\|_{L^3(\Omega_{R_m})} \, dt$$

$$\le C \|\rho_n - \rho\|_{L^\infty((0,T_0); L^2(\Omega_{R_m}))} C \int_0^{T_0} \|u^n\|^2_{H^1(\Omega_{R_m})} \, dt \to 0$$

as $n \to \infty$. Here we applied the Sobolev embedding $H^1(\Omega_{R_m}) \hookrightarrow L^p(\Omega_{R_m})$ for $2 < p \le 6$ and the facts that $\rho^n \to \rho$ strongly in $C([0,T_0]; L^2(\Omega_{R_m}))$ and that $u^n \subset L^\infty((0,T_0); H^2(\Omega_{R_m})^3)$ is uniformly bounded. The integral I_2 can be estimated by

$$|I_2| \le C \|\rho\|_{L^\infty((0,T_0); L^\infty(\Omega))} \int_0^{T_0} \|u^n - u\|_{L^2(\Omega_{R_m})} \|\nabla u^n\|_{L^2(\Omega_{R_m})} \, dt$$

for some $C > 0$ and hence as above we can conclude that it also converges to 0 as $n \to \infty$. Similarly, for I_3 we obtain

$$|I_3| \le C \|\rho\|_{L^\infty((0,T_0); L^\infty(\Omega))} \int_0^{T_0} \|u\|_{L^2(\Omega)} \|\nabla (u^n - u)\|_{L^2(\Omega_{R_m})} \, dt \to 0,$$

as $n \to \infty$. By using similar arguments, we also show that

$$\int_0^{T_0} \int_\Omega \rho^n (Mx \cdot \nabla u^n - Mu^n) \cdot \Phi_m \, dx \, dt \to \int_0^{T_0} \int_\Omega \rho(Mx \cdot \nabla u - Mu) \cdot \Phi_m \, dx \, dt,$$

$$\int_0^{T_0} \int_\Omega \rho^n [(b \cdot \nabla u^n) + (u^n \cdot \nabla b)] \cdot \Phi_m \, dx \, dt \to \int_0^{T_0} \int_\Omega \rho[(b \cdot \nabla u) + (u \cdot \nabla b)] \cdot \Phi_m \, dx \, dt$$

as $n \to \infty$. Now we send $n \to \infty$ in (5.88) and we see that

$$\int_0^{T_0} \int_\Omega (\rho u_t + \rho u \cdot \nabla u + \rho b \cdot \nabla u + \rho u \cdot \nabla b$$

$$-\rho Mx \cdot \nabla u + \rho Mu - \Delta u) \cdot \Phi_m \, dx \, dt = \int_0^{T_0} \int_\Omega (\rho f + \Delta b) \cdot \Phi_m \, dx \, dt. \tag{5.90}$$

Analogously, we send $n \to \infty$ in (5.87) and we obtain

$$\int_0^{T_0} \int_\Omega (\rho_t + u \cdot \nabla \rho + b \cdot \nabla \rho - Mx \cdot \nabla \rho) \cdot \Psi_m \, \mathrm{d}x \, \mathrm{d}t = 0. \tag{5.91}$$

Note that the functions of the form Ψ_m and Φ_m with $m \in \mathbb{N}$ are dense in $L^2((0, T_0); L^2(\Omega))$ and $L^2((0, T_0); L^2_\sigma(\Omega))$, respectively. If we set

$$\nabla \mathrm{p} := -(\mathrm{Id} - \mathbb{P}) \left(\rho u_t + \rho u \cdot \nabla u + \rho b \cdot \nabla u + \rho u \cdot \nabla b - \rho Mx \cdot \nabla u + \rho Mu - \Delta u - \rho f - \Delta b \right),$$

then it follows from (5.90) and (5.91) that (u, p, ρ) is indeed a strong solution to (5.5) on $(0, T_0)$.

Step 2: Uniqueness. Assume that there are two strong solutions $(u^{(i)}, \mathrm{p}^{(i)}, \rho^{(i)})$, $i = 1, 2$, on some interval $(0, T^*)$ which satisfy the properties stated in Theorem 5.4. Set

$$\tilde{u} := u^{(1)} - u^{(2)}, \qquad \tilde{\rho} := \rho^{(1)} - \rho^{(2)}, \qquad \tilde{\mathrm{p}} := \mathrm{p}^{(1)} - \mathrm{p}^{(2)}$$

and

$$G := \tilde{\rho} f - \tilde{\rho} u_t^{(2)} - \tilde{\rho} u^{(2)} \cdot \nabla u^{(2)} - \rho^{(1)} \tilde{u} \cdot \nabla u^{(2)} - \tilde{\rho} b \cdot \nabla u^{(2)} - \tilde{\rho} u^{(2)} \cdot \nabla b + \tilde{\rho} Mx \cdot \nabla u^{(2)} - \tilde{\rho} Mu^{(2)}.$$

Then a straightforward computation shows that $(\tilde{u}, \tilde{\mathrm{p}}, \tilde{\rho})$ satisfies

$$\begin{cases}
\tilde{\rho}_t + (u^{(1)} + b - Mx) \cdot \nabla \tilde{\rho} = -\tilde{u} \cdot \nabla \rho^{(2)} & \text{in } (0, T^*) \times \Omega, \\
\rho^{(1)} \tilde{u}_t + \nabla \tilde{\mathrm{p}} - \Delta \tilde{u} + \rho^{(1)} \left(u^{(1)} \cdot \nabla \tilde{u} + b \cdot \nabla \tilde{u} \right. & \\
\quad \left. + \tilde{u} \cdot \nabla b - Mx \cdot \nabla \tilde{u} + M\tilde{u} \right) = G & \text{in } (0, T^*) \times \Omega, \\
\operatorname{div} \tilde{u} = 0 & \text{in } (0, T^*) \times \Omega, \\
\tilde{u} = 0 & \text{on } (0, T^*) \times \Gamma, \\
(\tilde{\rho}, \tilde{u})|_{t=0} = (0, 0) & \text{in } \Omega,
\end{cases} \tag{5.92}$$

in the strong sense. By using the properties of $(u^{(i)}, \rho^{(i)})$, $i = 1, 2$, stated in Theorem 5.4 we conclude that

$$\left\| \sqrt{\rho^{(1)}(t)} \tilde{u}(t) - \sqrt{\rho^{(1)}(s)} \tilde{u}(s) \right\|_{L^2(\Omega)} \leq \int_0^t \left\| \partial_t \left(\sqrt{\rho^{(1)}(\tau)} u(\tau) \right) \right\|_{L^2(\Omega)} \mathrm{d}\tau \leq C(t - s)^{\frac{1}{2}}$$

for $t, s \in [0, T^*]$ and a constant $C > 0$. Consequently, we see that $\rho^{(1)} \tilde{u} \in C([0, T^*]; L^2(\Omega)^3)$.

Now we use energy estimates, similarly to the ones in Section 5.5, and Gronwall's inequality to obtain $\tilde{u} \equiv 0$ and $\tilde{\rho} \equiv 0$ as functions in $L^\infty((0, T^*); L^2_\sigma(\Omega))$ and $L^\infty((0, T^*); L^2(\Omega))$, respectively. To make this more precise we multiply the momentum equation in (5.92) with $\tilde{u}$, we integrate with respect to t and we do integration by parts with respect to the space variable x. Then we obtain

$$\int_\Omega \rho^{(1)}(t) \frac{|\tilde{u}(t)|^2}{2} \, \mathrm{d}x + \int_0^t \int_\Omega |\nabla \tilde{u}(s)|^2 \, \mathrm{d}x \, \mathrm{d}s$$
$$= -\int_0^t \int_\Omega (\rho^{(1)} \tilde{u} \cdot \nabla b) \cdot \tilde{u} \, \mathrm{d}x \, \mathrm{d}s + \int_0^t \int_\Omega G \cdot \tilde{u} \, \mathrm{d}x \, \mathrm{d}s$$

for all $t \in [0, T^*]$. Here again some integral vanished because $\tilde{u} \in C([0, T^*]; H^1_{0,\sigma}(\Omega))$. Moreover, it is easy to see that

$$\int_\Omega G \cdot \tilde{u} \, \mathrm{d}x$$

$$\leq C\|\tilde{\rho}\|_{L^2(\Omega)}\|f\|_{L^3(\Omega)}\|\tilde{u}\|_{L^6(\Omega)} + C\|\tilde{u}\|_{L^6(\Omega)}\|\tilde{\rho}\|_{L^2(\Omega)}\|u_t^{(2)}\|_{L^3(\Omega)}$$

$$+ C\|\tilde{u}\|_{L^6(\Omega)}\|\tilde{\rho}\|_{L^2(\Omega)}\|u^{(2)}\|_{L^6(\Omega)}\|\nabla u^{(2)}\|_{L^6(\Omega)}$$

$$+ C\left\|\sqrt{\rho^{(1)}}\tilde{u}\right\|_{L^2(\Omega)}^2 \|\nabla u^{(2)}\|_{L^2(\Omega)} + C\|\tilde{u}\|_{L^6(\Omega)}\|\tilde{\rho}\|_{L^2(\Omega)}\|\nabla u^{(2)}\|_{L^3(\Omega)} + \|\tilde{u}\|_{L^6(\Omega)}\|\tilde{\rho}\|_{L^2(\Omega)}$$

$$+ C\|\tilde{u}\|_{L^6(\Omega)}\|\tilde{\rho}\|_{L^2(\Omega)}\|Mx \cdot \nabla u^{(2)} - Mu^{(2)}\|_{L^3(\Omega)}$$

for a constant $C := C(N_0) > 0$. Hence, by using Young's inequality and the usual Sobolev embeddings, we conclude that for any $\varepsilon > 0$ there exists a constant $C := C(\bar{\rho}, \varepsilon) > 0$ such that

$$\int_\Omega G \cdot \tilde{u} \, \mathrm{d}x \leq \varepsilon\|\nabla\tilde{u}\|_{L^2(\Omega)}^2 + C\left\|\sqrt{\rho^{(1)}}\tilde{u}\right\|_{L^2(\Omega)}^2 \|\nabla u^{(2)}\|_{L^2(\Omega)}$$

$$+ C\left(\|u_t^{(2)}\|_{H^1(\Omega)}^2 + \|u^{(2)}\|_{H^2(\Omega)}^4 + \|u^{(2)}\|_{H^2(\Omega)}^2 + \|Mx \cdot \nabla u^{(2)} - Mu^{(2)}\|_{L^3(\Omega)}^2 + 1\right)\|\tilde{\rho}\|_{L^2(\Omega)}^2.$$

For simplicity we set

$$k := \|u_t^{(2)}\|_{H^1(\Omega)}^2 + \|u^{(2)}\|_{H^2(\Omega)}^4 + \|u^{(2)}\|_{H^2(\Omega)}^2 + \|\nabla u^{(2)}\|_{L^2(\Omega)}$$

$$+ \|Mx \cdot \nabla u^{(2)} - Mu^{(2)}\|_{L^6(\Omega)}^2 + \|Mx \cdot \nabla u^{(2)} - Mu^{(2)}\|_{L^2(\Omega)}^2 + 1.$$

By choosing $\varepsilon > 0$ in the estimate above sufficiently small we obtain

$$\left\|\sqrt{\rho^{(1)}(t)}\tilde{u}(t)\right\|_{L^2(\Omega)}^2 + \int_0^t \|\nabla\tilde{u}(s)\|_{L^2(\Omega)}^2 \, \mathrm{d}s$$

$$\leq C \int_0^t k(s)\left(\left\|\sqrt{\rho^{(1)}(t)}\tilde{u}(t)\right\|_{L^2(\Omega)}^2 + \|\tilde{\rho}\|_{L^2(\Omega)}^2\right) \mathrm{d}s$$

for some $C := C(N_0) > 0$. Similarly, by testing the transport equation in (5.92) with $\tilde{\rho}$ we obtain

$$\|\tilde{\rho}(t)\|_{L^2(\Omega)}^2 \leq \frac{1}{2}\int_0^t \|\nabla\tilde{u}\|_{L^2(\Omega)}^2 \, \mathrm{d}s + \int_0^t \|\rho^{(2)}\|_{H^2(\Omega)}^2 \|\tilde{\rho}\|_{L^2(\Omega)}^2 \, \mathrm{d}s$$

for all $t \in [0, T^*]$. Hence, we conclude that

$$\left\|\sqrt{\rho^{(1)}(t)}\tilde{u}(t)\right\|_{L^2(\Omega)}^2 + \|\tilde{\rho}(t)\|_{L^2(\Omega)}^2$$

$$\leq C\int_0^t \left(k(s) + \|\rho^{(2)}(s)\|_{H^2(\Omega)}^2\right)\left(\left\|\sqrt{\rho^{(1)}(t)}\tilde{u}(t)\right\|_{L^2(\Omega)}^2 + \|\tilde{\rho}\|_{L^2(\Omega)}^2\right) \mathrm{d}s \tag{5.93}$$

for all $t \in [0, T^*]$. Since $(u^{(2)}, \mathrm{p}^{(2)}, \rho^{(2)})$ satisfy the properties stated in Theorem 5.4 it is clear that the function $s \mapsto k(s) + \|\rho^{(2)}(s)\|_{H^2(\Omega)}^2$ is integrable on $(0, T^*)$. Hence, by applying

Gronwall's inequality, it follows from (5.93) that

$$\left\|\sqrt{\rho^{(1)}(t)}\,\tilde{u}(t)\right\|_{L^2(\Omega)}^2 + \|\tilde{\rho}(t)\|_{L^2(\Omega)}^2 \leq 0$$

for all $t \in [0, T^*]$. From this it follows that $\tilde{u} \equiv 0$ and $\tilde{\rho} \equiv 0$ as functions in $L^\infty((0, T^*); L^2_\sigma(\Omega))$ and $L^\infty((0, T^*); L^2(\Omega))$, respectively. This yields uniqueness and the proof of the main result is complete. $\qquad\square$

5.7 An example from geophysics

In the study of geophysical flows the external force is given by gravity, which is represented by a unit vector in $\mathbb{R}^3$ multiplied by the constant acceleration due to gravity of earth. As this external force is not an L^2-function over an exterior domain, it does not fit into Assumption 5.2, and therefore this situation is not covered by Theorem 5.4. As the presence of gravity is an interesting and physically relevant case in geophysics, we shall briefly discuss this situation and sketch how the methods from the previous sections can be modified to this setting.

Again, let $\mathcal{O} \subset \mathbb{R}^3$ be a compact set with boundary $\Gamma := \partial\mathcal{O}$ and assume that $\Omega := \mathbb{R}^3 \setminus \mathcal{O}$ is a domain of class C^3. We shall suppose that $\mathcal{O}$ is rotating around the y_3-axis with angular velocity $\omega = (0, 0, \omega_0)^\top$, where $\omega_0 \in \mathbb{R}$. As before, $M \in \mathbb{R}^{3 \times 3}$ denotes the matrix that describes the linear map $y \mapsto \omega \times y$ and $\Omega(t) := \{y(t) = e^{tM}x : x \in \Omega\}$. We shall suppose that the exterior of the rotating obstacle is filled with an inhomogeneous (density-dependent) fluid that is under the influence of gravity. For that purpose, let g denote the constant acceleration due to gravity and $e_3 = (0, 0, 1)^\top$ the standard unit vector in y_3-direction. In this section we consider the following equations:

$$\begin{cases} \tilde{\rho}_t + v \cdot \nabla\tilde{\rho} &= 0 & \text{for } t \in \mathbb{R}_+, \ y \in \Omega(t), \\ \tilde{\rho}v_t + \tilde{\rho}v \cdot \nabla v + \nabla\tilde{q} - \Delta v &= -\tilde{\rho}ge_3 & \text{for } t \in \mathbb{R}_+, \ y \in \Omega(t), \\ \operatorname{div} v &= 0 & \text{for } t \in \mathbb{R}_+, \ y \in \Omega(t), \\ v(t, y) &= My & \text{for } t \in \mathbb{R}_+, \ y \in \Gamma(t), \\ \lim_{|y| \to \infty} \tilde{\rho}(t, y) &= 1 & \text{for } t \in \mathbb{R}_+, \\ (\tilde{\rho}, v)|_{t=0} &= (\rho_0, u_0) & \text{for } y \in \Omega. \end{cases}$$

As above (see (5.2)) set

$$x = e^{-tM}y, \quad w(t, x) = e^{-tM}v(t, y), \quad q(t, x) = \tilde{q}(t, y), \quad \rho(t, x) = \tilde{\rho}(t, y).$$

Since ω is parallel to e_3, it holds that $e^{-tM}e_3 = e_3$, and thus, we obtain the following system:

$$\begin{cases} \rho_t + w \cdot \nabla\rho - Mx \cdot \nabla\rho &= 0 & \text{in } \mathbb{R}_+ \times \Omega, \\ \rho w_t + \rho w \cdot \nabla w - \rho Mx \cdot \nabla w + \rho Mw + \nabla q - \Delta w &= -\rho ge_3 & \text{in } \mathbb{R}_+ \times \Omega, \\ \operatorname{div} w &= 0 & \text{in } \mathbb{R}_+ \times \Omega, \\ w &= Mx & \text{on } \mathbb{R}_+ \times \Gamma, \\ \lim_{|x| \to \infty} \rho(t, x) &= 1 & \text{in } \mathbb{R}_+, \\ (\rho, w)|_{t=0} &= (\rho_0, u_0) & \text{in } \Omega. \end{cases}$$

We use the boundary extension b in (5.4) to rewrite these equations in the following form:

$$\begin{cases} \rho_t + u \cdot \nabla\rho + b \cdot \nabla\rho - Mx \cdot \nabla\rho = 0 & \text{in } \mathbb{R}_+ \times \Omega, \\ \left.\begin{array}{r} \rho u_t + \rho u \cdot \nabla u + \rho b \cdot \nabla u + \rho u \cdot \nabla b \\ - \rho Mx \cdot \nabla u + \rho Mu + \nabla q - \Delta u \end{array}\right\} = \rho\tilde{f} - \rho g e_3 + \Delta b & \text{in } \mathbb{R}_+ \times \Omega, \\ \operatorname{div} u = 0 & \text{in } \mathbb{R}_+ \times \Omega, \\ u = 0 & \text{on } \mathbb{R}_+ \times \Gamma, \\ \lim_{|x|\to\infty} \rho(t,x) = 1 & \text{in } \mathbb{R}_+, \\ (\rho, u)|_{t=0} = (\rho_0, a) & \text{in } \Omega, \end{cases} \quad (5.94)$$

where $a := u_0 - b$ and $\tilde{f} := Mx \cdot \nabla b - Mb - b \cdot \nabla b$.

Note that this equation is just (5.5) with $F = -ge_3$. However, this particular choice of F does not fit into the setting of Theorem 5.4 as it is not an L^2-function over Ω. Moreover, note that also $-\rho g e_3$ is not an L^2-function, due to the condition $\lim_{|x|\to\infty} \rho(t,x) = 1$. When considering this equation over a bounded domain, say $\Omega_R = \Omega \cap B_R$ for large $R > 1$, then $F = -ge_3$ is certainly an H^1-function and therefore fits into the setting of the previous sections. In particular, one can prove existence of a strong solution (ρ^R, u^R) to (5.94) on $\Omega_R = \Omega \cap B_R$, analogously to Proposition 5.15. Only for proving a priori estimates uniformly in $R > 1$ one needs extra considerations and a simple trick. Note that $-\rho g e_3 = -(\rho - 1)ge_3 - ge_3$. Since $-ge_3 = -\nabla(gx_3)$, this term can be absorbed into the pressure. Therefore, we can consider the following system of equations:

$$\begin{cases} \rho_t + u \cdot \nabla\rho + b \cdot \nabla\rho - Mx \cdot \nabla\rho = 0 & \text{in } \mathbb{R}_+ \times \Omega, \\ \left.\begin{array}{r} \rho u_t + \rho u \cdot \nabla u + \rho b \cdot \nabla u + \rho u \cdot \nabla b \\ - \rho Mx \cdot \nabla u + \rho Mu + \nabla p - \Delta u \end{array}\right\} = \rho\tilde{f} - (\rho-1) g e_3 + \Delta b & \text{in } \mathbb{R}_+ \times \Omega, \\ \operatorname{div} u = 0 & \text{in } \mathbb{R}_+ \times \Omega, \\ u = 0 & \text{on } \mathbb{R}_+ \times \Gamma, \\ \lim_{|x|\to\infty} \rho(t,x) = 1 & \text{in } \mathbb{R}_+, \\ (\rho, u)|_{t=0} = (\rho_0, a) & \text{in } \Omega, \end{cases} \quad (5.95)$$

This equation can now be treated by the L^2-techniques from the previous sections, and we can state the following modification of Theorem 5.4.

Theorem 5.45. *Assume that the conditions on (ρ_0, u_0) stated in Assumption 5.2 hold. Then there exists a $T_0 > 0$ and a unique strong solution (u, p, ρ) to (5.95) defined on $(0, T_0)$ satisfying*

$$u \in C([0,T_0]; H^1_{0,\sigma}(\Omega)) \cap L^\infty((0,T_0); \mathcal{D}(L_{\Omega,2})) \cap L^2((0,T_0); \mathcal{D}(L_{\Omega,6})),$$

$$u_t \in L^\infty((0,T_0); L^2_\sigma(\Omega)) \cap L^2((0,T_0); H^1_{0,\sigma}(\Omega)),$$

$$\nabla p \in L^\infty((0,T_0); L^2(\Omega)),$$

$$\rho \in L^\infty((0,T_0) \times \Omega),$$

$$\rho - 1 \in C([0,T_0]; L^2(\Omega)) \cap L^\infty((0,T_0); H^2(\Omega)),$$

$$\rho_t \in L^\infty((0,T_0); L^2(\Omega)),$$

$$\underline{\rho} \le \rho(t,x) \le \bar{\rho} \qquad \text{for almost all } (t,x) \in (0,T_0) \times \Omega.$$

The proof of Theorem 5.45 follows the lines of the proof of Theorem 5.4. We will not repeat all the details from Sections 5.4–5.6. In particular, as said before, the only point which needs extra considerations is to derive estimates on $\Omega_R = \Omega \cap B_R$ that do not depend on $R > 1$. The additional term which needs to be treated here is $-(\rho - 1)ge_3$. In the following we sketch how the proofs of the a priori estimates in Section 5.5.2 have to be modified at crucial points. By ρ_0^R we denote the initial density ρ_0 restricted to Ω_R. Let $(u, \rho) := (u^R, \rho^R)$ be a strong solution to the problem (5.95) on $(0, T_0) \times \Omega_R$ which has the regularity stated in 5.15.

The treatment of $-(\rho - 1)ge_3$ is based on the following observation: Since ρ solves the transport equation (5.14) in the strong sense with initial value ρ_0, we conclude that $\rho - 1$ solves (5.14) with initial value $\rho_0 - 1$. Thus, by Lemma 5.26 we know

$$\|\rho(t) - 1\|_{L^2(\Omega_R)} = \|\rho_0^R - 1\|_{L^2(\Omega_R)}$$

for all $t \in [0, T_0]$. By Assumption 5.2, $\|\rho_0^R - 1\|_{L^2(\Omega_R)} \leq \|\rho_0 - 1\|_{L^2(\Omega)}$ is finite. Thus, we control $\sup_{t \in (0, T_0)} \|\rho(t) - 1\|_{L^2(\Omega_R)}$ by a constant $C > 0$ just depending on the given data.

In the proof of Lemma 5.37 equation (5.56) needs to be modified by the term

$$\int_{\Omega_R} (\rho - 1)ge_3 \cdot u \, dx.$$

Since $1/\rho$ is uniformly bounded, this expression is easily estimated by

$$\int_{\Omega_R} (\rho - 1)ge_3 \cdot u \, dx \leq C \int_{\Omega_R} \rho |u|^2 \, dx + C \int_{\Omega_R} |\rho - 1|^2 \, dx$$

for some constant $C > 0$, just depending on the given data. Hence, by the observation above, the second term on the right-hand side is estimated by a constant just depending on the given data. Now, one can closely follow the proof of Lemma 5.37.

Next, we look at the proof of Lemma 5.38. There we have the additional term

$$\int_{\Omega_R} (\rho - 1)ge_3 \cdot (u_t - Mx \cdot \nabla u + Mx) \, dx.$$

By Young's inequality and since $1/\rho$ is uniformly bounded, we see that

$$\int_{\Omega_R} (\rho - 1)ge_3 \cdot (u_t - Mx \cdot \nabla u + Mx) \, dx$$

$$\leq \frac{1}{10} \int_{\Omega_R} \rho |u_t - Mx \cdot \nabla u + Mx)|^2 \, dx + C \int_{\Omega_R} |\rho - 1|^2 \, dx$$

for a constant $C > 0$, just depending on the given data. As above, the second term on the right-hand side is controlled by a constant just depending on the data, and therefore we can follow closely the proof of Lemma 5.38.

Finally, let us say a few words about the proof of Lemma 5.41. There the additional expression

$$\int_{\Omega_R} \rho_t \mathrm{g} e_3 \cdot u_t \, \mathrm{d}x$$

appears. We estimate

$$\int_{\Omega_R} \rho_t \mathrm{g} e_3 \cdot u_t \, \mathrm{d}x \le \mathrm{g} \Psi_R(t) \| u_t \|_{L^2(\Omega_R)},$$

where Ψ_R is the function introduced at the beginning of the proof of Lemma 5.41. Moreover, in this proof we also use that

$$\| (\rho(t) - 1) \mathrm{g} e_3 \|_{L^6(\Omega_R)}^2 \le C \| \rho(t) - 1 \|_{H^1(\Omega_R)}^2 \le C \Psi_R(t)$$

for a constant $C > 0$, independent of R, and all $t \in [0, T_0]$. Now, one can follow the remaining steps of Lemma 5.41.

Supplement: Method of characteristics for transport equations

This supplement is intended to give some details on the method of characteristics for linear transport equations which was applied in Lemma 5.24. Although this is a classical method this supplement is included for self-containdness.

Consider the linear transport equation

$$\begin{cases} \rho_t + u \cdot \nabla\rho &= 0 \quad \text{in } [0,T] \times \Omega, \\ \rho|_{t=0} &= \rho_0 \quad \text{in } \Omega. \end{cases} \tag{5.96}$$

Lemma 5.46. *Let $\Omega \subset \mathbb{R}^d$, $d \in \mathbb{N}$, be a bounded domain of class C^1 and $T > 0$. Moreover, let $j \in \mathbb{N}$, $mp > d > (m-1)p$ and*

$$u \in C^j([0,T]; W^{j+m,p}(\Omega)^d \cap W_0^{1,p}(\Omega)^d).$$

Then for an initial value $\rho_0 \in C^j(\overline{\Omega})$, problem (5.96) admits a unique, classical solution $\rho \in C^j([0,T] \times \overline{\Omega}_R)$ given by

$$\rho(t,x) = \rho_0(U(0,t,x)),$$

where $U \in C^j([0,T] \times [0,T] \times \overline{\Omega}_R)^d$ is the unique solution to the initial value problem

$$\begin{cases} \partial_t U(t,s,x) &= u(t,U(t,s,x)), \quad 0 \le t \le T, \\ U(s,s,x) &= x \in \overline{\Omega}, \qquad\qquad 0 \le s \le T. \end{cases} \tag{5.97}$$

The proof is based on the method of characteristics. See e.g. [NS04, Theorem 2.10] for a similar proof in the case $d = 1$.

Proof. The Sobolev embedding $W^{j+m,p}(\Omega) \hookrightarrow C^j(\overline{\Omega})$ yields $u \in C^j([0,T] \times \overline{\Omega})^d$. We use the extension operator $E : W^{j+m,p}(\Omega) \to W^{j+m,p}(\mathbb{R}^d)$ to extend u to a function $\widetilde{u} \in C([0,T]; W^{j+m,p}(\mathbb{R}^d)^d)$. By the Sobolev embedding $W^{j+m,p}(\mathbb{R}^d) \hookrightarrow C^j(\mathbb{R}^d)$ we conclude that $\widetilde{u} \in C^j([0,T] \times \mathbb{R}^d)^d$. The classical Picard-Lindelöf theorem yields that for fixed $0 \le s \le T$ there exists a time $\tau > 0$ and a unique solution $U(\cdot,s,x) \in C^1([T_0,T_1])^d$ of

$$\begin{cases} \partial_t U(t,s,x) &= \widetilde{u}(t,U(t,s,x)), \quad T_0 \le t \le T_1, \\ U(s,s,x) &= x \in \overline{\Omega}, \qquad\qquad 0 \le s \le T, \end{cases} \tag{5.98}$$

where $T_0 := \max\{0, s - \tau\}$ and $T_1 := \min\{T, s + \tau\}$. If $x \in \partial\Omega$, then the unique solution is given by $U(\cdot,s,x) = x$ since $\widetilde{u}(t,x) = 0$ for all $(t,x) \in [0,T_0] \times \partial\Omega$. Thus, it is clear that for any $x \in \overline{\Omega}$ the solution $U(\cdot,t,x)$ of (5.98) stays in $\overline{\Omega}$ for all $t \in [T_0,T_1]$, and therefore we can replace $\widetilde{u}$ in (5.98) by u. By repeating this argument we can show that $U(\cdot,s,x)$ solves (5.97) on $[0,T]$. Since $u \in C^j([0,T] \times \overline{\Omega})$, classical arguments for ordinary differential equations yield that $U \in C^j([0,T] \times [0,T] \times \overline{\Omega})$.

Next we define $\rho \in C^j([0,T] \times \overline{\Omega})$ by

$$\rho(t,x) := \rho_0(U(0,t,x)) \tag{5.99}$$

and we show that this function is indeed the unique solution to (5.96). For $s \in [0,T]$ and $x \in \overline{\Omega}$ fixed we define $C(\cdot; s, x) : [0,T] \to \mathbb{R}^{d \times d}$ by

$$C(t; s, x) := \nabla u(t, U(t, s, x)), \qquad t \in [0,T].$$

Differentiating equation (5.98) with respect to s and x_i, $i = 1, \ldots, d$, respectively, we see that $\partial_s U(\cdot, s, x)$ and $\partial_{x_i} U(\cdot, s, x)$, $i = 1, \ldots, d$, solve the ordinary linear differential equation

$$\frac{\mathrm{d}}{\mathrm{d}t} V(t) = C(t; s, x) V(t), \qquad t \in [0,T]. \tag{5.100}$$

Note that $\partial_t \partial_{x_i} U$, $i = 1, \ldots, d$ and $\partial_t \partial_s U$ exist by standard arguments for ordinary differential equations. Moreover, since $\partial_{x_i} U(s, s, x) = e_i$, where e_i is the i-th unit vector in $\mathbb{R}^d$, it follows that $\partial_{x_1} U(\cdot, s, x), \ldots, \partial_{x_d} U(\cdot, s, x)$ are linearly independent and form a fundamental system of solutions to (5.100). Hence, we conclude that there exist a vector $c(s, x) \in \mathbb{R}^d$, depending only on s and x, such that

$$\partial_s U(\cdot, s, x) = \sum_{i=1}^{d} c_i(t, x) \partial_{x_i} U(\cdot, s, x).$$

Looking at the initial condition in (5.97) yields that

$$c(s, x) = \sum_{i=1}^{d} c_i(s, x) \partial_{x_i} U(s, s, x) = \partial_s U(s, s, x) = -\partial_t U(s, s, x) = -u(s, x),$$

and thus we see that

$$\partial_s U(\cdot, s, x) + \sum_{i=1}^{d} u_i(s, x) \partial_{x_i} U(\cdot, s, x) = 0$$

for all $s \in [0,T]$, $x \in \overline{\Omega}$. By using formula (5.99) we then obtain

$$\partial_t \rho(t, x) + u(t, x) \cdot \nabla \rho(t, x) = \left(\partial_s U(0, t, x) + \sum_{i=1}^{d} u_i(t, x) \partial_{x_i} U(0, t, x) \right) \cdot \nabla \rho_0(U(0, t, x)) = 0.$$

Hence, the function defined in (5.99) is indeed a solution to (5.14). To show the uniqueness we assume that there exists another solution $\tilde{\rho}$. Then $\partial_t(\tilde{\rho}(t, U(t, 0, x))) = 0$, i.e. $\tilde{\rho}$ is constant along the characteristic $U(t, 0, x)$. From that we conclude that

$$\tilde{\rho}(t, U(t, 0, x)) = \rho_0(x) = \rho(t, U(t, 0, x)).$$

This directly yields the uniqueness. $\qquad \square$

Bibliography

[ABHN01]   W. Arendt, C. J. K. Batty, M. Hieber, and F. Neubrander, *Vector-valued Laplace transforms and Cauchy problems*, Monographs in Mathematics, vol. 96, Birkhäuser Verlag, Basel, 2001.

[ADN59]   S. Agmon, A. Douglis, and L. Nirenberg, *Estimates near the boundary for solutions of elliptic partial differential equations satisfying general boundary conditions. I*, Comm. Pure Appl. Math. **12** (1959), 623–727.

[AF03]   R. A. Adams and J. J. F. Fournier, *Sobolev spaces*, second ed., Pure and Applied Mathematics (Amsterdam), vol. 140, Elsevier/Academic Press, Amsterdam, 2003.

[AKM90]   S. N. Antontsev, A. V. Kazhikhov, and V. N. Monakhov, *Boundary value problems in mechanics of nonhomogeneous fluids*, Studies in Mathematics and its Applications, vol. 22, North-Holland Publishing Co., Amsterdam, 1990, Translated from the Russian.

[Ama95]   H. Amann, *Linear and Quasilinear Parabolic Problems. Vol. I*, Monographs in Mathematics, vol. 89, Birkhäuser Boston Inc., Boston, MA, 1995.

[Ama00]   ———, *On the strong solvability of the Navier-Stokes equations*, J. Math. Fluid Mech. **2** (2000), no. 1, 16–98.

[Are94]   W. Arendt, *Gaussian estimates and interpolation of the spectrum in L^p*, Differential Integral Equations **7** (1994), no. 5-6, 1153–1168.

[Are04]   ———, *Semigroups and evolution equations: Functional calculus, regularity and kernel estimates.*, Handbook of Differential Equations, Evolutionary equations, Vol. I (Dafermos, C. M. (ed.) et al.), Elsevier/North-Holland, Amsterdam, 2004, pp. 1–85.

[Bat99]   G. K. Batchelor, *An introduction to fluid dynamics*, paperback ed., Cambridge Mathematical Library, Cambridge University Press, Cambridge, 1999.

[Bog79]   M. E. Bogovskiĭ, *Solution of the first boundary value problem for an equation of continuity of an incompressible medium*, Dokl. Akad. Nauk SSSR **248** (1979), no. 5, 1037–1040.

[Bog86]   ———, *Decomposition of $L_p(\Omega; \mathbf{R}^n)$ into a direct sum of subspaces of solenoidal and potential vector fields*, Dokl. Akad. Nauk SSSR **286** (1986), no. 4, 781–786.

[Bor92]   W. Borchers, *Zur Stabilität und Faktorisierungsmethode für die Navier-Stokes-Gleichungen inkompressibler viskoser Flüssigkeiten*, Habiliationschrift, Universität Paderborn, 1992.

[BRMFC03] J. L. Boldrini, M. A. Rojas-Medar, and E. Fernández-Cara, *Semi-Galerkin approximation and strong solutions to the equations of the nonhomogeneous asymmetric fluids*, J. Math. Pures Appl. (9) **82** (2003), no. 11, 1499–1525.

[BS87] W. Borchers and H. Sohr, *On the semigroup of the Stokes operator for exterior domains in L^q-spaces*, Math. Z. **196** (1987), no. 3, 415–425.

[BV93] W. Borchers and W. Varnhorn, *On the boundedness of the Stokes semigroup in two-dimensional exterior domains*, Math. Z. **213** (1993), no. 2, 275–299.

[Cat61] L. Cattabriga, *Su un problema al contorno relativo al sistema di equazioni di Stokes*, Rend. Sem. Mat. Univ. Padova **31** (1961), 308–340.

[CK03] H. J. Choe and H. Kim, *Strong solutions of the Navier-Stokes equations for nonhomogeneous incompressible fluids*, Comm. Partial Differential Equations **28** (2003), no. 5-6, 1183–1201.

[CK04] Y. Cho and H. Kim, *Unique solvability for the density-dependent Navier-Stokes equations*, Nonlinear Anal. **59** (2004), no. 4, 465–489.

[CK06] _____, *Existence results for viscous polytropic fluids with vacuum*, J. Differential Equations **228** (2006), no. 2, 377–411.

[CM93] Alexandre J. Chorin and Jerrold E. Marsden, *A mathematical introduction to fluid mechanics. 3. ed.*, Texts in Applied Mathematics, Springer-Verlag, New York, 1993.

[CM95] M. Cannone and Y. Meyer, *Littlewood-Paley decomposition and Navier-Stokes equations*, Methods Appl. Anal. **2** (1995), no. 3, 307–319.

[CM97] Z. Chen and T. Miyakawa, *Decay properties of weak solutions to a perturbed Navier-Stokes system in $\mathbf{R}^n$*, Adv. Math. Sci. Appl. **7** (1997), no. 2, 741–770.

[CM04] F. Crispo and P. Maremonti, *An interpolation inequality in exterior domains*, Rend. Sem. Mat. Univ. Padova **112** (2004), 11–39.

[Dan03] R. Danchin, *Density-dependent incompressible viscous fluids in critical spaces*, Proc. Roy. Soc. Edinburgh Sect. A **133** (2003), no. 6, 1311–1334.

[Dan04] _____, *Local and global well-posedness results for flows of inhomogeneous viscous fluids*, Adv. Differential Equations **9** (2004), no. 3-4, 353–386.

[Dan06] _____, *Density-dependent incompressible fluids in bounded domains*, J. Math. Fluid Mech. **8** (2006), no. 3, 333–381.

[DGH09] E. Dintelmann, M. Geissert, and M. Hieber, *Strong L^p-solutions to the Navier-Stokes flow past rotating obstacles: The case of several obstacles and time dependent velocity*, Trans. Amer. Math Soc. **361** (2009), 653–669.

[DHP01] W. Desch, M. Hieber, and J. Prüss, *L^p-theory of the Stokes equation in a half space*, J. Evol. Equ. **1** (2001), no. 1, 115–142.

[DHP03]   R. Denk, M. Hieber, and J. Prüss, *R-boundedness, Fourier multipliers and problems of elliptic and parabolic type*, Mem. Amer. Math. Soc. **166** (2003), no. 788.

[DKS98]   W. Dan, T. Kobayashi, and Y. Shibata, *On the local energy decay approach to some fluid flow in an exterior domain*, Recent topics on mathematical theory of viscous incompressible fluid (Tsukuba, 1996), Lecture Notes Numer. Appl. Anal., vol. 16, Kinokuniya, Tokyo, 1998, pp. 1–51.

[DL89]   R. J. DiPerna and P.-L. Lions, *Ordinary differential equations, transport theory and Sobolev spaces*, Invent. Math. **98** (1989), no. 3, 511–547.

[Dor93]   G. Dore, *L^p regularity for abstract differential equations*, Functional analysis and related topics, 1991 (Kyoto), Lecture Notes in Math., vol. 1540, Springer, Berlin, 1993, pp. 25–38.

[DPL07]   G. Da Prato and A. Lunardi, *Ornstein-Uhlenbeck operators with time periodic coefficients*, J. Evol. Equ. **7** (2007), no. 4, 587–614.

[DS99a]   W. Dan and Y. Shibata, *On the L_q-L_r estimates of the Stokes semigroup in a two-dimensional exterior domain*, J. Math. Soc. Japan **51** (1999), no. 1, 181–207.

[DS99b]   ———, *Remark on the L_q-L_∞ estimate of the Stokes semigroup in a 2-dimensional exterior domain*, Pacific J. Math. **189** (1999), no. 2, 223–239.

[EN00]   K.-J. Engel and R. Nagel, *One-parameter semigroups for linear evolution equations*, Graduate Texts in Mathematics, vol. 194, Springer-Verlag, New York, 2000.

[ES04]   Y. Enomoto and Y. Shibata, *Local energy decay of solutions to the Oseen equation in the exterior domains*, Indiana Univ. Math. J. **53** (2004), no. 5, 1291–1330.

[ES05]   ———, *On the rate of decay of the Oseen semigroup in exterior domains and its application to Navier-Stokes equation*, J. Math. Fluid Mech. **7** (2005), no. 3, 339–367.

[Far06]   R. Farwig, *An L^q-analysis of viscous fluid flow past a rotating obstacle*, Tohoku Math. J. (2) **58** (2006), no. 1, 129–147.

[Fef00]   C.L. Fefferman, *Existence and smoothness of the Navier-Stokes equations*, 2000, available online under http://www.claymath.org/millennium/.

[Fei04]   E. Feireisl, *Dynamics of viscous compressible fluids*, Oxford Lecture Series in Mathematics and its Applications, vol. 26, Oxford University Press, Oxford, 2004.

[FH07]   R. Farwig and T. Hishida, *Stationary Navier-Stokes flow around a rotating obstacle.*, Funkc. Ekvacioj, Ser. Int. **50** (2007), no. 3, 371–403.

[FHM04]   R. Farwig, T. Hishida, and D. Müller, L^q-theory of a singular "winding" integral operator arising from fluid dynamics, Pacific J. Math. **215** (2004), no. 2, 297–312.

[FHZ10]   D. Fang, M. Hieber, and T. Zhang, Density-dependent incompressible viscous fluid flow subject to linearly growing data, Preprint, 2010.

[FJR72]   E. B. Fabes, B. F. Jones, and N. M. Rivière, The initial value problem for the Navier-Stokes equations with data in L^p, Arch. Rational Mech. Anal. **45** (1972), 222–240.

[FK64]   H. Fujita and T. Kato, On the Navier-Stokes initial value problem. I, Arch. Rational Mech. Anal. **16** (1964), 269–315.

[FN07]   R. Farwig and J. Neustupa, On the spectrum of a Stokes-type operator arising from flow around a rotating body, Manuscripta Math. **122** (2007), no. 4, 419–437.

[Gag59]   E. Gagliardo, Ulteriori proprietà di alcune classi di funzioni in più variabili, Ricerche Mat. **8** (1959), 24–51.

[Gal94a] G. P. Galdi, An introduction to the mathematical theory of the Navier-Stokes equations. Vol. I, Springer Tracts in Natural Philosophy, vol. 38, Springer-Verlag, New York, 1994.

[Gal94b] _____, An introduction to the mathematical theory of the Navier-Stokes equations. Vol. II, Springer Tracts in Natural Philosophy, vol. 39, Springer-Verlag, New York, 1994.

[Gal02]   _____, On the motion of a rigid body in a viscous liquid: a mathematical analysis with applications, Handbook of mathematical fluid dynamics, Vol. I (S. J. Friedlander and D. serre, eds.), North-Holland, Amsterdam, 2002, pp. 653–791.

[Gal03]   _____, Steady flow of a Navier-Stokes fluid around a rotating obstacle., J. Elasticity **71** (2003), no. 1-3, 1–31.

[GH11]   M. Geissert and T Hansel, A non-autonomous model problem for the Oseen-Navier-Stokes flow with rotating effects, J. Math. Soc. Japan **63** (2011), no. 3, 1027 – 1037.

[GHH06a] M. Geissert, H. Heck, and M. Hieber, L^p-theory of the Navier-Stokes flow in the exterior of a moving or rotating obstacle, J. Reine Angew. Math. **596** (2006), 45–62.

[GHH06b] M. Geißert, H. Heck, and M. Hieber, On the equation div $u = g$ and Bogovskiĭ's operator in Sobolev spaces of negative order, Partial differential equations and functional analysis, Oper. Theory Adv. Appl., vol. 168, Birkhäuser, Basel, 2006, pp. 113–121.

[GHH⁺10] M. Geissert, M. Hess, M. Hieber, C. Schwarz, and K. Stavrakidis, *Maximal L^p-L^q-estimates for the Stokes equation: a short proof of Solonnikov's theorem*, J. Math. Fluid Mech. **12** (2010), no. 1, 47–60.

[GHHS10] M. Geissert, H. Heck, M. Hieber, and O. Sawada, *Weak Neumann implies Stokes*, Preprint, 2010.

[GHHW05] M. Geissert, H. Heck, M. Hieber, and I. Wood, *The Ornstein-Uhlenbeck semigroup in exterior domains*, Arch. Math. (Basel) **85** (2005), no. 6, 554–562.

[Gig81] Y. Giga, *Analyticity of the semigroup generated by the Stokes operator in L_r spaces*, Math. Z. **178** (1981), no. 3, 297–329.

[Gig85] _____, *Domains of fractional powers of the Stokes operator in L_r spaces*, Arch. Rational Mech. Anal. **89** (1985), no. 3, 251–265.

[Gig86] _____, *Solutions for semilinear parabolic equations in L^p and regularity of weak solutions of the Navier-Stokes system*, J. Differential Equations **62** (1986), no. 2, 186–212.

[GK11] G. P. Galdi and M. Kyed, *A simple proof of L^q-estimates for the steady-state Oseen and Stokes equations in a rotating frame. Part I: Strong Solutions*, Preprint, 2011.

[GL08] Matthias Geissert and Alessandra Lunardi, *Invariant measures and maximal L^2 regularity for nonautonomous Ornstein-Uhlenbeck equations*, J. Lond. Math. Soc. (2) **77** (2008), no. 3, 719–740.

[Gri85] P. Grisvard, *Elliptic problems in nonsmooth domains*, Monographs and Studies in Mathematics, vol. 24, Pitman (Advanced Publishing Program), Boston, MA, 1985.

[GS02] G. P. Galdi and A. L. Silvestre, *Strong solutions to the problem of motion of a rigid body in a Navier-Stokes liquid under the action of prescribed forces and torques*, Nonlinear Problems in Mathematical Physics and Related Topics I, Int. Math. Ser. (N. Y.), Kluwer/Plenum, New York, 2002, pp. 121–144.

[GS05] _____, *Strong solutions to the Navier-Stokes equations around a rotating obstacle*, Arch. Ration. Mech. Anal. **176** (2005), no. 3, 331–350.

[GS06] _____, *Existence of time-periodic solutions to the Navier-Stokes equations around a moving body*, Pacific J. Math. **223** (2006), no. 2, 251–267.

[Haa06] M. Haase, *The functional calculus for sectorial operators*, Operator Theory: Advances and Applications, vol. 169, Birkhäuser Verlag, Basel, 2006.

[Han11] T. Hansel, *On the Navier-Stokes equations with rotating effect and prescribed outflow velocity*, J. Math. Fluid Mech. **13** (2011), no. 3, 405–419.

[HDR10] R. Haller-Dintelmann and J. Rehberg, *Coercivity for elliptic operators and positivity of solutions on Lipschitz domains*, Arch. Math. (Basel) **95** (2010), no. 5, 457–468.

[HDW05]   R. Haller-Dintelmann and J. Wiedl, *Kolmogorov kernel estimates for the Ornstein-Uhlenbeck operator*, Ann. Sc. Norm. Super. Pisa Cl. Sci. (5) **4** (2005), no. 4, 729–748.

[Hey80]   J. G. Heywood, *The Navier-Stokes equations: on the existence, regularity and decay of solutions*, Indiana Univ. Math. J. **29** (1980), no. 5, 639–681.

[His99a]   T. Hishida, *An existence theorem for the Navier-Stokes flow in the exterior of a rotating obstacle*, Arch. Ration. Mech. Anal. **150** (1999), no. 4, 307–348.

[His99b]   ――――, *The Stokes operator with rotation effect in exterior domains*, Analysis (Munich) **19** (1999), no. 1, 51–67.

[His00]   ――――, L^2 *theory for the operator* $\Delta + (k \times x) \cdot \nabla$ *in exterior domains*, Nihonkai Math. J. **11** (2000), no. 2, 103–135.

[His01]   ――――, *On the Navier-Stokes flow around a rigid body with a prescribed rotation*, Proceedings of the Third World Congress of Nonlinear Analysts, Part 6 (Catania, 2000), vol. 47, 2001, pp. 4217–4231.

[Hop51]   E. Hopf, *Über die Anfangswertaufgabe für die hydrodynamischen Grundgleichungen*, Math. Nachr. **4** (1951), 213–231.

[HR10]   T. Hansel and A. Rhandi, *The Oseen-Navier-Stokes flow in the exterior of a rotating obstacle: The non-autonomous case*, submitted, 2010.

[HR11]   ――――, *Non-autonomous Ornstein-Uhlenbeck equations in exterior domains*, Adv. Differential Equations **16** (2011), no. 3-4, 201–220.

[HS05]   M. Hieber and O. Sawada, *The Navier-Stokes equations in* $\mathbb{R}^n$ *with linearly growing initial data*, Arch. Ration. Mech. Anal. **175** (2005), no. 2, 269–285.

[HS09]   T. Hishida and Y. Shibata, L_p-L_q *estimate of the Stokes operator and Navier-Stokes flows in the exterior of a rotating obstacle*, Arch. Ration. Mech. Anal. **193** (2009), no. 2, 339–421.

[Iwa89]   H. Iwashita, L_q-L_r *estimates for solutions of the nonstationary Stokes equations in an exterior domain and the Navier-Stokes initial value problems in* L_q *spaces*, Math. Ann. **285** (1989), no. 2, 265–288.

[Kat84]   T. Kato, *Strong* L^p-*solutions of the Navier-Stokes equation in* $\mathbf{R}^m$, *with applications to weak solutions*, Math. Z. **187** (1984), no. 4, 471–480.

[Kat95]   ――――, *Perturbation theory for linear operators*, Classics in Mathematics, Springer-Verlag, Berlin, 1995, Reprint of the 1980 edition.

[Kim87]   J.U. Kim, *Weak solutions of an initial-boundary value problem for an incompressible viscous fluid with nonnegative density*, SIAM J. Math. Anal. **18** (1987), no. 1, 89–96.

[KS98]   T. Kobayashi and Y. Shibata, *On the Oseen equation in the three-dimensional exterior domains*, Math. Ann. **310** (1998), no. 1, 1–45.

[Lad69]   O. A. Ladyzhenskaya, *The mathematical theory of viscous incompressible flow*, Gordon and Breach Science Publishers, New York, 1969.

[LB07]   Luca Lorenzi and Marcello Bertoldi, *Analytical methods for Markov semigroups*, Pure and Applied Mathematics (Boca Raton), vol. 283, Chapman & Hall/CRC, Boca Raton, FL, 2007.

[Ler34]   J. Leray, *Sur le mouvement d'un liquide visqueux emplissant l'espace*, Acta Math. **63** (1934), no. 1, 193–248.

[Lio96]   P.-L. Lions, *Mathematical topics in fluid mechanics. Vol. 1*, Oxford Lecture Series in Mathematics and its Applications, vol. 3, The Clarendon Press Oxford University Press, New York, 1996.

[LL59]   L. D. Landau and E. M. Lifshitz, *Fluid mechanics*, Translated from the Russian by J. B. Sykes and W. H. Reid. Course of Theoretical Physics, Vol. 6, Pergamon Press, London, 1959.

[LS75]   O. A. Ladyženskaja and V. A. Solonnikov, *The unique solvability of an initial-boundary value problem for viscous incompressible inhomogeneous fluids*, Zap. Naučn. Sem. Leningrad. Otdel. Mat. Inst. Steklov. (LOMI) **52** (1975), 52–109, 218–219.

[Lun95]   A. Lunardi, *Analytic semigroups and optimal regularity in parabolic problems*, Progress in Nonlinear Differential Equations and their Applications, 16, Birkhäuser Verlag, Basel, 1995.

[Lun09]   ———, *Interpolation theory*, second ed., Lecture Notes. Scuola Normale Superiore di Pisa (New Series), Edizioni della Normale, Pisa, 2009.

[MB86]   V. N. Maslennikova and M. E. Bogovskiĭ, *Elliptic boundary value problems in unbounded domains with noncompact and nonsmooth boundaries*, Rend. Sem. Mat. Fis. Milano **56** (1986), 125–138.

[McC81]   M. McCracken, *The resolvent problem for the Stokes equations on halfspace in L_p*, SIAM J. Math. Anal. **12** (1981), no. 2, 201–228.

[Met01]   G. Metafune, *L^p-spectrum of Ornstein-Uhlenbeck operators*, Ann. Scuola Norm. Sup. Pisa Cl. Sci. (4) **30** (2001), no. 1, 97–124.

[MPF91]   D. S. Mitrinović, J. E. Pečarić, and A. M. Fink, *Inequalities involving functions and their integrals and derivatives*, Mathematics and its Applications (East European Series), vol. 53, Kluwer Academic Publishers Group, Dordrecht, 1991.

[MPRS02]   G. Metafune, J. Prüss, A. Rhandi, and R. Schnaubelt, *The domain of the Ornstein-Uhlenbeck operator on an L^p-space with invariant measure*, Ann. Sc. Norm. Super. Pisa Cl. Sci. (5) **1** (2002), no. 2, 471–485.

[MS90]   P. Maremonti and V. A. Solonnikov, *An estimate for the solutions of a Stokes system in exterior domains*, Zap. Nauchn. Sem. Leningrad. Otdel. Mat. Inst. Steklov. (LOMI) **180** (1990), 105–120, 181.

[MS97]   _____, *On nonstationary Stokes problem in exterior domains*, Ann. Scuola Norm. Sup. Pisa Cl. Sci. (4) **24** (1997), no. 3, 395–449.

[Nav27]   C.L. Navier, *Memoire sur les lois mouvement des fluides*, Acad. Sci. de France **6** (1827), 389–440.

[Nic96]   G. Nickel, *On evolution semigroups and wellposedness of nonautonomous Cauchy problems*, Ph.D. thesis, Universität Tübingen, 1996.

[Nir59]   L. Nirenberg, *On elliptic partial differential equations*, Ann. Scuola Norm. Sup. Pisa (3) **13** (1959), 115–162.

[NS03]   A. Noll and J. Saal, H^∞-*calculus for the Stokes operator on* L_q-*spaces*, Math. Z. **244** (2003), no. 3, 651–688.

[NS04]   A. Novotný and I. Straškraba, *Introduction to the mathematical theory of compressible flow*, Oxford Lecture Series in Mathematics and its Applications, vol. 27, Oxford University Press, Oxford, 2004.

[Oka84]   H. Okamoto, *On the equation of nonstationary stratified fluid motion: uniqueness and existence of the solutions*, J. Fac. Sci. Univ. Tokyo Sect. IA Math. **30** (1984), no. 3, 615–643.

[Paz83]   A. Pazy, *Semigroups of linear operators and applications to partial differential equations*, Applied Mathematical Sciences, vol. 44, Springer-Verlag, New York, 1983.

[Rud73]   W. Rudin, *Functional analysis*, McGraw-Hill Book Co., New York, 1973.

[Sch02]   R. Schnaubelt, *Well-posedness and asymptotic behaviour of non-autonomous linear evolution equations*, Evolution equations, semigroups and functional analysis (Milano, 2000), Progr. Nonlinear Differential Equations Appl., vol. 50, Birkhäuser, Basel, 2002, pp. 311–338.

[Shi08]   Y. Shibata, *On the Oseen semigroup with rotating effect*, Functional analysis and evolution equations, Birkhäuser, Basel, 2008, pp. 595–611.

[Shi10]   _____, *On a* C^0 *semigroup associated with a modified Oseen equation with rotating effect*, Advances in mathematical fluid mechanics, Springer, Berlin, 2010, pp. 513–551.

[Sil04]   A. L. Silvestre, *On the existence of steady flows of a Navier-Stokes liquid around a moving rigid body*, Math. Methods Appl. Sci. **27** (2004), no. 12, 1399–1409.

[Sim87]   J. Simon, *Compact sets in the space* $L^p(0, T; B)$, Ann. Mat. Pura Appl. (4) **146** (1987), 65–96.

[Soh01]   H. Sohr, *The Navier-Stokes equations*, Birkhäuser Advanced Texts: Basler Lehrbücher, Birkhäuser Verlag, Basel, 2001.

[Sol77]   V.A. Solonnikov, *Estimates for solutions of nonstationary Navier-Stokes equations.*, J. Sov. Math. **8** (1977), 467–529 (English).

[SS07a]   Y. Shibata and R. Shimada, *On a generalized resolvent estimate for the Stokes system with Robin boundary condition*, J. Math. Soc. Japan **59** (2007), no. 2, 469–519.

[SS07b]   Y. Shibata and S. Shimizu, *Decay properties of the Stokes semigroup in exterior domains with Neumann boundary condition*, J. Math. Soc. Japan **59** (2007), no. 1, 1–34.

[Ste70]   E. M. Stein, *Singular integrals and differentiability properties of functions*, Princeton Mathematical Series, No. 30, Princeton University Press, Princeton, N.J., 1970.

[Sto45]   G.G. Stokes, *On the theories of the internal friction of fluids in motion*, Trans. Cambridge Phil. Soc. **8** (1845), 287–319.

[Tem01]   R. Temam, *Navier-Stokes equations*, AMS Chelsea Publishing, Providence, RI, 2001, Reprint of the 1984 edition.

[Tri95]   H. Triebel, *Interpolation theory, function spaces, differential operators*, second ed., Johann Ambrosius Barth, Heidelberg, 1995.

[Uka87]   S. Ukai, *A solution formula for the Stokes equation in* $\mathbf{R}^n_+$, Comm. Pure Appl. Math. **40** (1987), no. 5, 611–621.

[Wie99]   M. Wiegner, *The Navier-Stokes equations: A never ending challenge?*, Jahresbericht der Deutschen Mathematiker-Vereinigung **101** (1999), 1–25.

[Wie07]   J. Wiedl, *Analysis of Ornstein-Uhlenbeck operators*, Ph.D. thesis, TU Darmstadt, 2007.

[Yia65]   C.-S. Yia, *Dynamics of nonhomogeneous fluids*, Macmillian, New-York/London, 1965.

[Zie89]   W. P. Ziemer, *Weakly differentiable functions*, Graduate Texts in Mathematics, vol. 120, Springer-Verlag, New York, 1989.